TOWARDS SYNERGISM

The Cosmic Significance of the Human Civilizational Project

Anthony E. Mansueto, Jr.

University Press of America, Inc.
Lanham • New York • London

Copyright © 1995 by
University Press of America,® Inc.
4720 Boston Way
Lanham, Maryland 20706

3 Henrietta Street
London, WC2E 8LU England

Library of Congress Cataloging-in-Publication Data

Mansueto, Anthony E.
Towards synergism : the cosmic significance of the human
civilizational project / Anthony E. Mansueto, Jr.
p.   cm.
Includes bibliographical references.
1. Civilization--Philosophy. 2. History--Philosophy.   I. Title.
CB19.M324   1995   901--dc20   95-31888   CIP

ISBN 0-7618-0083-2 (cloth: alk: ppr.)
ISBN 0-7618-0084-0 (pbk: alk:ppr.)

Dedication

This work is dedicated to Antonio Cassano, who was sold into slavery as a child, who taught himself to read though he was beaten for trying, who joined the Communist Party during the Great Depression, organizing workers in Chicago, Illinois, Eire, Pennsylvania, and Bowling Green, Kentucky, who worked as an "irrigation systems expert what they call ditch dig" until he was eighty two years old, and who died the year the Soviet Union collapsed, as hopeful as ever about humanity's future. Your struggle has not been in vain.

TABLE OF CONTENTS

God ... gave me true understanding of things as they are: a knowledge of the structure of the cosmos and the operation of the elements; the beginning and end of epochs and their middle course; the alternating solstices and changing seasons; the cycles of the years and the constellations; the nature of living creatures and behavior of wild beasts; the violent force of winds and human thought, the varieties of plants and the virtues of roots. I learnt it all, hidden or manifest, for I was taught by wisdom, by her whose skill made all things. Wisdom 7: 15a, 17-22

Acknowledgements

This work draws on the accumulated wisdom of many millennia of human civilization, and in particular of many millennia of struggle to unleash humanity's latent potential for participation in the cosmohistorical evolutionary process. My greatest debt is to the billions of workers and peasants whose labor has made human civilization possible——and to the prophets and philosophers, the priests and pastors, and the political leaders who have made straight our pathway towards Omega.

I owe a very special gratitude to those who have helped to make me, and thus this work, what they are:

to my grandfather, Salvatore, who first drew me into the mysteries of humanity's participation in the cosmohistorical evolutionary process, and to all my ancestors in all their many lines,

to my mother and father, whose devotion to the beautiful and the good, together with their constant support and companionship, nourished my vision, and to my brother John, a hunter and craftsman, who has shown me the beauty of commitment to family and tradition,

to Francois Houtart, who has done so much, often so silently, to nurture the dialogue and collaboration between Catholics and Marxists which frames the larger intellectual and political context out of which this work emerged,

to my teachers, both those who have inspired my vision from afar, especially Plato, Aristotle, Joachim of Fiore, Amalric of Bena, David of Dinant, Siger of Brabant, Thomas Aquinas, Baruch Spinoza, G.W.F. Hegel, Karl Marx, Frederick Engels, Jacques Maritain, Samir Amin, Alexander Bogdanov, George Lukacs, Antonio Gramsci, Errol Harris, Ernesto Cardenal, R. Buckminster Fuller, and Mary Daly, and those closer at hand, who drew out the logical implications and internal contradictions of my ideas, and urged me on towards a higher synthesis, even when they themselves could not follow, especially

Karen Crocker, with whom I climbed the stairs of the Great Observatory at Chichen Itza, who knew my love of the night sky and my equally great love for the villages of the Yucatan and who urged me on in my effort to chart humanity's path to the stars: here is part, at least, of the road map,

Truxton Hare, who questioned my belief in the meaningfulness of the universe, and to whom this book is a partial answer,

Karl Morrison, who introduced me to great chain of being, Fred Carstensen (it has everything to do with the price of beans in China), and Mietek Szporer (an antidote to your despair),

Norman Gottwald, whose application of sociological methods to the study of the scriptures provided one of the key building blocks of this study,

Maria Valiani, Josephine Torres, Tony Hinajosa, and Antonio Cassano and all the other old *partigiani* from whom I learned so much about the last great stage of humanity's struggle,

Rumi and Nancy, and my other comrades in the Communist Workers Party and the New Democratic Movement, from whom I learned to respect and fear Maoism, and

Marcia Kannry, in whom Israel really has become a beacon for the nations,

to my students, in dialogue with whom most of the ideas in this book were developed, especially

the members of the Justice and Peace Commission of the Catholic Diocese of Dallas (especially Ann Hambleton), and the people of the Catholic Diocese of Dallas, especially the members of the Catedral Santuario de Guadalupe, and Blessed Sacrament, St. Cecilia, Our Lady of Lourdes, Our Lady of Perpetual Help, and St. Michael's parishes, and

the students in my classes at the Carlow College, the University of Pittsburgh, Robert Morris College, the Community College of Allegheny County, the University of Dallas (Terry, Cheryl, Bozena ...), the College of Santa Fe and Santa Fe Community College, Harper College and the *Universidad Autonoma de Ciudad Juarez*,

Acknowledgements

to my friends, who have been there when the struggle was more prolonged than popular, and to those who haven't always been there, and have thus helped to teach me the meaning of going against the tide, especially

John (are you still searching?), Lee, Martha, Kathy, Bonnie, Leah, Brian, Mary, Peter, Mary Ann, Eric, Eddie, Susan, Kent, Sheila (let justice run down like the water ...), Mike (whither Shikasta?), Ted and Angela, Ruth (*una manciata di more*), Don, Carolyn, Steve (*accompañando el pueblo*), Ginnie (*il seme sotto la neve*), Jenny, Deb, Carol, Marlene, and Delia, Bob, and Jim,

to my adversaries, whose opposition only and always makes me aware of my limitations, and thus of what it means to be whole,

especially Frank Tipler and Ernie Cortes, in whom I saw my own shadow just in time,

to Gloria Vosburg, my wife's mother and a special friend, whose wisdom and love remind me daily of the enormous potential which is latent in humanity, and to Harry Vosburg, who helps to guide his community in a time of confusion and fragmentation,

and finally, to Maggie (the Magipens), who has helped me overcome the residual effects of the alienation engendered by the marketplace, who reintroduced me to Aquinas and Fromm, and introduced me for the first time to Eckhart and Fuller, and whose constant movement and growth in promotion of energetic natural solidarity has given me, these past five years, a constant vision of what humanity can become.

I would also like to thank those who have provided the direct and indirect financial and institutional support which made this work possible: Mansueto-BASI, the Mansueto and Vosburg families, the Catholic Diocese of Dallas, especially Rev. Msgr. Robert Rehkemper, St. Ansgar Catholic Church, especially Revs. John Tapper and Jerome Riordan, Carlow College, the University of Pittsburgh, the Community College of Allegheny County, the University of Dallas, the College of Santa Fe, Santa Fe Community College, Harper College, the *Universidad Autonoma de Ciudad*

Juarez, especially Rodolfo Rincones, and the members and friends of the Foundation for Social Progress.

Preface

People often ask me how I have come to believe what I do. My first impulse is always to present them with the *argument* for my system, for what I have come to call *synergism*, beginning with an explanation of the scientific foundations, and then proceeding to lay out the system step by step, moving from logic to cosmology, cosmology to axiology, and axiology to strategy, tactics, and organization. This book will lay out a very important part of the argument for my system——my philosophy of history. More specifically, it will explore the question of the cosmic significance of the human civilizational project.

But inevitably, people tell me that as important as it is, the argument is not enough. They want to hear the story behind the argument. This is a request which I have been reluctant to fulfill, because of my conviction that my work must stand on its own merits, and convince or fail to convince on the basis of reason alone. Still, the story may be useful as a point of entry for people who are less theoretically inclined, and may give even those with a taste for theory a sense of the interests which my system has evolved to fulfill.

***

When I was young, my grandfather used to take me on long walks through the Bushwick-Ridgewood section of Brooklyn where he lived. He was a thin man, and had seemed frail for as long as I can remember, but he could walk for hours on end. We would wander on warm summer evenings past the long rows of decaying brownstones, stopping from time to time to talk with one his old friends——most had long since died or moved away——or to buy bread, cheese, wine and oil in one of the little delicatessens which lined the main streets. We would wait under the elevated tracks for the clatter of the trains, and climb up on the railroad trestle where my father had played when he was a young boy. And we would talk.

It was a profoundly mysterious world that I entered when I was with him, deeply at odds with the world of suburban tract homes, regimented grammar schools, and sanitized post—conciliar Catholicism in which I was growing up. It was a world filled with a dark and impenetrable sadness but also with a terrible hope which I longed more than anything to understand. He would speak of his childhood, and of his native Trapani——of how his mother had died when he was young, and of how his father had him apprenticed to a blacksmith "which was the same thing as be a slave." His master didn't feed him, and he was forced to fend for himself, stealing fish from the nets of unwary fishermen, and roasting them with garlic, lemon and olive oil on the hot anvil in the foundry. He spoke of leaving Trapani in the midst of an epidemic and of being held in quarantine for months, of living "seventeen man to one room" on the Lower East Side of Manhattan, and of how he learned to be a fabric cutter, bought a house and raised a family. He spoke of raising canaries——"I used to lov-a so much when the little ones they come alive," of making wine, stealing olive oil from supermarkets, and of how we should come to visit him more often.

At other times he would take me into the basement of his house where he had a small workbench, and he would show me the various tools he had made. He seemed to be able almost magically to turn any thing one gave him into almost anything else. His capacity for labor was like a magical force which reorganized and transformed everything with which it came into contact.

Woven into this nostalgic tapestry——which otherwise was not too different from the sad reminiscing of any octogenarian——were threads of a history the real dimensions of which I would only gradually come to understand. My grandfather would speak of his days as a militant in the International Ladies' Garment Workers Union, of how "the work they busta their ass and geta noth' in return," because "the boss he is a no good scound' what he lives ona other people work. So we maka the union." He would speak of the days of the "popular front" struggles with an exhilaration which can only be called religious. I remember, somewhat later, during the 1972 general election, a kind of family council, at which my father asked my grandfather who he was planning support for President. My grandfather answered,

McGov' *non e vera socialista, pero* Nix' he is a crook. I *vota* McGov'.

The old Liberal-Democratic machine of Representative Emmanuel Cellar was in decay, so he organized a group of 12 and 13 year old kids, most of them African American, who called themselves "Samuel's Friends" to work the election, and carried the district handily. (My grandfather's name was Salvatore. His friends called him Sam. The Black kids, not knowing the name Salvatore, assumed that his real name was Samuel).

Like most Sicilian men of his generation, my grandfather was radically anticlerical, and would have nothing to do even with the "reformed" post-conciliar church in which my father had become something of a leader. Once when we were passing a church he stopped, turned towards me as though in a fury and said

A priest he is a no good. He is a womanize and steals the people's money.

He pulled out the ice pick which he always carried with him "for killa the scabs" and drew it across his throat saying

We fix-a good. We kill-a. Except-a the friar. They are on-a side of the poor.

Still, he was profoundly religious, and it is his spirituality, rather than that of my mother or my father, which I seem to have inherited. Once, I showed him a picture of Our Lady of Sorrows, *Maria Addolorata*, being carried in procession on Good Friday his native Trapani. He looked closely at the picture, his face overwhelmed with an unspeakable sadness and said.

I know that-a wom'. She is-a very sad. The police they kill-a her son. So sad she is, we worship her for it. I am not sure, but her son I think-a they call-a him Jes'."

It was only later that I discovered that my grandfather's religion represented a survival of the very ancient Mediterranean cult of the *Magna Mater*, whose various forms, each of which articulated a distinct dimension of the aspirations, sufferings, and struggles of the peoples of the Mediterranean Basin, had been subsumed, albeit with significant patriarchal distortions, in the cult of the Virgin Mother. Isis, the Great Cosmic Librarian, storehouse of Wisdom, Demeter, the Grain Mother, symbol of the peasants' labor, staging a cosmic strike to liberate her daughter, Persephone, symbol of the ripe, harvested grain, from the forces of Pluto, the god of wealth——they were all there for him, just beneath the surface, in that image of *Maria*, of Miriam (the name means rebellion) searching for her captured son. And even though it was only later, after decades of study, that I came to understand the significance of this ancient cult, from the time I was a child, it was always the image of Mary——a woman clothed in the sun, her skirt covered with stars like the night sky——rather than that of Jesus which was most powerful for me.

Layered over the cult of the *Magna Mater* was another, distinct form of religiosity, the true nature of which also took me years of study to uncover. "Except-a the friar ..." My grandfather was also an heir to the tradition of Joachim of Fiore, the Calabrese Abbot who taught that the divine nature was fully present not in any one of the three persons of the trinity, but only in the community which exists between them, and that,

furthermore, this divine nature is progressively revealed, even *realized*
throughout history. The Age of the Father, what we would call tribal
society, gives way to the age of the Son, rule by priests and kings, the
tributary social order, which in turn gives way to the Age of the Holy
Spirit, when all things will be held in common. Joachim was the first
thinker to offer a progressive vision of human history as an active
participation in the life of God, and the first to argue that communism was
both possible and necessary if humanity was to realize its place in God's
plan for the universe. His vision captured the imagination of many
members of the mendicant orders. Not surprising, many were
excommunicated or worse. But others held out in the remote mountain
villages of Sicily and the *Mezzogiorno* passing the tradition on to lay
confraternities, which then became a key constituency for the scientific
socialism of the nineteenth and twentieth centuries. Thus "... the friar,
they are on the side of the poor."

My grandfather's stories spoke to me of an aspiration for a better
world, of a deep sadness but also of a terrible hope, of a struggle which I
longed more than anything to understand——no, to enter into, to become
a part of, and to carry on.

My father held a similar vision——but articulated it in very different
language. He turned down an appointment to Annapolis, which my
grandfather's cunning had extracted from a Brooklyn ward heeler, to fight
fascism in Europe. When he returned from the war he studied engineering
under the GI bill, at Princeton, Columbia, and New York University, and
went on, together with a few of his colleagues, to establish a construction
management firm which, for a time, was the leader in its field.
Construction management, for those who are not familiar to the discipline,
is essentially the application of the principles of modern scientific
management to the construction industry——historically one of the
technologically most backward, and politically most corrupt sectors of the
economy. The firm grew, in large part, because of the massive expansion
of the public sector during the 1960's. My father was a builder of schools,
of hospitals which, under the Hill Burton Act were required to dedicate a
portion of their services to low-income communities, and of vast
installations for the National Aeronautics and Space Administration.

My father's role in all of this was not that of innovator and organizer,
so much as bearer of the vision. He was able to transform the scientific
rationality of his colleagues into a compelling hope for a better tomorrow,
and to connect that vision to diverse and often suspicious
constituencies—federal bureaucrats and general contractors, labor leaders

and international investors, and thereby helped to make the vision a reality. By the end of the 1960's he was bringing home blue prints for "new cities" he planned to build, cities which would be integrated from the very day of their foundation, which would guarantee decent housing, education, and health care for working class families, which would have public transportation systems that rendered the automobile superfluous. I would spend hours just contemplating the plans for those projects, which seemed to me to be a kind of technological New Jerusalem, a fulfillment of the human potential for productivity, power, knowledge and solidarity.

The anticommunist hysteria of the 1950's made the socialist option an impossibility for my father's generation. Like most second generation Italian Americans he was a loyal Democrat, and during the 1950's drifted back towards the Catholic Church. There was, however, more than a little of the socialist utopia in those plans for new cities which we would pore over on cold winter evenings. And his Catholicism, while it preserved elements of the older devotionalism, was first and foremost a religion of salvation through ethical conduct and service to the community.

My mother came from a very different tradition. She was born in the town of Coeburn, in Virginia, high in the Appalachian mountains. Her father, who came from a German settler family (Groseclose) which had received a land grant in Wythe County signed by none other than Patrick Henry himself, owned and drove an oil truck. Her mother seems to have had somewhat obscure ties to the aristocratic families of the Tidewater (she was a Bush and a Richmond), but came from a branch of the family which had fallen on hard times. Nominally Methodist, their ties to the church seem to have been tenuous at best. They were clearly not evangelicals of the type associated today with the "Bible Belt." On the contrary, whenever we would visit my aunt, who still lived in Southwestern Virginia, talk would always turn eventually to matters occult, and the older women would recite stories of ghosts and poltergeists, and mysterious lights on remote mountain hillsides. When I was young, these stories frightened me, not so much because of their content, but rather because of the hushed and mysterious tones in which they were told, as though there was some Great Sin being committed just in the telling. Today, I often wonder if some elements in my mother's family were not heirs of the Craft, who, however, had internalized the Christian denunciations of their cult as evil and demonic.

In any case, my mother devoted her youth to music, and obtained a scholarship to Barnard. She hoped to become a concert pianist, but like so many women in the 1950s, abandoned her career to seek wholeness and salvation through motherhood. I am, in so many ways, the product of her careful craftsmanship. For my first five years, she was my constant companion. The strength of our bond gave me a sense of deeply rooted connectedness, and of my own value, which would enable me to stand alone, to sacrifice relationships, and even to engage in intense conflict when circumstances required it, as they so often would in years to come. She spent endless hours reading to me, opening up imaginary worlds which became the seedbeds of vision. She brought me books of every kind, which allowed me to journey endlessly through space and time. And she gave me space to be alone ——to reflect on and ponder everything she had given me, gradually digesting it, making it my own.

My mother converted to Catholicism when she married my father. Like my father, her religion was basically one of ethical conduct. But it was also a religion of beauty. She loved the grandeur of the ancient liturgy, and when the reforms of the Second Vatican Council came, she mourned its passing. I would come home from school to the sounds of Chopin's "Revolutionary Etude" and "Fantasie Impromptu."

Together, my father and mother and brother and I travelled. We journeyed to Rome and Athens. I followed Socrates' path through the Agora and crawled though the catacombs of the old imperial capital. I climbed the steps of the Great Observatory at Chichen Itza and entered the sanctuary of Bangkok's Great Emerald Buddha. I grew up feeling that each of these civilizations was my own, part of one great civilizational project, a project that extended back millennia, and which was about to open up into a bright and glorious new future for humanity. And I learned to appreciate the contributions of ordinary people to that project, people whose names would never be remembered: the people whose labor built the Parthenon and the Great Observatory, the Basilicas and the Pagodas, as well as the people who designed them. Wandering through the narrow streets of the old Arab quarter in Palermo, or floating down the *klongs* of Bangkok, I came to know the tremendous beauty and energy of humanity. Having seen what I had seen by the age of nine, I will never doubt humanity's capacity for growth and development.

We travelled, but there were also quiet times at home. We lived, until I was thirteen, on the far eastern edge of New York's Long Island suburbs, at the point where the suburbs gradually gave way to the countryside. There were long summer afternoons spent reading, or playing "Star Trek," or even softball in the court outside our house. I had a friend, by the name of John Bastianelli. We would take out books from the library about starships and space travel and UFOs and debate the relative merits of photon and ion based propulsion systems. We would go on "expeditions" through the woods and fields, collecting "specimens," and "analyzing" them (usually with disastrous results) using our chemistry sets. There were deep dark summer nights when we would spend endless hours gazing through my telescope, wondering what, and who, was out there, hoping against hope itself that someday we might be among the first to make "Contact," or pretending, in the way only a child can pretend, that we already had. For me, the boundaries of the civilization extended much farther than Bangkok's *Wat Po* or the jungles of the *Peten*. I already believed, already *knew* that humanity was part of something much larger than itself, a participant in a great cosmic project in which our own civilization played a vitally important role.

These were the years of the New Frontier and the Great Society ——years when it really seemed that scientific innovation and civic responsibility would at long last triumph over humanity's millennia of ignorance and egoism. The aspirations of the period were perhaps best captured by the popular science fiction series *Star Trek*, which depicted a society in which not only had poverty and hunger been vanquished and racial and national contradictions overcome, but everyone performed useful, challenging labor. The only hierarchies were those based on ability, and even these were remarkably fluid. James T. Kirk, the 34 year old starship captain, was a kind of cosmic Jack Kennedy, grouping around himself the expertise he needed to complement his own decisive (if sometimes reckless) leadership. For those of us who grew up in the 1960s, society on the model of *Star Trek* seemed not only possible but inevitable. It was only a matter of time.

The election of 1968 was something of a disaster for our family. The new Republican Administration ushered in a long period of reduced investment in infrastructure, education, research, and development. And my father's firm, which had always made contributions to progressive Democrats, was placed on the Administration's blacklist. Entire projects, including the one new city to progress beyond the visionary stage, were impounded. Now in my teens, I watched helplessly as the market gradually destroyed what my father's hard work had built up. It was this, more than anything, that set me on the road towards socialism.

But other things were changing as well. Intensely religious when I was very young, around the age of eleven or twelve I became alienated from the Church, partly because of the irrationality of its dogmas, at least as the parish clergy were able to explain them, conflicted with the scientific world view I was absorbing at school, and partly because the hierarchy seemed to be contributing nothing *new* to the human civilizational project. Even though I was only beginning to understand the tradition I had inherited from my grandfather, it was increasingly apparent to me that any attempt to rigorously draw out the conclusions of the values my parents had taught me pointed clearly beyond the irrationality and individualism of the marketplace, towards what had historically been called of socialism.

There was very little, however, in my own questioning, to propel me towards a full blown atheism. I would, in all probability, have settled into a kind of anticlerical deism, had it not been for the influence of my teachers at the Latin School of Chicago. The school presented me, for the first time, with a profoundly secular milieu. The teachers often expressed antireligious views. They never, to be sure, took on religion directly. Rather, they simply conveyed in passing that sentiment, characteristic of the secular intelligentsia, that "no reasonable person could ever believe in such things." And "such things" included, apparently, not only the Father, the Son, and the Holy Spirit, but also God as such, and the ultimate meaningfulness and rationality of the world at large.

The various different departments of the school conspired with each other unconsciously in this work. The department of mathematics eschewed a synthetic approach to mathematics based on geometric intuition and deductive proof in favor of the artificial and humanly contrived "rationality" of analytic geometry, to which we were introduced by the tenth grade. The science teachers railed endlessly against the "teleological world view," and lobbied constantly in favor of the idea that life, reason, and human culture were ultimately the product of the random interactions of elementary particles, expressing no necessity higher than the fundamental laws of physics and chemistry. The literature department, meanwhile, pushed Nietzsche and the existentialists, insisting that humanity could find no meaning in a world governed by the laws which physics discovered, but must, rather, create this meaning for itself. The school's adolescent literary intelligentsia took as its own the Sartrean slogan "Life begins on the other side of despair."

To some extent, of course, this kind of sentiment was simply an intellectually pretentious expression of the adolescent search for identity. But it would be a mistake to underestimate the effect such an intellectual assault can have on a person's intuitive capacity to grasp the underlying relationality, organization, and meaningfulness of the world. I suspect that there are very many who never recover from being told by someone in authority that the universe is ultimately without meaning.

Fortunately, my own intuitive powers remained strong. In my heart of hearts I never doubted that there was a logic to nature and thus a meaning of some sort to the larger cosmos which during my childhood had seemed so alluring and beautiful. But I needed to reconcile my intuition regarding the meaningfulness of the universe with the claims of scientific rationality, and to discover a more complex and satisfying way to participate in the self-organizing activity of the cosmos than that offered by the Roman Catholic Church.

Resolution came partly through my political activity, which began during this period, and partly through my philosophical and theological studies. I was elected Vice President of the Student Council at my Junior High School, at age 12, as an open Communist, and the next year, when we moved to Chicago, became active in electoral politics, working first for George McGovern, and then for a number of progressive Democratic candidates for the Illinois state Assembly, learning the art of politics by struggling toe to toe with some of Mayor Richard J. Daley's most experienced precinct captains.

When it was not campaign season, I spent my free days exploring every nook and cranny of the city, beginning first with the neighborhoods along the lakefront, then the West, Northwest, and Southwest sides. What I found was a universe infinitely richer, and infinitely more complex, than anything I had ever imagined. I watched the sun rise over the frozen lake and I trudged through the snow to watch it set behind the onion domes of the Orthodox Churches of the Ukrainian Village. I peered into the wholesale butcher meat markets along Randolph Street at Haymarket and the fishmarkets along Fulton street. I walked down Halsted Street to Bridgeport, the home of "The Mayor," and looked into the storefront offices of the insurance agents and real estate agents and small time lawyers who seemed to form the matrix out of which the city's political caste emerged. I walked past Cabrini Green, and began to wonder for the first time, about the "rationality" of modern urban planning. Surely it didn't make sense to pile all those poor people up on one place, and then abandon them. It was like a concentration camp.

The city was like one vast organism, a complex unity which was able to comprehend and integrate an incredible diversity of elements, each in its own way contributing something unique to the creativity, the power, and beauty of the whole. And on election day all these elements came together as the city struggled with itself to find a higher unity, a more profound synthesis.

By night I would read revolutionary novels——Malraux's *La conditione humaine,* Gorky's *Mother*——and began to find in the political struggle that sense of meaning and direction, and that experience of transcendent solidarity, which the Roman Catholic Church could no longer provide. Later, at the University of Chicago, I attended Paul Ricoeur's lectures on "Ideology and Utopia," and began to steep myself in the writings of Hegel and Marx, Lukacs and Althusser. But I also developed, thanks to the

efforts of a medieval historian by the name of Karl Morrison, a sense of
the larger philosophical tradition of which these revolutionary thinkers are
the heirs. So I pored over Plato and Aristotle, Augustine and Aquinas,
Teilhard and Maritain. Somewhere, between the arguments of the
Platonists and the Aristotelians, the Thomists and the Hegelians, I found
satisfying rational arguments for my own intuitive grasp of the relational,
holistic, self-organizing and teleological character of the universe. And
Hegel specifically was able to demonstrate to me convincingly that the
logic of the city, in which I had immersed my self since my family moved
to Chicago in 1971, was not in any sense opposed to the logic of the
cosmos, to the orderly motion of the stars or the steady cycle of the
seasons, but was, in fact, simply a higher, more complex expression of this
same cosmic order, humanity's own gift to the universe, what we humans
gave back to God in exchange for the precious gift of creative initiative and
spiritual freedom.

It was during this same period that I began my tumultuous romance with
dialectical materialism. It was already clear to me (the efforts of my
neoliberal professors at the University of Chicago notwithstanding) that the
marketplace was not an adequate mechanism for centralizing and allocating
society's surplus product. On the one hand, the marketplace transforms all
activities into merely a means of advancing individual consumer interests,
undermining the social fabric and eroding humanity's ability to know and
love the Whole. On the other hand, the marketplace has no access to
information regarding the impact of various activities on the development
of human social capacities, but systematically redeploys resources away
from investment in infrastructure, education, research, and development,
and squanders them on luxury consumption. My father's own career was
testimony of this. He spent the 1960s and early 1970s, a period when
market forces were being held in check by progressive Democratic
administrations, building schools and hospitals, as well as the vehicle
assembly building for the Saturn V rocket. He spent the 1980s, when
market forces were unleashed, building luxury hotels. All this Marx and
his successors documented and explained.

At the same time, dialectical materialism seemed to have lost the sense,
which I treasured in Hegel, that human history was part of a great cosmic
evolutionary process, the process of God's own, gradual, self-realization,
the sense that we are here for a reason, to add something to the universe,
in favor of an ultimately more somber view of humanity as an island of
meaning and order in an ultimately meaningless and chaotic universe. This
sense of ontological rootlessness troubled me, and even in the late 1970s
I had a sense that it had something to do with the historical failures of the

communist movement.   But I had not matured philosophically to the point that I was ready to attempt a new synthesis.

In 1977 I moved to New Haven, and spent two years at Yale University, studying the Hebrew Scriptures, the New Testament, Athanasius and Cyril of Alexandria, Luther, Calvin, Edwards, and Tillich, while continuing my militance in community struggles and my private study of the Marxist tradition, under the tutelage of a Polish emigre literary critic.   This was a very difficult time in many ways.   My two years at Yale represented my first systematic exposure to Protestant theology. Unlike my professors at the University of Chicago, most of whom at least tolerated my Hegelianism, the Yale theologians taught that reason could at best pose theological questions, but that the answers always came from faith, and that the human drive towards holism, far from being authentic love of God, was simply a kind of refined selfishness.   Christian love was always self-effacing and self-sacrificial.   The cross, which defied all reason, was the norm for Christian belief and action.

At the conscious level I resisted this onslaught, for the first time embracing my Catholic heritage as a bulwark against Protestant pessimism. But the exposure to Protestant theology affected me in subtle ways that I have only gradually come to understand——and only gradually overcome.

There was more at work here than the intellectual content of Protestantism, which, with the exception of some of Edwards' theology, I never found very compelling.   First of all, I was very lonely.   I wanted very much to have a life partner, to find a woman with whom I could share everything, from my most intimate secrets to my most difficult theoretical quandaries, someone who understood all of the passions which moved me, from sexual desire to the desire for warmth and acceptance to my insatiable drive for power.   I had good female friends, to be sure, but the kind of life partnership that I sought eluded me.   It seemed like all of the women who were interested in anything serious were already in committed relationships, or had no interest in me, and that the rest just wanted a "good time."

My prospects for any kind of career also seemed increasingly dim.   It was already apparent in the late 1970s that the academic job market was tight and getting tighter.   It was also already apparent that the kind of philosophy I wanted to write was very much out of fashion.   It was clear that I was unlikely to find the kind of academic sponsorship which is necessary for an academic career.   I nonetheless applied to the doctoral program in Ethics and Society back at Chicago, figuring it would at least be intellectually more rigorous and stimulating than what I had found at Yale, but was turned down.   I frankly had no idea what I was going do after I became a "Master of Arts in Religion."

In 1979 I returned to Chicago. Those were heady days for the left. The economic stagnation of the 1970's seemed at long last to be opening up into a real crisis, and municipal governments all around the country were in disarray. In just over one year, the city of Chicago faced strikes by teachers, police, and firemen, turned the public hospital over to Hyatt Corporation, and raised transit fares by 66%. To many of us it seemed there was a real opening for the left. Then in 1980 Ronald Reagan was elected president, and when his program was announced it seemed that we might be able to draw together ——under communist leadership no less——a broad front of resistance to austerity and militarism. And so I threw myself into the struggle, building a small community organization in West Englewood, preaching on the virtues of socialism in a tiny Lutheran Church, and serving on the steering committees of several ill-fated "united front" and "popular front" formations. I was even named "Secretary of Transportation" in former Chicago Alderman Dick Simpson's "shadow government" for the State of Illinois! By August, however, it became apparent that we had no base outside the ethnic minority communities, where despair and social disintegration made organizing almost impossible. Our base organizations atrophied as the real implications of Reagan's victory began to take hold of the popular consciousness, and our popular fronts disintegrated in a frenzy of sectarian strife.

Both my personal loneliness and the difficulty I experienced building a stable institutional base, whether in the academic arena or in the "popular movements" I interpreted as part of the "hegemonic individualism of capitalist society," as indeed they were. But here was the danger. As I became increasingly isolated, and increasingly angry, I began to call on the "ideal" of self-sacrificial Christian love both as a weapon with which to attack bourgeois individualism, and as "bridge to the formation of socialist consciousness." I was internalizing more than I realized of the Protestant theological problematic with its Christocentrism and its cruci/fixation.

The result was an intellectual and spiritual detour, which led me into a deeper and more explicit identification with both Christianity and Marxism than might otherwise have been the case. My energies were turned away, at least for the time being, from the task of achieving a new philosophical synthesis which transcended the limitations of existing paradigms, and focused instead on elaborating a consistent Christian Marxism. On the one hand this meant demonstrating theologically that only communism could realize in practice the historic values of the Christian tradition, and that communism was therefore obligatory for every Christian, and, on the other hand arguing to the communist movement, on strategic grounds, that creation of a mass socialist movement presupposed a breach with the hegemonic individualism of bourgeois society, and that this was possible only on basis of what amounted to a religious conversion, whether this was understood in Christian terms or in language drawn from some other tradition. I became absorbed in the sociohistorical study of the Christian tradition, drawing on the biblical sociology of Norman Gottwald and Marvin Chaney, on social historical studies of various theological controversies, etc., attempting to build up an argument that, at least in its origins (which for me were always the Yahwistic origins of ancient Israel, not Christian origins in the narrower sense) Christianity was first and foremost a movement towards creation of a classless and communal social order.

At the same time, I developed a highly original argument, based on Durkheim's sociology, that atheism was first and foremost a form of bourgeois ideology. Durkheim points out that, whatever deeper ontological realities they may disclose, religious symbols are a kind of "collective representation" of the social order. Atheism, the absolute denial of the sacred, is thus nothing other than a denial of society itself, and thus provides an "ontological ground," if one can call it that, for bourgeois individualism. These two strains of investigation came together in an emerging strategic orientation centered on using existing religious solidarities as a "bridge to the formation of socialist consciousness." Politically, this put me close to the Catholic left of Latin America, though I was more explicit than they were both in my commitment to the communist movement, and in my insistence that atheism was an insuperable obstacle to communist organizing and socialist construction. More so than liberation theology in the narrower sense, I loved the writings of Ernesto Cardenal, whose poems articulated a vision of cosmohistorical evolution which resonated with my own deepest convictions, and, like my own vision, stretched the boundaries of Christian Marxism, pointing beyond it to a higher synthesis.

During roughly this same period, I became involved in a study of Italian immigrant and ethnic history at the University of Illinois, and conducted some sixty interviews with Italian Americans, only to discover that a significant minority were anticlerical Christian socialists like my grandfather. I returned to Sicily for a time and conducted more interviews, finding that the tradition there was alive and well.

In 1981 I moved to San Francisco, where I continued my studies, pursuing a doctorate in Religion and Society at the Graduate Theological Union. During my three years in San Francisco I was exposed to nearly every dimension of the communist movement as it existed in the United

States at that time——party building, united front, and popular front, solidarity work, electoral politics and "theoretical struggle." I joined what was then the Communist Workers Party, a mostly Asian and African American group of Maoist origins, which was undergoing a period of self-examination, but was suspended within a matter of weeks for "violations of democratic centralism," and soon resigned. This did not, however, end the relationship. The party continued, largely of its own accord, what amounted to a cross between an "investigation" of my "case" and an open-ended theoretical dialogue. My work was never published in party journals because, I was told, my critiques of atheism would be regarded as "too controversial." But they were widely circulated among the party leadership, and seemed to have some influence. In 1985, the party dissolved and reconstituted itself as the New Democratic Movement. The new strategy called positioning cadre in key institutions, and struggling for "cultural hegemony." There was a significant overture to the religious left, religious traditions being recognized as one of many bridges to the formation of what was now called "postindustrial society," along with new technologies, democratic struggles etc. The party did not, however, accept my critique of atheism, and indeed, sympathy for this aspect of my position seemed to be stronger among the more orthodox Maoists, particularly those of Asian immigrant origin, than among partisans of the new thinking, who had absorbed more than a little of North American "new left" culture, with its incipient postmodernism. In 1986, now in Pittsburgh and teaching at Carlow College, I somewhat reluctantly rejoined the new formation. But the "new line," as it was called, never caught on, and the organization disintegrated within a matter of two or three years, leaving little more than a mailing list and an unemployed former general secretary.

During this period I had pulled together the results of my research on Italian Americans in Chicago, which I presented as my doctoral dissertation, and had extracted from it an article laying out systematically my analysis of "atheism as bourgeois ideology," which was finally published in 1988 in the international journal *Social Compass*. Publishing, however, turned out to be very difficult. The academic reviewers who evaluated my dissertation for publication just didn't buy my argument that there had been a mass socialist movement in the Italian immigrant communities, much less my analysis of the implications of that movement for socialist strategy. A similar response met most of my other submissions. My holistic and teleological philosophical and theological commitments were deeply at odds with the "postmodernist" consensus prevalent in leftist academia, and my very explicit communism was hardly welcome in most theological circles. And frankly the problem of defining the conditions for the emergence of a mass socialist movement, and the construction of a classless and communal social order, were not really regarded as a legitimate topic of academic inquiry. And of course difficulty in publishing meant difficult in securing a permanent academic position. I became part of the academy's "reserve army," earning just enough to get by.

During this period I also became painfully aware that I actually had much less in common with much of the Catholic left than I had believed. Discussions with "liberation theologians," and leftist Catholic organizers alike ended with denunciations of my "rationalism" and "theoreticism." I found myself put off by their anti-intellectualism and what seemed like a mindless populism and a paternalistic romanticization of "the poor." But still, it was hard for me to put a finger on the precise nature of these differences.

All this came to a head in a rather unexpected way, through an encounter with one of my students at the University of Pittsburgh, where I was teaching American Religious History. Her name was Marcia Kannry. Marcia had recently resigned as Regional Executive Director of the National Jewish Fund in Pittsburgh in order to finish her B.A.——something which had been put off when she left the U.S. in her early twenties to live the Zionist dream on a Kibbutz in Israel. An accomplished fundraiser, she had gradually risen through the ranks of the Zionist hierarchy to a position of middling responsibility, but she needed a degree to go further. And she was authentically interested in knowledge. She had a fine, if somewhat haphazard and unsystematic mind, her talent for incisive criticism over-reaching her capacity for constructive thought. Her greatest virtue, however, was a passionate commitment to the truth, and a willingness to sacrifice relationships, and even civility, for what she believed.

One day after a discussion of Virgilio Elizondo's *Galilean Journey*, Marcia stayed after class, and asked me to join her for a cup of coffee. No sooner had we sat down than she launched into what initially seemed like a mad tirade, accusing me of distributing antisemitic literature in the course. As I sorted through this storm of words, however, the outlines of a remarkably clear argument gradually emerged. Christology, she argued, *all Christology*, is inherently antisemitic. The claim that Jesus is the promised messiah, through whom alone redemption is possible, both negates Jewish national claims (historically bound up with the expectation of an inner-worldly political messiah) and Jewish claims regarding their own specific road to redemption, through fulfillment of the law. She then went on to show how, more specifically, liberation theologians like

Elizondo, by stressing Jesus' struggles with the "Jewish power structure" of his day, simply added insult to injury, reproducing ancient antisemitic stereotypes.

Marcia's argument opened up a whole new world for me. I had always been somewhat uncomfortable with what I saw as the growing antisemitism of the left. One of the attractions of the Communist Workers Party had been its rejection of the absurd UN declaration that "Zionism is racism," preferring instead to regard Zionism as a type of "bourgeois nationalism," something which, in Maoist terms, put it among the progressive forces. Marcia was able to introduce me to forms of Zionism which went far beyond bourgeois nationalism, and which were genuinely dedicated to building a society which fully developed human social capacities.

More important, from my point view, however, was what Marcia's argument said about Christianity. The truth is, I had never really known what to do with old *Gesu*. The historical figure, dimly discernable through the accounts of the Gospels, had always been singularly uncompelling to me. And the crucified and risen Christ of the Pauline corpus presented insuperable theological problems for any attempt to elaborate a consistent Christian Marxism. If salvation comes about through the death and resurrection of Jesus, then it *does not* come about through the struggle for a classless and communal social order, which is at best relegated to a secondary task which draws out the implications of the salvific act. Progressive "reinterpretations" had always smacked of intellectual dishonesty.

But there was more. I began to re-examine my whole pre-occupation with self-sacrificial Christian love. Whatever the social basis for, and tactical usefulness of such a doctrine in a context of armed struggle for national liberation (in, for example, Central America in the 1970s and 1980s) it had little place in the struggle to build a new society which taps into the full range of human social capacities. What we want from people is not necrophilic self-sacrifice, but productivity, power, knowledge, creative and nurturing love ——things which add something to the dynamic, organized complexity of the cosmos, rather than taking something away. Witness the difficulty of the Sandinistas and their allies on the Catholic left in making the transition from armed struggle to socialist construction.

In June of 1988 I was asked by the Catholic Bishop of Dallas to serve as Director of his Diocesan Justice and Peace Commission. I found myself in a paradoxical position. I now had the kind of institutional positioning which I needed to implement the strategy I had developed during the 1980s——using religious traditions as a bridge to the formation of socialist consciousness——albeit under the very difficult conditions presented by a city which serves as a kind of headquarters for the religious right. But I was already beginning to move beyond the Christian Marxist theory which provided the larger framework for that theory.

My starting point was simply to join to the original formulation of the strategy the need to criticize the "otherworldliness" which was implicit in the Christological formulae, more or less explicitly, depending on what political conditions would permit. After early moves to put the diocese solidly behind the struggle of national minority communities for increased representation on the city council, which helped to build for me a lasting strategic reserve of support within the Latino community, and a parallel move to "reorganize" the commission and neutralize what up until then had been a Christian pacifist majority, I concentrated on two principal initiatives. First, I secured an invitation for the Industrial Areas Foundation to come to Dallas to build a congregation based community organization. My hope was that the active involvement of the parishes in the public arena would create a basis in experience for rethinking the political valence of the Christian tradition. I played the leading role in building the sponsoring committee for this effort, recruiting over 50 congregations, making Dallas Interfaith one of the larger organizations in the network, and indeed in the country. Second, I built a multilayer network of study groups, the purpose of which was to transform the Christian faith of the participants into a bridge to the formation of socialist consciousness. My hope was to recruit the most advanced elements to a leadership core, which could continue my work, while I moved on to other cities to replicate it, gradually building a national organization.

The first stages of this plan were successfully completed, but serious difficulties soon emerged. The first had to do with the Industrial Areas Foundation. When I began organizing in Dallas I took more or less at face value the IAF's claim to be uninterested in larger questions of political theory or theology. I assumed that it simply organized people to improve their communities, never advancing beyond what communists call "trade union consciousness" to a critique of the underlying structure of capitalist society, but also innocent of any political-theological agenda of its own. I assumed it would be possible to relate to it as communists have always related to spontaneous mass organizations, building up a concentration and gradually asserting leadership, while respecting its distinctive role and its autonomy. What I found was something very different, an organization with its own distinctive political theology and its own long term political strategy. This perspective is best described as a kind of updated, postmodern Augustinianism. Human beings are motivated by self-interest.

Some have broader and more complex self-interests than others; it is these who are able to understand and tap into the interests of others, and who thus have the potential to become leaders, and who can contribute in some measure to the welfare of the community. But there is no rational basis for judgements of value, and certainly no underlying, rationally accessible, ontological ground for the moral order. On the contrary, authentic vision and values are accessible only through faith. Thus the central role of the churches in the IAF's political strategy. Of course, the ecclesiastical hierarchy remains the only legitimate interpreter of the deposit of faith. The result is an irrationalist doctrine pregnant with authoritarian potential.

At the same time, the Catholic hierarchy was moving farther and farther to the right, or more specifically, towards is own variant of neo-Augustinian, crypto-Protestant Christolatry. The dominant expression of the trend is the *Communio* theology currently in vogue in the Vatican. In 1990, the Catholic Bishop of Dallas, Thomas Tschoepe, himself rather conservative, but also very unobtrusive, retired, and was replaced by Charles Grahmann, a friend of Cardinal Ratzinger's, and someone determined to make his mark on Dallas. But there were also more extreme expressions of this trend. Opus Dei had made the University of Dallas one of its principal concentrations in the U.S. And all this took place in a milieu already dominated by evangelical Protestantism. The work of political-theological transformation proved very difficult.

Finally, my years in Dallas were also the years of the final collapse of the international communist movement. My longstanding criticisms of the ruling Communist Parties notwithstanding, this was a crushing blow. It is nearly impossible to recruit people to a movement which is, at best, facing a prolonged period of rethinking. And it was clear to me that more was at issue than just rethinking. Communism as I had known it was over. It was time for something new.

And so, by the summer of 1990 it was already apparent to me that I would not be able to realize my plans in Dallas. I began trying to consolidate a small core, and preparing for a defensive struggle which would ultimately lead to a protracted strategic retreat. And I was becoming more and more deeply convinced of the necessity to rethink fundamentally the whole socialist project.

Earlier that year I had met Maggie Vosburg, who was serving as Assistant Chaplain at the University of Dallas. She almost immediately became my principal deputy in the political theological struggle, and she contributed immensely to the work of retheorization which was soon to begin. She also became an authentic life partner, the kind of life partner I had been seeking for years, and had given up hope of ever finding.

Maggie brought to our partnership fine critical capacities. She soon went to work exposing the damaging effects of neo-Augustinianism and crypto-Protestantism everywhere, not only in our opponents, but in our own thinking as well. She helped me to see that these doctrines were not only politically backward, because of their pessimism and otherworldliness, but were also spiritually damaging, and represented a kind of practical atheism, a denial of the divine drive towards holism, the self-organizing dynamism, and teleological centeredness present in matter itself. She brought a powerful sense of the goodness, the underlying rationality, and the tremendous beauty of the universe. She helped me to gradually recover and deepen my own sense of connectedness to the cosmos, never wholly lost, but certainly strained and fractured by years of personal isolation and political frustration, and thus helped to create in me the spiritual conditions for the theoretical achievements of the past three years. Her own training was rigorously Thomistic, acquired at the University of Dallas before it fell into the hands of the extreme right. These Thomistic foundations are rounded out by an apparently eclectic but internally consistent set of influences which range from creation spirituality of Matthew Fox to the synergetics of Buckminster Fuller, from Erich Fromm to Mary Daly, and from New Zealand novelist Keri Hulme to German psychotherapist Alice Miller.

We were married in 1991, and left Dallas almost immediately. The next three years were a time of wandering for us——a time of exile. We lived in New York, Washington, Albuquerque——even in a remote mountain cabin on the high road from Los Alamos to Chaco Canyon——before finally returning home to Chicago in 1993. We have had to fight to earn just enough to survive, and have found it nearly impossible to connect with people, either personally or politically. But it has also been a time of immensely productive reflection and retheorization. In some ways, for me at least, this has meant picking up where I left off, back in 1977, before my "grand detour" through Christian Marxism——a return to philosophy and above all to Hegel. But I have come armed with new weapons. I have the benefit of some seventeen years of study, seventeen years of teaching and organizing experience. Having *personally* followed the dialectic along *both* of the paths which along which it developed *historically*, after the collapse of Hegel's synthesis——objective, religious idealism, and dialectical and historical materialism——I have been able to comprehend in a way very few others have both the achievements and the limitations of these twin traditions. And along the way I have deepened profoundly my understanding of the actual process of cosmic self-organization. Partly this has been the result of formal study.

I have made a real effort to grasp the contributions of such new scientific disciplines as unified field theory, complex systems theory, and postdarwinian evolutionary biology, which have more or less demolished the old atomistic paradigm and laid the scientific foundations for a new philosophy which recognizes the universe as a relational, holistic, self-organizing, and teleological system. Partly it has been the result of years of organizing, which have dispelled any illusions I may have had regarding the possibilities for *either* the "spontaneous self-organization of the masses," *or* the imposition of organization and direction from the outside. Human organizations develop through a complex dialectical process in which the most advanced elements pose questions which draw out the implications and internal contradictions of the existing structure, thus driving the whole towards a higher synthesis.

The result is what, for the lack of a better term, I call "synergism," though in many ways it simply represents just another stage in the development of the dialectical tradition which stretches back through Marx and Hegel, Spinoza and Aquinas, Aristotle, Plato, Socrates and beyond, to humanity's first (probably woman's first) intuitions of Wisdom, of the underlying order of the universe, the *Magna Mater* which gives birth to all things. At the core of synergism is the conviction that being itself is organization, and that this complex, self-organizing dynamic realizes itself in an infinite scale of increasingly complex and more highly integrated forms: the metric field, matter itself, chemistry, biology, and our own social form of matter. And it does not end there. The universe is ever struggling to become more creative, more powerful, more intensely beautiful and more profoundly rational, more powerfully loving with that nurturing love which always and everywhere gives life. The *telos* or aim of this process, is nothing other than God herself. Human society is a laboratory for the creation of dynamic, organized complexity, a key link in a grand cosmohistorical project.

Synergism conserves from objective, religious idealism the conviction that there is an underlying, organizing principle which makes the universe necessary——which guarantees that there is cosmos rather than chaos, and of which everything is ultimately an expression. This is the ontological ground which is missing in Marxism. But synergism also conserves elements of the dialectal materialist tradition. First, I hold that the organizing principal of the cosmos does not act from without, so much as from within matter itself, which, even in its simplest form, the metric field, is endowed with organization and with a drive towards increasing complexity and integration. Second, development takes place through the interaction of particular systems with each other. It is relationship which draws out the implicit holism of every particular, and drives each to realize the totality of which it is an integral part. If God acts "from outside" or "from beyond" it is always as final rather than as efficient cause, her incredible beauty, wisdom, and goodness constituting an ever present lure for growth and development.

I have, furthermore, conserved from historical materialism the critique of the marketplace, which I believe continues to constitute an obstacle to the full development of human social capacities, undermining the integrity of the ecosystem and the social fabric, and holding back efforts to centralize the resources necessary for investment in infrastructure, education, research, and development. Our aim, of course, is not centralized planning, which can never tap into the rich diversity of interests and talents characteristic of a complex society, but rather some new structure, the nature of which we are only beginning to understand, which uses the dialectic of the organizing process itself to identify and tap into existing interests and talents, while catalyzing the development of new aspirations and capacities in a way neither market nor plan have been able to do.

Synergism is a political-theological system, where communism was merely political-economic. Communism sought to harvest the fruits of human labor in order to catalyze an ongoing process of social development. Synergism goes further, and seeks to harvest the fruits of human social development, the fruits of the human civilizational project, for the cosmic evolutionary process, for the realization of God's plan for the universe.

I believe that there is a constituency for this vision, or rather a range of constituencies. The first, and most important constituency, is among those highly talented elements in society who have the ability and drive to innovate, or to carry innovations into practice, in every sphere of human endeavor. These are the people who are on the cutting edge of the arts, the sciences, and philosophy——people who are painting, writing epic poems and symphonies, working out the unified field theory, or drawing out the philosophical implications of the new science. But they are also the people who want to go to Mars, or to Alpha Centauri or who want to build a global telecommunications system. They are the people who want to reform education in the inner cities and tap into the incredible potential which is languishing in our ghettos. They are the people who want to invest their energies in raising children who are loving and creative, powerful and wise. All these elements in society, who authentically make up the most advanced section of the working class, find their efforts frustrated by the marketplace, as well as by ideological problematics which devalue creativity and misunderstand its critical role in the cosmohistorical evolutionary process. They constitute the core of our constituency, the people for whom synergism is an organic expression of their own life experience.

The second constituency is made of those who, with great efforts, are laboring to conserve what remains of the ecosystem and the social fabric, and of humanity's intuitive sense of the underlying unity and meaningfulness of the cosmos. These are the true conservatives: mothers and fathers and grandparents, elders in village communities around the world, true pastors who, the limitations of the theology which they have inherited notwithstanding, have tried to shelter their people from the storm, preserving the nurturing communal matrix out of which creativity emerges. This will be a more reluctant constituency: reluctant to acknowledge that their traditions are not sufficient, by themselves, to carry humanity forward, and reluctant perhaps to see their resources taxed to support innovation, when in the past "innovation" meant simply a new way to rip them off. But they are vitally important to our enterprise, and worth the effort to win over, because they above all have conserved humanity's Archaic connection with its ontological ground, the primordial matrix, mother of all things. They are our principal strategic reserve.

The third constituency is composed of those elements of capital which are currently engaged in the work of drawing humanity together into a single, integrated global civilization, partly through the mechanism of the world market, but partly also through the creation of international organizations, governmental and nongovernmental. These "one-worlders," as their right-wing critics call them, are not only an important force for world peace and for global ecosystem integrity. They are also capable and creative organizers, who are mobilizing immense energies, often in the service of innovative and progressive projects. They will soon discover that the world market, which made their existence possible in the first place, is a frustrating obstacle to their organizing activities. This constituency will be the most difficult to win. Entrepreneurs and political leaders operating at the very summits of world power are not likely to believe they need us, or if they do, they are likely to see us as an interesting addition to their collection of "think tanks." But even this sort of tactical arrangement can be profitable.

The road ahead is difficult indeed. We face long years of continuing to develop our vision. We must build relationships in all three of the different constituencies I have identified, train the people we recruit, position them or help them to understand the strategic importance of the positions they already hold, and mentor them as they work out the implications of the synergistic vision in their own field of endeavor. But

I am hopeful, as hopeful now as I was when I was a child, on that warm December night twenty four years ago, looking out from the steps of the Great Observatory at Chichen Itza, the villages and corn fields of the Yucatan below me, the stars high above, my heart racing, lured by God's incredible beauty to chance my life on the conviction that what has been is only the beginning and that we humans, individually and as a species, are capable of so much more. I do not regret that decision, nor do I regret the often difficult and crooked path which I have taken. I know that I, that we, our species, our cosmos, are, all appearances to the contrary, still moving forward.

Chicago
12 December 1994

CHAPTER ONE:
Introduction

I. Statement of the Problem

We humans have a relentless drive towards wholeness and completion——a drive which is apparent in everything we do. We join together in intimate union——and produce a new whole, the child. Our labor produces technological artifacts, which are nothing if not complex totalities. We form organizations which allow us to accomplish more together than a single individual ever could alone. Our minds struggle endlessly to grasp the cosmos as an organized totality. Religion *is* this drive towards holism, become conscious of itself as an end in its own right. The Latin *religere* means to re-connect, and *salvare*, to save, originally meant to make whole. It is the principal thesis of this work that humanity's drive towards holism is not simply an isolated phenomenon in a universe otherwise characterized by chaos and fragmentation. It is, rather, an advanced and complex manifestation of a dynamic which we find at work throughout nature, from the simplest structures of the metric field, space and time, through the increasingly complex and more highly organized forms of physical, chemical and biological matter. Through our constant struggle to create ever more complex forms of organization, we participate in the creative activity of the cosmos itself, and contribute to the cosmohistorical evolutionary process. This process, furthermore, does not end with humanity, but rather transcends the social form of matter, continuing into infinity, as the universe continues to grow and develop towards the Good which draws all things to itself.

From this standpoint, humanity appears to be poised on the threshold of a new era. Recent scientific developments have made it possible for human beings to comprehend, and to realize, their vocation in the cosmos in a qualitatively more complex fashion. Unified field theories, complex systems theory, postdarwinian evolutionary biology, dialectical sociology, and anthropic cosmology, have laid the groundwork for a new understanding of the universe. For the first time in its history, humanity is in a position to grasp being as relationship, structure, and organization, to comprehend the cosmos as an organized totality developing towards higher and higher levels of complexity and integration——and thus to recognize human civilization as an active participant in the cosmohistorical evolutionary process——a center for the creation of dynamic, organized complexity. The new science has, furthermore, opened up the road to the development of new energy sources, new ways of approaching the production process, new ways of centralizing and allocating resources, building and exercising power, and organizing our experience of the world. We have reached, or are very close to reaching, the level of development necessary for deployment of fully renewable energy sources and production processes which use almost exclusively renewable natural resources. We are capable of producing goods and services sufficient to lift the vast majority of the population out of poverty, of automating many of the less creative forms of labor, and of engaging the creativity of each and every person in the service of the human civilizational project. New means of transportation and new media of social communication have drawn peoples and nations together into a tightly knit global civilization. New forms of institutional organizing make it possible to tap into the talents and interests of ever larger numbers of people, building organizations more powerful than any the planet has known. Diverse cultural trends across a broad ideological spectrum, from those rooted in traditional religious communities (Bellah 1985, 1991) to elements of the New Age movement (McKenna 1994) are struggling to develop a spirituality centered on recognition of the unity and interdependence of the cosmos and on an ethic of social responsibility and service to the common good. We humans have taken our first, tentative, steps beyond the earth. We are ready to begin our journey to the stars.

And yet these possibilities remain largely unrealized. Indeed, far from realizing the potential latent in the new science, we seem to be caught in a downward spiral of chaos and disintegration. The integrity of the ecosystem and the social fabric continue to erode. Widespread recognition of the ecological crisis notwithstanding, we continue to treat the planet's biosphere as if it were nothing more than raw material from which to fashion instruments for our own self-gratification. Right and left vie for control of the "social agenda," while families disintegrate and millions of children are deprived of the nourishment and nurture they need in order to become creative, powerful, wise, and loving participants in the human civilizational project. It becomes more difficult with each passing year to centralize the resources necessary for investment in infrastructure, education, research, and development. Producers of fossil fuels and of the outmoded internal combustion engines which consume them continue to receive generous state subsidies, while only pennies are allocated for the development of solar, wind, biomass, or even fusion technologies or for the development of urban mass transit, high speed rail, and other ecologically sound and socially progressive transportation systems. Our schools are perpetually underfunded. Funding for basic research continues to decline, as projects which have no military or commercial value are scuttled and research priorities rewritten to meet market imperatives. And perhaps most troubling, very few seem to care. Political participation continues to decline. People are increasingly isolated from each other and increasingly disconnected from the institutions through which their social capacities are organized and developed. The people are more deeply mired than ever in a senseless, nihilistic consumerism.

Clearly something is seriously wrong with the structure of our society. Indeed, sometime during the past quarter century, the development of our society seems to have ground suddenly to a halt. Fashionable neoliberal doctrine tells us that free markets provide an optimum environment for scientific and technological innovation, economic growth, and development (Hayek 1988). But the evidence just isn't there. There is too much unrealized potential, and too much luxury consumption.

For the past 150 years humanity has looked to socialism for an analysis of the contradictions of industrial capitalist society and a strategy for the construction of a mode of social organization which would make possible the full development of human social capacities. But the past several years have witnessed an all-sided crisis of socialism, which has left the working class everywhere in retreat. The socialist systems of the Soviet bloc, despite their extraordinary ability to centralize resources for investment in technological and economic development, education and research, could not contain the rising consumer expectations generated by the persistence of commodity relations——nor could they organize the diverse talents and interests necessary to the next steps in the human civilizational project. The Communist led national liberation movements, which only a decade ago seemed to be on the verge of a global victory, are rapidly losing influence to reactionary tendencies such as fundamentalist Islam. The Communist Parties of the advanced industrialized countries are declining even in their historic stronghold in southern Europe, while social democracy has just barely held its own against victorious neoliberalism and the resurgent right. Even where leftist parties have come to power, as in some of the countries of the former Soviet bloc, it is on a platform written to the specifications of the neoliberal orthodoxy. And the crisis is not confined to the socialist tradition narrowly understood. Christian Democrats, as well as Arab, Asian and African religious socialists and secular nationalists have also suffered defeats, or have retreated into a moderate neoliberalism.

As a result, the only forces left to contest the order of the marketplace do so on the basis of a plethora of competing gender, ethnic, and religious particularisms. On the right, religious fundamentalists attempt to reestablish a social order which has been undermined by the penetration of market relations by an appeal to the divine law as laid out in one or another religious text: the Hebrew or Christian scriptures, the Koran, various Hindu scriptures, etc. On the left a broad array of constituencies——"greens," women, the oppressed peoples of the Third World——assert their claims on the basis not of a universal moral law, but rather of the unique specificity of their own identity and/experience, which others, including potential allies, can never share. Some of these particularist ideologies do, in fact, conserve authentic insights regarding the underlying order of the cosmos and powerful aspirations for the development of human social capacities. This is certainly true of the ecological movement, which calls attention to the interconnectedness of the natural world (Sheldrake 1989), certain forms of radical feminism, which have generated qualitatively new insights regarding teleology and self-organization (Daly 1984), and those ethnic particularisms which conserve strong communitarian religious traditions. But in the absence of a rationalizing theory which can unlock the univeralistic kernel contained within the particularistic hull, it is all but impossible to construct a united, global opposition to the disintegration unleashed by capital, or to promote a rigorous understanding of the real social conditions under which the aspiration for universal human development might actually be realized. And much of the new opposition is frankly chauvinistic or even fascistic in character, or else rejects on principle the notion of social progress. This is true not only of the religious right, but of those elements in the ecological movement which reject human technology out of hand, postmodernist "feminisms" which reject the whole philosophical tradition, and ethnic identity movements which confuse respect for cultural diversity with resistance to global integration.

There is, to be sure, a large and vitally important center, which more or less rejects both postmodern nihilism and fundamentalist extremism. This centrist bloc reflects a broad range of interests, including high technology capital, the religious institutions, state and "non-profit" bureaucracies, and various sectors of the intelligentsia, working class, and peasantry. It includes people from diverse political tendencies. In the advanced capitalist countries it is composed primarily of moderate liberals such as Clinton and Gore, "communitarians" and Christian Democrats, radical democrats and social democrats, together with (in Europe at least) "reform" communists who continue to uphold the vision of *perestroika*. In the former Soviet bloc, this center includes a broad spectrum of forces which grew out of the disintegration of the movement towards *perestroika*, and which now embrace a wide range of positions from moderate nationalist liberalism and Orthodox Social Christianity, through social democracy, to a moderate socialist restorationism. In the Third World, the center includes a broad array of religious and secular socialists, as well as emerging *indigenista* currents which seek to recover political-theological projects buried by the European conquests.

The various elements within the center, however, whether taken individually or as a bloc, face serious obstacles. First and foremost, their political perspectives are defined largely by a recognition of the limitations of both market and bureaucratic structures and of the dangers of both unbridled nihilism and national or religious fundamentalism. They lack a profound and unified understanding of the nature and significance of the human civilizational project. They lack a really coherent analysis of the current crisis of and the next steps in that project, and they lack a clearly defined strategic perspective. The forces of the center are carrying out an important holding action against the process of social disintegration which is engulfing the planet. And they include many creative individuals and organizations which need to be part of the dialogue about our planet's future. But they do not have solutions. And they will not lead.

Clearly it is necessary to rethink fundamentally humanity's place in the cosmos, and to re-envision the next steps in the human civilizational project. This is a formidable task, especially at the present conjuncture. For the political crisis of the progressive forces has been accompanied by a theoretical crisis of unprecedented proportions. The apparent failure of dialectical and historical materialism to deliver on its claim to understand the "line of march, conditions, and ultimate general result" (Marx 1848/1978) of the historical process, to know "the solution to the riddle of history," (Marx 1844/1978), has called into question all theories which attempt to grasp the underlying structure of the cosmos, and to use this knowledge to organize and direct social activity (Laclau and Mouffe 1985, Unger 1987). The past quarter century has been a period of growing skepticism regarding such totalizing rationalist "metanarratives" (Lyotard 1984). And, in theory as in practice, the crisis has not been confined to the socialist tradition. Rationalistic, holistic philosophies outside the dialectical materialist tradition, such as Neothomism, process philosophy, general systems theory, synergetics, and allied doctrines have also lost ground. And yet it is precisely this failure to comprehend the cosmos as an organized, meaningful totality, and to grasp the critical significance of humanity's role in the universe, which makes it so difficult to break the impasse and unleash the potential latent in recent scientific and technological developments. More is at issue, therefore, than simply correcting certain errors in social theory or developing alternatives to failed political strategies or public policies. Rather, we need to reestablish the logical-ontological, cosmological and axiological foundations of the human civilizational project. Only on such a foundation can we attempt to chart a course for the future of humanity.

The present work forms an integral and very specific part of this larger project. It attempts, given certain ontological, cosmological, and axiological assumptions, which are explained and justified briefly in the first chapter, to

1) demonstrate that human civilization constitutes an integral part of a larger process of cosmic evolution from less to more highly organized forms, a kind of laboratory for the development of dynamic, organized complexity,

2) characterize just how this process of development has, and continues to take place, by presenting a synoptic view of the history of human civilization, and

3) outline, at least in broad terms, a vision and a strategy for the future.

The book is, in other words, a philosophy of history, grounded in an analysis of the logical-ontological implications of recent developments in the natural sciences, informed and richly illustrated by comparative historical and sociological research, and issuing in a preliminary statement of political-theological strategy.

II. Method

An inquiry of this sort poses difficult methodological problems. On the one hand, resolution of the questions we have posed presupposes extensive, in-depth, empirical study. It is only by observing the universe generally, and human society in particular, that we can determine whether or not they constitute a meaningful, self-organizing totality, and ascertain just how they develop and towards what end. At the same time, the organization, development, and ultimate end of the cosmos are not themselves directly accessible to empirical observation. Resolution of the questions we have posed *presupposes* the results of the empirically based special sciences, which serve, as it were, as the raw material for philosophical thought. But it also involves a further process of philosophical synthesis.

This study thus draws extensively on a wide range of works in the natural and social sciences. My starting point in exploring recent developments in the natural sciences developments was the excellent series of studies by Paul Davies (1988, 1991). More rigorous treatments still accessible to nonspecialists include Barrow and Tipler (1986), Bennet (1988), Pines (1988), Campbell (1989), Zurek (1989), Waddington (1957), Lenat (1980), Denton (1985), Sheldrake (1981, 1989), Margulis (1991), and Wesson (1991). These studies provide the natural-scientific groundwork for the philosophical argument developed in the first chapter.

My understanding of the nature and development of human society has been influenced most powerfully by the historical materialist tradition and by certain currents in feminist theory. My developing understanding of the socialization process owes much to the work of Erich Fromm (1943), psychoanalytic feminists such as Chodorow (1978) and Rubin (1975), the antipedagogical work of Alice Miller (1990), and the as yet unpublished work of my wife, Maggie Mansueto. Samir Amin's work on precapitalist social formations (1980, 1989), and Gerhard Lenski's neoevolutionary theory (1982) have been very important to the development of my synoptic view of human history and my account of the tension between communitarian and militaristic tendencies in precapitalist societies. Debates between historical materialists and contemporary neoliberals (Hayek 1973, 1988), as well as debates within the historical materialist tradition between Trotskyists (Mandel 1968, 1978) and dependency theorists (Emmanuel 1972, Amin 1978, 1980, 1982) have had a profound impact on my understanding of the impact of market systems and the socialist state on social development. Studies by Bellah (1985, 1991), Fromm (1941), Lasch (1977, 1879, 1981, 1990), and Sennet (1976) have influenced my understanding of market and bureaucratic structures on the human capacity for relationship building and on the socialization process. Emile Durkheim's *Formes elementaire* (1911) remains my point of departure for the analysis of religious phenomena. My understanding of the dynamics of social movements owes a great deal to the work of Antonio Gramsci and to studies by Hobsbawm (1958), Wolf (1969), and Lancaster (1988).

I have drawn on case studies examining societies in most, but not all, of the principal cultural areas on the planet, and from all of the principal historical epochs. The reader can consult the bibliography for a complete list of the sources which I have used. Among the studies most influential in shaping my analysis I would cite the studies of Aztec civilization by Brundage (1985) and van Zantwijk (1985), Norman Gottwald's study of ancient Israel (1979), Merlin Stone's study of ancient Mediterranean civilization (1976), de Ste. Croix's study of ancient Greece (1981), studies of early Christianity by Horsely (1985), Theissen (1982), and Dymitris Kyrtatis (1987), studies of feudalism by Perry Andersen (1974), of the development of capitalism in North America by Moore (1966), Lockrdige (1970), Boyer and Nissanbaum (1974), Heimart (1966), Crowley (1974) Hatch (1977), Geissler (1981), Greven (1977) and Howe (1979), and Davis (1986), as well as numerous studies of the socialist experiment in North America (Burbank 1976, Glazer 1961, Green 1978, Mansueto 1985), Latin America (Hodges 1986, Lancaster 1988), Europe (Sewell 1980, del Carria 1966, Hobsbawm 1959, Renda 1977, Spriano 1967), Russia (Venturi 1966, Radkey 1958, 1962, Lewin 1968, and Bettelheim 1976), the Middle East (Amin 1978b, Avineri 1981, Cohen 1987), East Asia (Sarkisyanz 1965, Wolf 1969, Hinton 1990), and Africa (Kelly 1991).

Viewing these studies synoptically it becomes possible to identify certain common patterns. It gradually becomes clear that there is an order to the cosmos generally and to human history in particular. One begins to get a sense of how organization emerges across the system as a whole and of the different ways in which the underlying organizing dynamic is expressed at different levels on the dialectical scale (physical, chemical, biological, social) and during different epochs of human history. The result is a philosophical synthesis of the results of the special sciences. In this sense, my procedure is rather like that outlined by Plato in his *Republic* (531c-d), where the gradual training of the mind in the disciplines of mathematics (arithmetic, geometry, astronomy, and music) prepares it for an intuitive grasp of the whole, with the one difference that I have found organization not only in the formal objects of the mathematical sciences, but also in the more complex, concrete, and more highly organized objects of physics, chemistry, biology, and particuarly sociology. A long immersion in the special sciences gradually awakens one's ability to intuit the organizing principle which lies behind particular forms of organization, making it possible to comprehend the universe as an organized totality and eventually to grasp the "community and interrelationship of all things," the principle of value which is the highest expression of the dialectic.

In carrying out this work of philosophical synthesis, I have drawn on the whole philosophical tradition, and most especially on the work of Plato, Aristotle, Aquinas, Joachim of Fiore, Spinoza, Hegel, and Marx. I owe to them both the general way in which I have framed the problem, and the way in which I have identified the various possible solutions. Several contemporary philosophers and theologians have also been important to my work. In attempting to sort out the philosophical implications of the new science, debates among Neothomists (Wetter 1958), neoliberals (Barrow and Tipler 1986), general systems theorists (von Bertalanffy 1968), process philosophers (Whitehead 1929), synergeticists (Fuller 1981, 1991) and dialecticians idealist (Harris 1991, 1992) and materialist (Zeman 1988), have been especially important. The work of feminist philosopher Mary Daly (1984), read late in the preparation of this book contributed important insights regarding teleology and self-organization. The debate around liberation theology, especially the debate between Segundo and Ratzinger (in Segundo 1985) and the work of Clodovis Boff (1986) and Hans urs von Balthasar (1968) have been particularly helpful in framing the question of the degree and nature of human participation in the self-organizing activity of the cosmos.

Ultimately, the test of any theory is in practice. The more complex and powerful our interpretation of the world, the better able we will be to contribute to the human civilizational project, and to the larger cosmohistorical evolutionary process of which it is a part. While my principal purpose in this work has been to show that the cosmos is an organized totality in which human civilization plays an important contributing role, and to characterize as clearly as possible the way in which development, especially social development, takes place, I have offered in the concluding chapter some broad strategic directions which flow from the sociological analysis and philosophical synthesis developed in the work as a whole.

My understanding of political-theological strategy derives from two principal sources. Debates between religious socialists and Christian democrats who emphasize the importance of spirituality, and materialist socialists who focus attention on the objective conditions for social transformation, debates within the socialist movement between social democrats and communists and representatives of what I have called the "third force," including populists, "theomachists" such as Bogdanov and Lunacharsky, (see Rowley 1987), and socialist humanists such as Gramsci, and finally, debates within the third force among advocates of "insurrectionalist," "proletarian," and "prolonged struggle" tendencies (Hodges 1986) have all been influential. Second, I have learned a great deal from participation in organizing efforts in working class and oppressed nationality communities in cities throuought the United States, and in particular from efforts lead by the Industrial Areas Foundation and the Communist Workers Party (later the New Democratic Movement). I learned directly from the accumulated experience of these organizations——and from active struggle with them over the limitations of their vision and strategy.

It would be a serious omission were I to fail to mention the numerous works of fiction and poetry which have helped to shape my vision of humanity and its place in the cosmos. The novels of Dostoyevsky (1880/1950), Malraux, Silone (1948, 1950, 1955, 1968) and Lessing (1979, 1980, 1983) and the poems of Ernesto Cardenal (1980, 1989, 1992), Roque Dalton, Giocanda Belli and Ricardo Morales Aviles (in Aldaraca, Baker, Rodriguez and Zimmerman 1980) have been especially powerful in this regard.

III. Outline

This study has been organized in such a way as to present systematically the development of matter generally, and of the social form of matter in particular, towards increasingly complex levels of organization, and to demonstrate the leading role of the social form of matter, and of human social movements, in that process. Chapter One, "Dialectic, Cosmos, and Society," traces briefly the development of holistic philosophy and its crisis in the wake of the scientific revolution of the seventeenth century. It analyzes the social basis, political valence, and internal contradictions of the nihilistic "postmodern" tendency which is currently dominant in social theory. The chapter then lays out the theoretical foundations for the book as a whole, analyzing the philosophical implications of such recent scientific developments as unified field theory, complex systems theory, postdarwinian evolutionary biology, and anthropic cosmology.

Chapter Two, "The Social Form of Matter," outlines a general theory of human society. It treats human activity as an organizing process integrating various dimensions (socialization, production, power, communication, artistic, scientific, and philosophical creativity, and spirituality) regulated by a definite social structure (kinship, relations of production, authority structure, ideology). At various points in time the existing structure of society becomes an obstacle to the further development of human organizing capacity. Movements form to reorganize society and once again unleash the development of human social capacities. The character of these movements, however, is constrained by the overall level of development of human organizing capacities.

Chapter Three, "The Mandate of Heaven," examines the early stages in the development of human society: the development of horticulture and agriculture and the organizing role of the village community. The village community provided the context in which emerging human society understood its participation in the self-organizing activity of the cosmos.

It is not surprising that in several languages the words for "village community " (the Russian mir) or "right order for the community" (the Hellenic *kosmos*), also mean peace and "universe" in the sense of the ordered whole of everything existing. The chapter then traces the emergence of warlord states which tapped into the surplus produced by the village communities to finance luxury consumption, warfare, and sacrificial cults which held back the further development of human social capacities. The great salvation religions emerged out of resistance to these states and represent an attempt to "restore the mandate of heaven"——to re-establish the cosmic harmony which had been shattered by exploitation and oppression. These movements grasped the cosmos as an organized totality, but, due to the low level of development of human organizing capacities, and the distorting impact of patriarchal and tributary structures, did not yet grasp humanity as a real participant in the self-organizing activity of the cosmos. The chapter examines movements in Europe, Asia, and the Americas.

Chapter Four, "The Mandate of the People," explores the significance of the industrial, democratic and scientific revolutions. For the first time humanity became an active participant in reorganizing not only the natural world (through industrial technology) but also the social world (through the democratic revolutions). Participation in the self-organizing activity of the cosmos no longer meant simply conforming to an order fixed by the mandate of heaven. It meant, rather, actually helping to design that order. Mediation of this organizing activity by the marketplace, however, transforms all activity into simply a means of realizing individual consumer interests, leading to loss of ecosystem integrity and the erosion of the social fabric. Market systems obstruct the centralization of resources necessary for investment in infrastructure, education, research, and development. They break down the social networks necessary to carry out complex political organizing tasks. And they gradually erode our ability to understand, and take an active interest in promoting, the self-organizing activity of the cosmos. Thus the current crisis.

Chapter Five, "The Mandate of History," explores the socialist experiment in both its religious or idealistic and materialist manifestations. The dialectical tradition grasped the cosmos as a relational, self-organizing, and teleological system, and recognized human history as a real participation in the cosmic evolutionary process. Dialectical philosophy, however, did not find adequate support in the natural science of the nineteenth century, dominated as it was by an atomistic paradigm. The result was a disintegration of the dialectical trend into idealist and materialist trends. Idealists looked for the source of order in an extramaterial organizing principle, and generally ended up opting for one or another form of religious restorationism. While conserving important values, objective or religious idealism did not realized the potential inherent in the original dialectical insight regarding the self organizing character of the cosmos, and human participation therein. As the marketplace has undermined humanity's faith in the underlying rational unity of the universe, objective idealism has disintegrated and given way to various forms of religious fundamentalism.

Materialists, on the other hand, tended increasingly to regard organization and meaning as something imposed on an otherwise chaotic universe by humanity, rather than something implicit in matter from the very beginning. This is the underlying reason for the atheism of the dialectical materialist trend. When they lost their grasp of the cosmos as an organized system, the socialist movements also lost their grasp of the organizing process in general. Practically this meant that they were unable to formulate an adequate strategy for tapping into the creative potential of the working classes, and for catalyzing their development to a level sufficient to support fully communist social relations. Indeed, the empirical working class seemed largely stalled at the level of a more or less sophisticated interest group or trade union politics. Some socialists (the social democratic trend) settled for organizing workers based on their existing understanding of their self interest.

The Communist movement, on the other hand, recognized, at least implicitly, the limitations of both of this approach. Lenin developed an effective strategy for state power, based on the ability of a Communist vanguard to respond to the largely presocialist demands of the worker and peasant masses, but was unable to meet the challenge of actually building a communist society. Socialist systems were generally more effective than the marketplace in centralizing the resources necessary for development, especially during the early stages of industrialization. For the most part, however, they have proven unable to tap into the complex networks, interests, talents, and aspirations characteristic of advanced industrial societies. And socialist systems for the most part left intact the market in labor power. People in socialist countries still work for a wage, in order to purchase consumer goods. This means that the consumerism characteristic of capitalist societies has affected the socialist countries as well, gradually eroding commitment to the construction of socialism and the development of human social capacities. As the marketplace has undermined humanity's faith in the underlying rational unity of the universe, dialectical materialism has disintegrated and given way to various forms of postmodern relativism.

There is a cluster of trends which looked beyond the limitations of both social democracy and Leninism, and which grasped socialism as the "negation of the negation," the recovery, at a higher level of differentiation and integration, of humanity's archaic holism. Populists, for example, tapped into peasant millenarianism and other popular collectivist traditions, only to find that these very traditions were themselves being rapidly undermined by the penetration of market relations. Theomachists like Bogdanov and Lunacharsky sought to raise the working class to communism through in-depth organizing and education. Socialist humanists joined elements of both populism and theomachy in a complex strategy for cultural hegemony. While these three tendencies, which together make up what I have called the "third force," made important contributions, not the least of which was to open up a dialogue with religious socialists which points beyond both idealism and materialism, towards a new synthesis, they were not by themselves able to escape the disintegrating effects of the marketplace, or the limitations of their own materialist standpoint, which left their socialist ethics without an adequate ontological ground, and thus vulnerable to the postmodernist assault.

Chapter Six, "The Next Steps in the Human Civilizational Project," argues that we now possess the scientific tools necessary to overcome the errors which led to the deformation of the socialist project. The results, however, involve much more than a mere rectification of the strategic errors of the socialist movements. Specifically, unified field theories, complex systems theory, postdarwinian evolutionary biology, and anthropic cosmology, in so far as they imply a dialectical, relational, self-organizing, and teleological cosmos, point beyond the atheism of the socialist movement to a new understanding of human history as a real participation in the life of God. The emerging "synergistic" theory provides the basis for the development of new technologies which tap into the self-organizing character of matter itself, of new methods of organizing and developing human social capacities, and a new spirituality centered on participation in the self-organizing activity of the cosmos. The full potential of these new developments cannot, however, be realized in the context of a market society which systematically undermines the integrity of the ecosystem and the social fabric and obstructs the centralization of resources necessary for investment in infrastructure, organizing, education, research, and development. The chapter analyses briefly the current situation, and outlines a strategy for the gradual, patient, and systematic reorganization of institutions in the light of the new synergistic theory.

The work concludes with a consideration of the "supersocial" levels of organization towards which humanity is evolving. The biological infrastructure on which the social form of matter depends cannot survive into the far future of the universe, and, in any case, imposes real limits on the development of our organizing capacities. If some of the bolder formulations of the anthropic cosmological principle are correct, then the very existence of the cosmos *requires* the continued development of intelligence until it embraces the universe as a whole, tapping into the infinite potential for the good, for the true, and for the beautiful, which it carried within itself from the very beginning, but which is fully realized only in the complex, synergistic unity we call God.

CHAPTER TWO:
Dialectic, Cosmos, and Society

I. Archaic Holism

Throughout most of its history, humanity has understood itself as part of an organized cosmic totality. It was above all the emergence of horticultural forces of production, which involved a complex symbiosis between humanity and the other species of the biosphere, and the birth of the village community, with its rich fabric of social relationships, which provided the matrix out of which emerged such concepts as order and organization, the relationship of whole and part, force and law, etc. Indeed, the Hellenic word *kosmos* means "right order for the community," in the sense of the traditional order of the village, and the Slavic *mir* means "village community." Both mean "universe," in the sense of the ordered totality of being (Bogdanov 1928/1980, Mandel 1968: 30-36, Wolf 1969: 58-63, Hayek 1973: 37)." These communitarian, predominantly matriarchal societies recognized the universe itself as one vast interconnected system, regulated by a "perfect pattern of creation" (Waters 1968) which was less something imposed on the world by a transcendent creator god than something implicit in each and every thing, and above all in the harmonious relationships of all things with each other. This view of the world found its most typical (and most profound) symbolic expression in the cult of the *Magna Mater* who is at once, in the form of Demeter or Tonantzi, the profoundly material goddess of the earth and of its fruits and, as Isis, Sophia, or Sussistinako, the goddess of wisdom, the latent pattern from which all complex organization emerges.[1]

Gradually, with the emergence of warfare, the archaic matriarchy gave way to a new, patriarchal regime (Engels 1891/1972), and the village community came under the sway of the warlord states and emerging petty

---

[1]. We should keep in mind that the word matter derives from the Latin *mater*, or mother. Originally matter referred simply to the **potential** for being, and thus for complex organization. It was only later, as patriarchy and the warlord state gained hold that this potentiality was transformed into simply a passive capacity to receive form from the outside --from the Father God, or his philosophical reflex: the Idea. In this sense, the communitarian worldview was profoundly materialist, not in the modern sense of denying spirituality, but in the archaic sense of locating the capacity for spirituality within, rather than outside, the self-organizing universe.

15

commodity production (Amin 1980).   Human society began to seem
increasingly out of harmony with the underlying structures of the universe.
    Humanity did not, to be sure, completely lose sight of the archaic
cosmic harmony.   Increasingly, however, this harmony seemed like
something ideal, from which the material world had fallen away, and which
could be recovered only with great effort.   This is the period of the birth
of the great salvation[2] religions, the time when prophets called the people
back to fidelity to the law (*Tao* or *Dharma, Torah* or *Dike*)[3] and
philosophers began to seek an objective, rational basis for judgements of
value (Mansueto 1992).   This quest to recover the archaic holism took
many forms.   Taoism, the Hopi religion, and resurgent cults of the *Magna
Mater* argued tenaciously that the underlying order of the universe was
immanent in matter itself——if only we would stop to look——and that
human society had to be brought back into harmony with the "way," with
the "perfect pattern of creation." Emerging Hinduism and Buddhism, on
the other hand, argued that the manifold diversity of the material world was
simply an illusion, and that unity and wholeness could be discovered only
in withdrawal from the world, in *atman* and ultimately in *brahman* or, in
the case of Buddhism, in selfless *nirvana*.
    The philosophical and religious traditions which emerged in the
Mediterranean basin took a middle path between these two extremes.
Hellenic philosophy sought wholeness in rational knowledge of the cosmic
order.   Socrates demonstrated that by drawing out the implications and
internal contradictions of human opinion it was possible to catalyze the
development of authentic knowledge which grasps the underlying unity and
order of the cosmos, and which disposes the soul towards the Good.   Plato
built on this insight, teaching that the study of mathematics (arithmetic,
geometry, astronomy, and music) would bring the soul back into harmony
with the cosmos, and prepare it to grasp the dialectical "community and
inter-relationship of all things" (Plato, *Republic* 531d) which was the
foundation of the transcendental values.   Aristotle captured even more
powerfully the self-organizing dynamism of the material world.   His grasp

---

    [2]. The word "salvation" derives from *salvare*, one meaning of which is to make
whole.

    [3]. All of these terms articulate humanity's immediate intuition of the underlying order
and harmony of the cosmos.   *Tao* means the "way" and was used by the Chinese to refer
to the ineluctable, and rationally inexpressible order of nature.   *Dharma* is a Sanskrit
word, important in both Hindu and Buddhist thought, which means law.   *Torah*, of
course, is the divine law decreed by *YHWH*.   *Dike* is a Hellenic word for justice.

of the immanent teleology of matter and form, potency and act, dynamia and *energeia*, product of his work in the new sciences of biology and psychology, represented humanity's first recognition of the lawful and progressive character of change——the first authentic theory of development.

Judaism, on the other hand, sought wholeness in the struggle to bring human society into harmony with the divine cosmic law. Born out of the struggle of an oppressed people of the eastern Mediterranean against the hegemony of the warlord state, Judaism taught that wholeness was to be found first and foremost in ethical conduct, and in construction of a just social order. Judaism endowed political struggle with a radically new theological significance and offered new hope that the world really could be made whole.

As these two traditions flowed together in the philosophical and theological systems of Philo, Plotinus, Alfarabi, Avicenna, Avveroes, Maimonides, Aquinas, and Eckhart, a new synthesis emerged which grasped the cosmos as an organized totality within which human civilization played a real and significant, albeit limited role. The structure of the cosmos was knowable, at least in part, by human reason, and knowledge of this structure, of the "community and interrelationship of all things," of their *arche* and *telos*, in turn, formed the basis for political action directed at the construction of a just social order——what Philo called the great cosmopolis. Full participation in the self-organizing activity of the cosmos, however, was impossible for finite humanity, and human beings found their greatest joy not in action but in contemplation, and sought fulfillment not in this life, but in the next.

These new religious and philosophical conceptions constituted a powerful force for the reorganization of human society and helped to reorient human civilization towards the divine cosmic law. It was humanity's rekindled love of that law which once again awakened the spirit of scientific inquiry and unleashed the development of human social capacities, issuing, ultimately, in the industrial, democratic, and scientific revolutions.

At the same time, real limitations remained. The most important of these is what has come to be known as idealism: the notion that order or organization come to matter from the outside.[4] Partly this is a result of

---

[4]. It is interesting to note that those philosophers and theologians who argued that the principle of organization is immanent in matter itself, such as Avveroes and the Latin Avveroists and later Spinoza, as well as those who interpreted idealism in a panentheistic direction (such as Eckhart), were all regarded with suspicion, and in some cases even

the low level of development of human organizing capacities. In agrarian societies human beings did not seem to be significant participants in the self-organizing activity of the cosmos. But this dynamic was reinforced by two aspects of the social structure. Patriarchy makes a sharp distinction between the active "male" principle (idea or spirit), which is supposed to contribute form, and the passive, female principle (matter) which is merely the capacity to receive form. The subjugation of women is the subjugation of matter generally, by a society which believes order must be imposed from the outside.[5] Second, in the tributary structures of most agrarian societies, only the priests and warlords "organized" in any meaningful sense, only they participated in the universal. The vast majority, the peasants, were simply the objects of their organizing activity, raw material in their hands, and worked in order to make their organizing activity possible. The organizing principle was thus regarded as radically distinct from matter and thus incapable of realization in the material world. To be good meant to be in harmony with a pre-existing order, not to be oneself an active creator of order. The mode of creating order which had become typical in the warlord state, furthermore, left its traces in the salvation religions which emerged to resist this state. Warfare or its sublimated form, self-sacrifice, rather than productive labor, seem to be at the core of the cosmic evolutionary process. Ultimately this was true even——perhaps especially——of the dynamic philosophical and theological systems which emerged from the Hellenistic-Jewish synthesis. Catholic Christianity and Medieval Islam, while conserving a powerful sense of the underlying unity of the cosmos, were both also characterized by very real necrophilic deformations: Islam by a cult of warfare and Christianity by a cult of self-sacrifice.

II. Protestantism, Liberalism, and Nihilism

Humanity's archaic sense of the underlying unity of the cosmos was undermined by a complex series of developments. The industrial,

---

declared heretical, by the religious hierarchies.

[5]. Mary Daly (1984) does an excellent job of showing how patriarchal philosophy transformed the concept of *dynameia* or *potentia* (potential) from a latent capacity for organizing activity into a purely passive "capacity" to receive form from the outside --and how the concept of form was, in the process distorted, so that it no longer implied an ability to be creative, powerful, wise, and loving, and instead meant submission to a dead, ritualized orderliness.

democratic, and scientific revolutions vastly increased the level of human organizing capacities, creating for the first time a sense that human society, and the cosmos with it, were actually developing, and making human beings real participants in the creation of order and organization. Science unlocked the processes which governed physical, chemical, biological, and social interactions. Industry used this knowledge to vastly increase human productive capacities, and the democratic revolutions demonstrated once and for all that far from being merely an imperfect reflection of an eternal divine order, human society is, in fact, a social product, as fully susceptible to rational reorganization as a piece of clay.

At the same time, the emergence of a market economy gradually destroyed the village community and undermined the social basis for humanity's grasp of the universe as an organized totality. As society itself became increasingly atomized, matter came to be regarded as a system of interacting particles, externally related to each other, the relationships between which were governed by the laws of mechanics. Gradually this paradigm was modified to make room for the more complex chemical interactions between atoms and molecules of the various elements, and the still more complex ecological interactions between organisms competing for scarce resources within a single ecological niche. Even so, the basic paradigm remained the same, and there was a tendency to regard these more complex interactions as ultimately reducible to the laws of mechanics.

These developments had complex and ambiguous philosophical implications. On the one hand, the new science liquidated the matter/form dualism which had crippled the philosophical tradition. What had historically been understood by form was simply the pattern of relationships between particles, or what is the same thing, the organization of matter. The most advanced philosophers of the seventeenth and eighteenth century recognized that there is only one Substance (Spinoza) or general system (Edwards), in which all particulars participate to the extent of their organization. This was an important advance which in many ways prefigured the achievements of the dialectical tradition. But the emerging mechanistic paradigm made it impossible to theorize organization as such, and especially the fully teleological level of organization which characterizes complex systems. Teleological arguments for the existence of God based on the notion of cosmic purposefulness thus gave way to eutaxiological arguments based on a sense of cosmic order. Some (Paley) regarded God as a kind of divine "watchmaker" who had fixed the initial conditions of the universe, and set its various parts into motion. Others (the evangelical tradition, and especially Edwards) insisted that the universe is purposeful but that God's purposes radically transcend human reason, and thus remained inscrutable to natural humanity, becoming accessible, and then only in a very limited way, through faith. Others still (Spinoza) rejected the notion of purposefulness altogether as an anthropomorphism, a reflection of humanity's limited capacity to comprehend the beauty of the "order without a purpose" which characterized the one substance in which all things lived and moved and had their being (Spinoza).

Gradually, however, this rationalistic synthesis broke down. It seemed increasingly that order and meaning were simply something imposed by the human mind on its perceptions of world, which was itself chaotic, or at least opaque to human reason, like the orthogonal coordinates of the Cartesian system which gave measure and direction to otherwise formless space (Descartes 1637/1975, 1641/1975). Or else it is something which emerges spontaneously from the interaction of individuals——the "fundamental particles" of human society, as they interact in the marketplace or come together in a social contract to "make" laws——a conception which was essentially foreign to the philosophical tradition. Philosophers such as Kant (Kant 1781/1969) were forced to appeal to the synthetic capacities of practical reason to provide the coherence which the universe itself no longer seemed able to supply, or which had, in any case, become hidden from the human mind. English and Scottish philosophers such as Mandeville and Smith (Smith 1776/1976) began to speak of vice as a catalyst for the progress of civilization, and of the "invisible hand" which insured that even as individuals pursued their own self interest, the common good was served, and human civilization advanced. Historically this view found expression in both the British materialist and empiricist tradition, which includes thinkers such as Hobbes, Mandeville, Locke, Hume, and Smith, and, in slightly different form, in the subjective idealist tradition represented by Berkeley.

Where the older Aristotelian science explained things by reference to their place in a systematic whole, and thus required some concept, however inadequate, of the nature of the whole, the new science simply provided increasingly more elegant mathematical descriptions of the laws of motion of various particles. In this sense Laplace was quite correct when, asked whether or not he believed in God, he replied that he had "no need of that hypothesis." God, after all, is simply a symbolic term for the whole, and once science has abandoned the struggle to grasp the whole, it has abandoned the search for God as well.

The eutaxiological accommodation to the new science was, in any case, dealt a sharp blow by the discovery of the second law of thermodynamics, which stated that all closed systems of particles (such as the universe was then believed to be) tend to develop over time towards an increasingly random state (Davies 1988: 19). In 1854 the German physicist Hermann von Helmholtz proclaimed that the universe was, in fact dying.

The remorseless rise in entropy that accompanies any natural process could only lead in the end, said Helmholtz, to the cessation of all interesting activity throughout the universe, as the entire cosmos slides irreversibly into a state of thermodynamic equilibrium. Every day the universe depletes its stock of available, potent energy, dissipating it into useless, waste heat (Davies 1988: 19).

Increasingly human perceptions not only of purpose but also of order in the cosmos came to be regarded as the product of anthropocentric wishful thinking, and the very real development of order and complexity within our very localized region of space and time simply a magnificent, but ultimately meaningless coincidence. Thus Bertrand Russell:

... all the labours of the ages, all the devotion, all the inspiration, all the noonday brightness of human genius are destined to extinction in the vast death of the solar system, and the whole temple of Man's achievements must inevitably be buried beneath the debris of a universe in ruins ... Only within the scaffolding of these truths, only on the firm foundation of unyielding despair, can the soul's habitation henceforth be safely built (Russell 1957: 107).

Within this context, humanity came to be regarded as a realm of fragile and localized meaning in an otherwise meaningless universe. Having despaired of discovering any meaning in the natural order revealed by the sciences, positivistic philosophy devoted itself to the task of formalizing the intellectual operations which govern scientific thought, while a whole series of subjectivist, "existentialist" thinkers, from Kierkegaard and Nietzsche through Heidegger and Sartre searched for some way in which human beings could render life meaningful after the "death of God."

III. The Dialectical Tradition

There were, however, a cluster of philosophical trends which resisted the nihilistic conclusions so readily embraced by positivists and existentialists alike. The most important of these was the dialectical tradition, in both its idealist and materialist variants, but important contributions have also been made by objective idealism (neo-Platonism, neo-Aristotelianism, neo-Thomism, vitalism) and "centrist" tendencies which stand between the dialectical and positivist traditions——process philosophy, general systems theory, etc.

Beginning with Hegel, dialectical philosophy sought to fashion a theory of development which at once recognized the cosmos as an organized totality, and comprehended better than the Scholastic tradition the relational, self-organizing character of being. For Hegel, the whole cosmos was simply the necessary self-objectification of the Idea, which gradually became conscious of itself through the long, slow, and often contradictory progress of natural evolution and human history. From this standpoint, the human civilizational project was nothing other than "God's march through history," the process by which God became conscious of himself. On the one hand, Hegel took seriously——perhaps even more seriously than any thinker before him——the idea that there was a single, underlying, organizing principle which manifested itself, which could be grasped by human reason, and which could thus explain the whole complex manifold of nature and history. At the same time, he recognized that complex organization as we understand it emerges only through a prolonged process of real natural and social interactions——through natural evolution and historical progress.

From the very beginning, however, the dialectical tradition faced serious obstacles. Nineteenth century science did not, by and large, support the idea that nature is a relational, holistic, self-organizing, and teleological system. Physical, chemical, and biological matter were understood in atomistic terms, and evolution was understood as the product of competition and natural selection. The result was a rapid reassertion of dualistic trends. If matter is not fully self-organizing and teleological, then an organizing principle must be sought outside of matter, in God or in the Idea——or else in the one form of matter which demonstrates indisputable self-organizing and teleological tendencies——i.e. in human labor power. The initial dialectical synthesis developed by Hegel broke down into idealist and materialist variants.

A. Objective Idealism

Faced with the somber picture of nature presented by nineteenth century science, a growing number of philosophers reverted to an idealist posture, seeking an "organizing principle" outside of matter. There are a number of variants of this strategy. Philosophers rooted in the scholastic tradition were best equipped to resist the positivistic onslaught. Beginning in the nineteenth century, and continuing up into the present period a whole series of Catholic scientists, philosophers, and theologians, the most important of whom are probably Teilhard de Chardin, Maritain, and Rahner, attempted to reinterpret the Thomistic tradition in a way which conserved its historic sense of the underlying, ontological unity and dynamism of the cosmos, while stressing the role of humanity as an active participant in the life of God, thus creating room for an authentic theory of development. Teilhard de Chardin, for example, posited the existence of something he called "radial energy" to account for the drive towards holism and complexity which, in his view, ran counter to the entropic trend identified by the physical sciences. Maritain focused his efforts in the political arena, reinterpreting the historic Thomistic doctrine that the state must serve the common good in the light of the democratic revolutions, and arguing for the creation of an effective, democratic global political authority and for a communitarian "third way" between capitalism and socialism. Rahner extended traditional Thomistic doctrines regarding the human capacity to know and love God, arguing that faith is as much a transcendental capacity of the human subject as it is something infused from outside.

Neo-Thomism made important contributions, laying the theoretical groundwork for the progressive political initiatives of Social Catholicism, and for a policy of dialogue and alliance with the dialectical materialist tradition and the communist movement. At the same time, while the basic categories of Neo-Thomistic thought were in fundamental tension with atomistic science, Neo-Thomism never produced an alternative scientific system. Because of this its ontology and cosmology remained unconvincing. Teilhard, for example, was never able to give the concept of "radial energy" physical content, and was thus never able to win the battle against atomistic-entropic pessimism. And "transcendental Thomism," at least in its more radical forms, tended to cede the entire natural world to atomism, locating the effective ground of meaning in the human subject.

Philosophers outside the Thomistic tradition, on the other hand, accepted more or less uncritically the positivistic understanding of reason and its limits. Some of these concluded that knowledge of the whole was accessible only outside of reason, through aesthetic or religious intuition. This is the approach taken by the whole line of Protestant thinkers which leads from Jacobi and Schelling to Tillich. It is also the approach taken by Russian Orthodox philosopher Vladimir Soloviev and by intuitivists in the Latin countries such as Bergson and Vasconcellos. According to this view, reason is irreducibly analytic, capable of grasping the nature of physical, chemical, biological, and social processes——and of identifying their internal limitations and contradictions, but not of fulfilling humanity's deeply rooted desire to know and love the whole. Thus the necessity of aesthetic or religious intuition. While this sort of "irrationalist" idealistic holism can certainly support a progressive politics——Tillich was a religious socialist, and Vasconcellos a left-wing populist——the tendency as a whole is pregnant with dangerous authoritarian potential. This is because it asks the individual, as a precondition for achieving "wholeness" or salvation, to yield to a superior power which s/he cannot understand and which alone can bring fulfillment.

Others attempted to interpret the results of the natural sciences in a way which allowed for an "organic" or "systems" perspective. The "process philosophy" of Alfred North Whitehead theorized the universe as a process of "concrescence," in which the eternal objects or potential entities implicit in God's primordial nature gradually realize themselves, becoming actual entities, giving birth to organisms, societies, etc. and ultimately to God in his "consequent nature (Whitehead 1929: 524)." Whitehead attempts to reconcile atomism with a "philosophy of organism," by making the plurality of particular entities potential in God as "eternal objects" or "conceptual feelings", and only actual as objectified in him. The resulting system does indeed significantly mitigate the effects of atomism, but the source of organization, as in the other idealist trends, is ultimately external to the particular entities, which, Whitehead insists, have their own independent existence.

General systems theory, of the kind developed by Ludwig von Bertalanffy, attempts to build a holistic theory on the foundations of atomistic science by looking for general organizing principles shared by systems at all levels of complexity (physical, chemical, biological, social), and then, on the basis of these principles, elaborating a "systems ontology," a "systems epistemology" and a "systems approach to values (von Bertalanffy 1968: xxi-xxii)." While emphasizing the limitations of "analytic-summative" methods however, and the impossibility grasping the behavior of complex, highly interdependent systems simply by understanding their parts, general systems theory comprehends only *behavioral* interdependence. There is little or no recognition that the nature and indeed the very existence of the elements of systems is determined by their relationships with each other, so that only the system itself is really real. "System" is an organizing principle external to the "material" elements of which it is composed. Once again, we have a doctrine which mitigates, but does not really transcend atomism.

In general, objective idealism represents an advance over Scholastic philosophy, in so far as it does provide for authentic progress and development. And unlike dialectical materialism, objective idealism provides both a coherent ontological ground for the process of development, and the basis, at least, for a coherent theory of value which avoids the dangers of relativism and nihilistic postmodernism. At the same time, objective idealism tends to regard the evolution of matter towards ever higher levels of organization as the result of the action of a principle outside of matter. By making organization something imposed on matter from the outside, objective idealism preserves the patriarchal, authoritarian tendency which entered philosophy in the bronze age. This approach to the problem of organization tends to stifle efforts to catalyze development of the potential latent in matter itself and belittles the role of real material (physical, chemical, biological, social) interactions in the development process, making all development ultimately a religious problem——a problem of relationship to the ontological ground. At best this leads to an insufficiently critical attitude towards social structures which hold back the development of human social capacities. Even the most progressive objective idealists (e.g. left wing Christian Democrats and religious socialists) have tended to rest content with moral denunciations of injustice, and have not really devoted much attention to the critical analysis of social structures or the elaboration of new strategies for development. At worst, especially in traditions in which respect for human reason is weak, or where this respect has been eroded by reification, it can lead to an authoritarian attempt to "restore order" to human society on the basis of an irrational submission to religious authority.

It is only with the scientific developments of the twentieth century——relativity, quantum mechanics, complex systems theory, and postdarwinian evolutionary biology that it became possible for philosophers working within the dialectical tradition to continue Hegel's project in a way which is at once faithful to the scientific evidence and authentically dialectical in nature. Errol Harris(1991, 1992), heir of the British Hegelian tradition, for example, regards the universe as an expression of a single, underlying "organizing principle" which realizes itself in an increasingly complex "dialectical scale of forms:" physical, chemical, biological, and social. At this point the boundaries between idealist and materialist dialectics begin to blur. The organizing principle is immanent in matter and realizes itself through concrete material interactions, as the cosmic system strives towards the synergistic integrity which is latent within itself.

B. Dialectical Materialism

It is precisely a focus on real material relationships which is the principal philosophical attraction of dialectical materialist theories of development. Unlike both positivistic atomism and idealist dualism, dialectical materialism recognizes that all matter constitutes a single interconnected system. Precisely because of this interconnectedness, matter is characterized by a capacity for "reflection." This means that material systems are affected by their interactions with other material systems, and conserve the effects of that interaction. Gradually, over a period of time, these quantitative interactions build up, so that, at a certain point, a transition takes place and the accumulated interactions effect a qualitative change in the systems involved. Over a period of time, systems thus evolve towards greater complexity and integration (Zeman 1988: 104-106).

Dialectical materialist philosophers have differed significantly over the precise formulation and relative importance of the various "laws of the dialectic" and over the way in which these general laws apply to the different forms of matter (physical, chemical, biological, social). Engels, for, example, emphasized the gradual accumulation of quantitative contradictions, which, he believed, would issue of their own accord in qualitative systemic change. Lenin, Stalin, and Mao, on the other hand emphasized the "unity and struggle of opposites" and what Mao called the "principle of contradiction"——i.e. the tendency of systems to develop through struggle and conflict. All of these thinkers upheld what they called a "dialectics of nature", and regarded all forms of matter as subject to these same basic laws. "Humanistic" socialists, finally, have tended to stress the principle of the "negation of the negation," of the recovery, at a higher level of differentiation and integration, of humanity's archaic, communitarian, holism——an emphasis which laid the groundwork for a very fruitful dialogue with objective idealism and the salvation religions.

The dialectical materialist approach to this problem has the merit of focusing attention on the role of concrete interactions in the actual process of development. The result has been a level of critical social analysis which we do not see in other traditions. It is to the dialectical materialist tradition which we owe not only the critique of capitalism specifically, but the whole enterprise of assessing various social structures from the standpoint of the their contribution to cosmohistorical development. At the same time, by rejecting out of hand the idea that there is an underlying, organizing principle immanent within matter, and that the working out of this principle through concrete relationships endows the cosmos with a teleological quality, dialectical materialism provides an inadequate logical/ontological ground for the process of development, and makes it appear that more complex forms of development are the result of ultimately random interactions, or, given the emergence of intelligence through such random interactions, the result of conscious human intervention (labor, political struggle, etc.). There are two dimensions to this problem. First, in the absence of some logical/ontological principle which *requires* there to be space-time, matter, chemistry, biological, society, etc., we have no real explanation for the fact that we find an organized universe rather than chaos and nothingness. In this sense, the Neothomist Gustav Wetter is correct when he says that dialectical materialism fails to meet the criterion of sufficient reason (Wetter 1958). Second, if the drive towards organized complexity is not written into matter itself then the ends towards which really complex systems (human beings, human organizations) strive are ultimately ungrounded. The result is either an authoritarian drive to impose order in a fundamentally chaotic world, or else a nihilistic relativism which denies meaning and value.

It is this contradiction which has led to the crisis of dialectical materialism, and of the socialist movements. The socialist tradition expended considerable energy trying to ground the development of socialist consciousness in the laws of nature[6] or at least in the laws of human history.[7] Generally such theories produced very poor practical results. Lenin, alone among the principal socialist leaders, recognized that socialist consciousness would never emergence spontaneously, but had to be "constructed" by the party, which led the proletariat through a series of political contradictions which would gradually expand their vision, and lay the groundwork for socialist construction. The difficulty, of course, is that this approach at least implicitly abandons the claim that socialism is grounded in the cosmohistorical evolutionary process. When an attempt was made to rescue the ontological foundations of socialism, this was done by endowing humanity, "the historical process," or even the party and its "leader" with the full burden of authority historically carried by God, with results which were philosophically absurd and historically tragic. When no

---

[6]. Engels, for example, argues that laws of the dialectic apply to both nature and history, and that socialist consciousness will emerge gradually through the growth of the working class and accumulation of contradictions between the working class and the bourgeoisie, which, at a certain point, will cease to be merely "quantitative" --e.g. conflicts over the division of surplus-- and will become qualitative --i.e. a conflict over the basic organization of human society.

[7]. Consider for example, Eduard Bernstein's attempt to derive the development of socialist consciousness from the gradual moral evolution of humanity.

such attempt was made the result was a rapid descent into nihilistic relativism.[8]

The realization of organization as a social product thus came almost inadvertently to mean the negation of organization as such. If the order, meaning, and purpose which we find in nature and society are simply the product of human organizing activity, (be it technological, economic, political, or cognitive), then how can we claim that it is the universe which is ordered and meaningful, and not merely our minds? If each culture organizes the universe differently, then how can we argue that there is even a universal human teleology? Moral norms derived from the logic of human social life thus began to look increasingly like mere expressions of the will of one or another social group——like alibis for individual or collective self-interest rather than authentic moral norms. This "realization" tended to undermine the moral authority of socialism, which was increasingly regarded as the meaning which humanity (or some section thereof) gave to its own history rather than as the expression of an underlying drive towards organized complexity written into the fabric of being itself. Humanistic socialism, which attempted to ground the "negation of the negation" in human practice, was not immune to this difficulty.

It is in this context that we must understand the emergence of "postmodern" skepticism regarding "totalizing rationalist metanarratives" (Lyotard 1984)——i.e. regarding the effort to grasp nature and/or history as a unified totality. "Postmodernism," as it has come to be called, is

---

[8]. The work of Luis Althusser represents a key link in this process. Althusser makes a distinction between "expressive" and "complex structured" totalities. Belief in the first, which he associates with Hegel and "socialist humanism," represents a kind of religious residue within dialectical materialism, a naive faith in the existence of a latent order which gradually becomes conscious and explicit as the working class matures. Complex structured totalities are, on the contrary, the result of conscious political intervention on the part of Communist organizations, which stitch together complex political alliances in their protracted struggle for hegemony. Certain questions, however, immediately come to mind. Why should these alliances be led by the working class and in particular by the Communist Party? Why should they aim towards socialism? These questions were eventually asked --and answered-- by Althusser's followers, Ernesto Laclau and Chantal Mouffe (1985), who, we will see, argue that there is no logical or historical necessity to ground either the leading role of the working class and its political party or the struggle for socialism. The result was a wholesale breach with the dialectical materialist tradition and a turn towards postmodernism. The resulting doctrine, which treats socialism as simply one political option among many, is highly pessimistic. Is it any wonder that Althusser's brightest student, Nicos Poulantzas killed himself, and that Althusser drowned his wife in the toilet?

defined by an "incredulity toward metanarratives" (Lyotard 1984: xxiv) which claim that science can grasp the underlying structure of reality and thereby contribute to the "dialectics of Spirit, the hermeneutics of meaning, the emancipation of the rational or the working subject, or the creation of wealth (xxiii)." Indeed, most postmodernists argue more or less strongly that it is precisely this claim to grasp the "standpoint of totality" which lies behind the "terrorism" and "totalitarianism" of the modern period (1984:82). Postmodernism proposes, through its "deconstructive" activity to relativize the grand narratives which legitimate the totalitarian project, and thus to free humanity from its "nostalgia for the whole."

Postmodernism has a double significance. On the one hand it represents a radical realization of human being as social being, constituted without remainder by the system of relationships between human beings, so that our knowledge itself is a product of historically limited social structures. At the same time, it is only a *negative* realization of human sociality. It has not yet realized sociality as itself the truth of being, the expression of an organizing principle which provides it with an ontological ground, in such a way that knowledge, by becoming itself the object of social scientific analysis, rather than simply dissolving in its own historicity, begins to see itself as an historical force which, by its very existence, transforms and realizes the organization of the reality which it knows.

This negativity is a product of the social location of postmodernist theory. On the one hand capitalism has developed to the point that the principal object of labor is less and less physical, chemical, and biological matter, and more and more the organization of institutions, and of signs. This represents a qualitative advance in the complexity of human society. At the same time this increasingly postindustrial labor process is mediated through the marketplace——i.e. through the spontaneous interaction of a multitude of individual preferences. The gradual development of human society to higher levels of organization thus appears as the creation of a multitude of products, social forms, systems of meaning——*lifestyles*—— to satisfy a myriad of different preferences, no one of which can lay claim to universality.

The difficulties of this problematic should be obvious, its current popularity notwithstanding. First of all, it fails to account for the evidence. For all their very real diversity, different human cultures have, for the most part, shared a common perception that the cosmos is an organized totality. And we will see that there is powerful new evidence from the physical and biological sciences that humanity's intuitive grasp of cosmic order, far from being a mere projection of human social or cognitive structures, in fact discloses profound truths about the universe itself. The fact that postmodern "multiculturalists" seem unable to take this perception seriously, suggests that behind their concern for "difference" lies a thinly veiled contempt for the religious traditions of the diverse peoples of this planet.

Second, postmodernism is internally inconsistent. If all knowledge and all values are themselves cultural products, if none can claim to grasp reality in a way superior to the others, then on what basis are we asked to accept postmodernism itself? On what basis, and from what vantage point are we to criticize terror and totalitarianism?

Finally, postmodernism, even when it offers powerful critiques of injustice and oppression, is unable to offer either a credible vision for the future or a credible strategy for change. In the absence of a clear vision of the cosmic significance of the human civilizational project, it is impossible to make any serious contribution to charting the future of humanity.

C. The Reemergence of Holistic Philosophy

It is only during the past decade or so, as objective idealism and dialectical materialism have entered a period of crisis, and the social sciences and philosophy have retreated into postmodern fundamentalism and nihilism that strong scientific evidence has begun to accumulate which fully and clearly supports the original insights of the dialectical tradition. Interestingly enough, the main impulse in this direction has come not from the social sciences (which ought to be sciences of holistic systems *par excellence*), but rather from the natural sciences, and especially from physical cosmology, complex systems theory, and postdarwinian evolutionary biology. In this section we will examine the results of the new science in some detail, assess the "neoliberal" or positivistic interpretation of this science, and then outline a new synergistic theory.

1. The New Sciences of Organization

The new science has transformed our understanding of every level of cosmic organization. Newtonian physics regarded space as a formless void in which particles moved over the course of time without thereby affecting or being affected by the space time continuum through which they moved. The theory of relativity, on the other hand, has demonstrated that space-time itself has a structure, a "curvature," so that far from being "void and without form," it is a kind of matrix, out of which more complex forms of organization emerge. Indeed, it is precisely the curvature of the space-time continuum which requires gravity, and therefore mass——and ultimately the whole complex of structures which make up the material world (Harris 1965).

Closely related to the relativistic theory of space-time and gravitation (if not yet fully unified with it) is the shift from particle theory to field theory. Newtonian mechanics regarded the universe as constituted by irreducible particles which interact with each other through the medium of various forces, such as gravity and electromagnetism. As physicists have carried on the search for increasingly more fundamental particles, as they have discovered new physical forces (the strong and weak nuclear forces) and as they have tried to develop a unified theory which explains the relationship among the various forces, they have been forced to abandon the atomism which characterized their original paradigm. Contemporary physics regards matter (including observed and theoretically postulated "particles") as *actually constituted by* gravitational, electromagnetic, and strong and weak nuclear fields. These four fundamental forces, furthermore, seem originally (during the earliest stages in the development of the universe) to have been identical with each other. It is only as the universe develops that they become distinct (Davies 1988: 12).

The full significance of this shift is best suggested by the EPR (Einstein, Podolsky, Rosen) nonlocality. Einstein

conceived of an experiment in which two particles interact and then separate to a great distance. Under these circumstances the quantum state of the combined system can be such that a measurement performed on one particle apparently affects the outcome of measurements on the other, distant particle ...

More precisely, it is found that independently performed measurements on widely separated particles yield correlated results. This itself is unsurprising because if the particles diverged from a common origin each will have retained an imprint of their encounter. The interesting point is the degree of correlation involved ...

... quantum mechanics predicts a significantly greater degree of correlation than can possible be accounted for by any theory that treats the particles as independently real and subject to locality. It is almost as if the two particles engage in a conspiracy to cooperate when measurements are performed on them independently, even when these measurements are performed simultaneously. The theory of relativity, however, forbids any sort of instant signalling or interaction to pass between the two particles.[9] There seems to be a mystery, therefore about how the conspiracy is established.

Resolution of this contradiction presupposes recognition that

> ... the two particles, though spatially separated are still part of a unitary quantum system with a single wave function ... it is simply not possible to separate the two particles physically, and to regard them as independently real entities ...

> ... the fate of any given particle is inseparably linked to the fate of the cosmos as a whole, not in the trivial sense that it may experience forces from its environment, but because its very reality is interwoven with that of the rest of the universe (Davies 1988: 177).

The new physics thus points to the irreducibly relational character of matter. The cosmos is not composed of interacting things, but is rather itself a complex network of relationships. Because of this, both the universe as a whole, and particular systems within it, are best described not by classical equations which regard the position of a particle as a function of its trajectory and velocity, but rather by "quantum wave functions" which define the relative probability of each possible state of the system. Since, furthermore, all elements of the system are interrelated with each other, and with the observer who is in fact part of the system, the state of quantum systems appears to be influenced by the persons who observe them.

This has led some interpreters to adopt a subjectivist epistemology, which in its more radical form tends towards solipsism. It is possible, for example, to conclude from quantum theory that we never really know anything about the "objective" state of physical systems. This seems unlikely, however, given the practical use we have been able to make of scientific knowledge in exercising power in relationship to the physical world. Only slightly more modestly, we might claim that any conscious being can collapse the wave function describing a system. This also seems unlikely, given the fact that different observers generally achieve identical

---

[9]. This is because, according to the theory of relativity, nothing can travel faster than the speed of light.

results when making the same measurement (Barrow and Tipler 1986: 468-469).[10]

A second interpretation of quantum mechanics——the Many Worlds Interpretation——suggests that wave functions don't collapse. According to this view, all logically possible universes in fact exist. As Errol Harris (1991: 9-15) has pointed out, however, the very concept of "many universes" is itself internally contradictory. Even if we allow that the cosmic totality is infinitely more complex than we might have imagined, and includes not only an enormous, perhaps infinite spatial expanse, but also infinitely many trajectories defined by the constant splitting of the universal quantum wave function, the very fact that these various trajectories branch off from a single wave function means that it nonetheless remains a unified cosmic whole. The realization of the manifold "logical possibilities" inherent in the cosmic wave function would be better theorized as the logical unfolding of a single organizing principle, which realizes itself across a (logically interrelated) complex of topologies. Adequate theoretical specification of this alternative would seem to presuppose complete unification of relativity with quantum mechanics.

Perhaps the most promising interpretation of quantum mechanics is that suggested by John Wheeler: that conscious beings collectively collapse the wave function describing the universe, and thus, in a very real sense collectively bring the universe into being (Barrow and Tipler 1986: 470). There seem to me to be two problems with this notion, which are, however, easily corrected. First, Wheeler's solution contains a residual subjectivism. It is the radically interconnected character of the cosmic system, and not some peculiar property of consciousness itself, which causes "observations" to affect the state of quantum systems. We should, therefore, remove the criterion that only conscious beings participate in collapsing the wave function. Every element in a system adds to its determination and thus participates in collapsing the wave function which describes it. Second, human beings seem to be able to affect only very small scale properties of the universe. It would require a form of organization far more developed than ourselves to affect the large scale properties of the universe. Barrow and Tipler refer to this as an "Ultimate Observer" who coordinates the separate observations of all of the lesser observers, and who thus brings the cosmos as a whole into being (Barrow and Tipler 1986: 470-471). It would be more appropriate to speak of an Ultimate Organizer. Since the process of cosmopoesis is not complete until

---

[10]. These two interpretations of quantum mechanics correspond roughly to the postmodernist solution to the problem of cultural relativism.

the final state, if there is a final state, or else continues indefinitely, this Ultimate Organizer must continue to exist as long as the universe endures. And since all systems must be organized at the level of the Ultimate Organizer in order to come fully in to being, complex organization must eventually become coextensive with the cosmos as a whole. Where would this Ultimate Organizer come from? It is the entire community of organizers acting throughout cosmic history which brings the final organizer into being. From this point of view human beings, and human society generally, would be participants in a creative process which nonetheless radically transcends the merely social form of matter.

The second dimension of the current scientific revolution is the growing recognition of the existence, in even the simplest forms of matter, of a drive towards self-organization. For over a century now discussions about the ultimate fate of the universe have been dominated by gloomy predictions of "heat-death," the inevitable result of the tendency of "all systems" to evolve towards increasingly random or "entropic" states. This generalization of the second law of thermodynamics into a theory of cosmic catastrophe has never made much sense. The tendency towards increased entropy is a tendency which characterizes *closed* systems of *particles* which interact with each other in a purely *external* manner. But it has never been definitively demonstrated that the universe is a closed system. And there is growing evidence that it is not a system of particles at all, but rather *of radically interdependent relationships* in which all *particular systems* are *internally related* to each other. If this is true, then there is no basis on which to generalize the law to cover the cosmos as a whole.

In recent years, moreover, there has been growing evidence that matter in fact tends to evolve in the direction of organized complexity. Nowhere is this process more apparent than in the non-linear dynamic systems which have become the focus of the new science of complex systems theory. (For a general overview of complex systems theory cf. Gleick 1987; for fuller treatments cf. Campbell 1989, Zurek 1990).

For most of the past 300 years, the natural sciences have concentrated on the analysis of relatively simple linear systems, which were characterized by a small number of variables, which seemed to be related to each other in a linear (i.e. proportional) manner. Electromagnetic fields, weak gravitational fields, the diffusion of gases, and the behavior of many materials under stress are all characterized by a high degree of linearity. To the extent that they have analyzed nonlinear systems at all, the natural sciences have generally resorted to linear approximations of these systems. The calculus itself, one of the principal tools of the natural sciences, is characterized by the reduction of curves to lines (differentiation) and of the area under curves to a series of rectangles (integration).

Complex systems theory has made three important discoveries. First, it is becoming increasingly clear that even the most complex systems, which are characterized by behavior which is "chaotic" and unpredictable, are in fact governed by relatively simple mathematical algorithms. In this sense, it appears, even chaos is deterministic. The difficulty in predicting the behavior of chaotic systems results from the fact that in nonlinear equations very small differences in initial parameters can lead to very large differences in final results.

> That a system governed by deterministic laws can exhibit effectively random behavior runs directly counter to our normal intuition ... we typically assume that if the initial conditions of two separate experiments are *almost* the same, the final conditions will be *almost* the same. For most smoothly behaved "normal" systems, this assumption is correct. But for certain nonlinear systems it is false, and deterministic chaos is the result. (Campbell 1989: 13).

While the resulting behavior is extraordinarily difficult to predict, it is, nonetheless, characterized by the emergence of complex, intricate patterns. As systems approach chaos, they tend to go through a phase of "period doubling" in which they oscillate between two values with constantly increasing frequency. Graphs plotting the behavior of chaotic systems are characterized by "self-similarity." Patterns which develop at a larger scale repeat themselves, with minute variations, at successively smaller scales, creating objects of extraordinary beauty.

This leads us to the second contribution of complex systems theory: a deeper understanding of the process by which coherent structures emerge in the midst of apparently chaotic behavior.

> From the Red Spot of Jupiter through clumps of electromagnetic radiation in turbulent plasmas to microscopic charge-density waves on the atomic scale, spatially localized, long-lived, wave like excitations abound in nonlinear systems. These nonlinear waves and structures reflect a surprising orderliness in the midst of complex behavior (Campbell 1989: 15).

Finally, complex systems theory has suggested that not only do natural systems behave in a way which is deterministic even when it is chaotic, and spontaneously form coherent structures, they also have the capacity to adapt to environmental conditions in such a way as to better maintain and reproduce their structures. This adaptive process in not confined to living systems.

> The process of pattern formation and selection occurs throughout nature in nonlinear phenomena ranging from electromagnetic waves in the ionosphere through mesocale textures in metallurgy to markings on sea shells and stripes on tigers (Campbell 1989: 18).

Complex systems theory has provided valuable tools for understanding the behavior of a tremendous range of natural systems, from stellar structure and the complex patterns of coastlines and mountain ranges to weather patterns and the beating of the human heart, from the operation of the immune system and fluctuation of animal populations in a given habitat to the behavior of the stock market and of complex economic systems as a whole. We are now able to explain why a butterfly in China might cause a hurricane in Florida, we are coming closer, at least, to the creation of artificial living systems (Langton 1989), and we have the first mathematical formalization of the kind of nonlinear patterns which social scientists have been studying for over a century (Gleick 1987).

On the fringes of complex systems theory scientists are beginning to explore some of the extraordinary pattern generating capacities of living matter. While complex biological processes can be reduced analytically to simpler chemical and mechanical processes, this does very little to explain how, for example the genetic information encoded in a strand of DNA can

> exercise a coordinating influence, both temporal and spatial, over the collective activity of billions of cells spread over what is, size for size, a vast area of three dimensional space (Davies 1988: 105).

Biologist Rupert Sheldrake has suggested that emerging structures create morphogenetic fields.

> The idea is that once a new type of form has come into existence, it sets up its own morphogenetic field, which then encourages the appearance of the same form elsewhere ...

> Morphogenetic fields are not ... restricted to living organisms. Crystals possess them too. That is why ... there have been cases where substances which have never previously been seen in crystalline form have apparently been known to start crystallizing in different places at more or less the same time (Davies 1988: 164).

While the theory of morphogenetic fields has not yet been adequately developed, the fact remains that, even as entropy has increased, matter has managed to assume increasingly complex forms, organizing itself in such a way that it can reproduce these forms, and, as we shall see, in the case of humanity, in such a way that it can begin to consciously reorganize physical, chemical, and biological matter in order to serve human purposes, and consciously increase the complexity of the ecosystem.

It might be objected, of course, that the emergence of complex systems is simply a counterpoint to a larger process of cosmic disintegration, a local dynamic governing an area more extensive, perhaps, than human society, but ultimately no match for large scale cosmological processes which will lead ultimately either to the endless expansion of the universe, so that communication and thus complex interactions become all but impossible, or else to the recollapse of the universe to a final singularity in which molecular matter, and thus all complex organization as we know it, will have disappeared. There is an emerging body of science, however, which suggests that this is not the case, that complex systems, including not only life but also intelligence, are in fact a constitutive dimension of the cosmos, and fully necessary for its existence. It appears that a whole series of fundamental physical constants——the masses of the elementary particles, and the strengths of the fundamental forces (gravity, electromagnetism, and the strong and weak nuclear forces) are fixed at just the values necessary for the emergence of intelligent life (Barrow and Tipler 1986). This is the so-called "anthropic cosmological principle."[11]

The fact that physical constants are fixed at precisely the levels necessary for the emergence of intelligent life can be interpreted in relatively modest terms——what has come to be called the Weak Anthropic Principle. According to the Weak Anthropic Principle, the values of physical and cosmological quantities are restricted by the requirement that "there exist sites where carbon-based life can evolve and by the requirement that the Universe be old enough for it have already done so (Barrow and Tipler 1986: 16)." This form of the Anthropic Principle is simply "a restatement ... of one of the most important and well-established principles of science: that it is essential to take into account the limitations

---

[11]. The term "anthropic" is, in fact, is something of a misnomer, as the principle has very little to do with humanity as such. In its weak forms, the principle might be reformulated to state that "the universe appears to be constituted in just such a way as to may possible the existence of viruses, or bacteria, or protozoa," and in its strong forms, which make intelligent organization really necessary to the existence of the universe, it in fact requires organizing capacities which far transcend the merely human.

of one's measuring apparatus when interpreting one's observations (Barrow and Tipler 1986: 23)." The values of the fundamental physical constants tell us nothing of logical, ontological, or cosmological significance about the necessity of complex organization. It is still just a coincidence that we are here, and there is nothing remarkable in the realization that we would be unable to observe a universe in which our existence was not possible.

Some cosmologists, however, argue that there is evidence for stronger interpretations of the Anthropic Principle: i.e. that "the Universe must have those properties which allow life to develop within it at some stage in history (Barrow and Tipler 1986: 21)." This interpretation they call the Strong Anthropic Principle. If, furthermore, intelligent life is necessary for the existence of the universe, but dies out before it has developed sufficiently to affect the universe on a cosmological scale, then it is difficult to see why it would have been necessary in the first place. Barrow and Tipler thus conclude that "intelligent information processing must come into existence in the Universe, and, once it comes into existence, it will never die out (Barrow and Tipler 1986: 23)." This conclusion is referred to as the "Final Anthropic Principle." According to this view, intelligent life will continue to develop towards the "Omega Point," at which it will embrace the cosmos as whole, and be capable of organizing all cosmic events. Recently Tipler (1994) has gone further, arguing that there will in fact be a universal resurrection as the universe approaches Omega, and that all intelligent beings will experience what to them subjectively appears to be eternal life.

Clearly something like the Final Anthropic Principle or the Omega Point Theory would be a valuable asset in re-establishing the ultimate meaningfulness of the universe, and we will argue that there are in fact grounds for affirming the necessity of intelligent organization, including intelligent organization which embraces the universe as a whole. There are, however, a number of very different ways of making such an argument, at least some of which turn out to be highly problematic. Let us examine Tipler's version in some detail, and see to what extent it adequately comprehends the results of the new science taken as a whole.

2. The Neoliberal Interpretation of the New Science

We begin by outlining Tipler's assumptions. Tipler takes as his starting point a high-technology variant of Berkeley's subjective idealism. The universe is a vast information processing system. Matter is the "hardware" component of the system, the laws of nature the "software." Drawing on the information theory developed by Shannon and Weaver (1949), Tipler argues that the organization of a system is its negative entropy, or the quantity of information encoded within it. "Life" is simply information encoded in such a way that it is conserved by natural selection. A system is intelligent if it meets the "Turing test," i.e. if a human operator interrogating it cannot distinguish its responses from those of a human being (Turing 1950). Intelligent life continues forever if

1) information processing continues indefinitely along at least one worldline γ all the way to the future c-boundary of the universe; that is, until the end of time.

2) the amount of information processed between now and this future c-boundary is infinite in the region of spacetime with which the worldline γ can communicate; that is the region inside the past light cone of γ.

3) the amount of information stored at any given time τ within this region diverges to infinity as τ approaches its future limit (this future limit of τ is finite in a closed universe, but infinite in an open one, if τ is measured in what physicists call "proper time") (Tipler 1994: 132-133).

The first condition simply states that there must be one cosmic history in which information processing continues forever. The second condition states that it must be possible for the results of all information processing to be communicated to world-line γ. This means that the universe must be free of "event horizons," i.e. regions with which an observer on world line γ cannot communicate. It also means that since an infinite amount of information is processed along this world line, an observer on this line will experience what amounts subjectively to eternal life. The third condition avoids the problem of an eternal return, i.e. an endless repetition of events as memory becomes saturated and new experience thus impossible.

Tipler then goes on to describe the physical conditions under which "eternal life" is possible. In accord with the as yet incompletely unified state of physics, he presents separate "classical" or "global general relativistic" and "quantum mechanical" theories. We take his "classical" theory first. Information processing is constrained by the first and second laws of thermodynamics. Specifically, the storage and processing of information requires the expenditure of energy, the amount required being inversely proportional to the temperature.

... it is possible to process and store an infinite amount of information between now and the final state of the universe only if the time integral of P/T is infinite, where P is the power used in the computation and T is the temperature (Tipler 1994: 135).

Eternal life thus becomes essentially a problem of finding an adequate energy source. Tipler proposes finding this source in the "gravitational shear" created as the universe collapses at different rates in different directions. This imposes a very specific set of constraints on the process of cosmic evolution. Only a very special type of universe, the so-called "Taub" universe, named after mathematician Abraham Taub, collapses in just precisely the way required. And even most Taub universes tend to "right" themselves, returning to more nearly spherical form. For information processing to continue forever, life must gain control of the entire universe, and force it to continue its Taub collapse in the same direction far longer than it would spontaneously (Tipler 1994: 137). Thus the requirement that intelligent life gain control of the universe as a whole, and control the rate and direction of its collapse, so as to create the enormous energies necessary to guarantee eternal life.

Meeting the second and third conditions outlined above requires, furthermore, that the universe be closed, because "open universes expand so fast in the far future that it becomes impossible for structures to form of sufficiently larger and larger size to store a diverging amount of information (1994: 140)." It also requires that "the future c-boundary of the universe consist of a single point ... the Omega Point (1994: 142)." Finally, in order to meet information storage requirements, "the density of particles must diverge to infinity as the energy goes to infinity, but nevertheless this density of states must diverge no faster than the cube of the energy (1994: 146)." Tipler identifies, in addition to these requirements, which he calls "weakly testable," a variety of other predictions which can be used to test his theory, including the requirement that the mass of the top quark be 185 +/- 20 GeV and that the mass of the Higgs boson must be 220 +/- 20GeV (1994: 146). Fermilab recently measured the top quark at just a little bit below this mass.

In order to understand Tipler's Quantum Omega Point Theory, it is necessary to understand some of the internal contradictions of current quantum cosmology. In general relativity the spatial metric h and the nongravitational fields F are taken as given on the underlying three-dimensional manifold S. Cosmologists then attempt to find a four-dimensional manifold M with a Lorentz metric g (the gravitational field) and nongravitational fields F such that M contains S as a submanifold, g restricted to S is the metric h, and K is the extrinsic curvature of S, or, to put the matter differently, K says how quickly h is changing along the fourth, "temporal" dimension (1994: 162). In quantum cosmology, on the other hand, the universe is represented by a wave function $\Psi(h,F,S)$, which determines the values of h and F on S (1994: 174-175). One feature of the system, however, remains arbitrary: the selection of the fixed three-dimensional manifold S. Hartle and Hawking have proposed to eliminate this contingency by allowing the wave function to be a function of any three-dimensional manifold. According to this view, the domain of Ψ includes all possible values of h, F, and S (1994: 178). The Hartle-Hawking formulation, however, still requires h to be spacelike on all three-dimensional manifolds S. This restriction brings the formulation into conflict with classical general relativity, which does not distinguish so sharply between space and time.

Tipler points out, however, that the requirement that h be spacelike derives from a subjectivist interpretation of quantum mechanics, which interprets the wave function as a probability amplitude at a given time. This, obviously, requires times to be sharply distinguished from space. Tipler, however, favors a Many-Worlds interpretation of quantum mechanics, according to which all possible values of the wave function exist mathematically, and all those which permit the existence of observers exist physically. This removes the need to distinguish between space and time, and thus the requirement that h be always spacelike. Tipler proposes instead to allow the domain of the wave function to include all four-dimensional manifolds which permit a Lorentz metric g. All such manifolds permit what is known as a foliation. They can, that is, be represented as a "stack" of three-dimensional manifolds S(t), each representing the topology of a possible universe at a different moment of time. Each foliation will have a metric h, which need not be space like, as well as nongravitational fields, induced by the enveloping spacetimes (M,g). Any (h,F,S) which cannot be represented this way has $\Psi=0$; it does not exist. Similarly, there will be many spacetimes which permit the same (h,F,S). Some of these may have a future c-boundary which is a single point——the Omega Point (1994: 174-181). Thus the "Omega Point Boundary condition on the universal wave function:

> The wave function of the universe is that wave function for which all phase paths terminate in a (future) Omega Point, with life continuing into the future forever along every phase path in which it evolves all the way to the Omega Point (1994: 181).

Now, the Four-Manifold Non-Classification Theorem states that there does not exist any algorithm which can list or classify all compact four-dimensional topological or differentiable manifolds without boundary, nor is it possible to tell if any two given manifolds are the same or different

(1994: 190). This means that it is impossible to derive the system as a whole from any one of its elements——a situation which, following William James, Tipler identifies with radical, ontological indeterminism (1994: 187). This means that the existence of life and intelligence, and the *decision* on the part of intelligent life to guide the universe towards Omega, is in fact logically and ontologically prior the universal wave function itself (1994: 183): "The wave function is generated by the self-consistency requirement that the laws of physics and the decisions of the living agents acting in the universe force the universe to evolve into the Omega Point (1994: 203)." Indeed, in so far as the equations of both general relativity and quantum mechanics are reversible, there is no scientific reason to assume that causality runs only in one direction: from the past, through the present, into the future. It might just as well be seen as running from the future, through the present, into the past. From this point of view it is God, the Omega Point, which, existing necessarily, brings the entire universe into existence and draws it to himself.

> At the instant the Omega point is reached, life will have gained control of *all* matter and forces not only in a single universe, but in all universes whose existence is logically possible; life will have spread into *all* spatial regions in all universes which could logically exist, and will have stored an infinite amount of information, including *all* bits of knowledge which it is logically possible to know. And this is the end (Barrow and Tipler 1986: 677).

The question arises, quite naturally, just how we are to reach Omega. The key link between actually existing carbon based life, and this nonmolecular intelligent living system are a "race" of intelligent, self-reproducing, interstellar probes (the so-called von Neumann probes). Tipler proposes launching a series of such interstellar probes in the expectation that as they evolve they will grasp the conditions for the long term survival of intelligent life in the cosmos, and eventually reorganize the universe on a cosmic scale in order to bring into being the nonmolecular life form(s) which can survive into the final stages of cosmic evolution.

Such probes would, of course, be extremely expensive. It thus becomes necessary to identify an optimum path of economic development. It is interesting to note that both Barrow and Tipler make extensive reference to the neoliberal economist F.A. Hayek in their work. Hayek, like Barrow and Tipler, identifies complex organization with negative entropy, or with the quantity of information which a system can encode. An economy is simply an information processing system. No centralized planning agency or redistributional structure can grasp the complexity of a highly interdependent, rapidly developing human system, and any attempt on the part of such agencies to plan the society will inevitably result in a loss of complexity and will hold back growth and development.

Certainly nobody has yet succeeded in deliberately arranging all the activities that go on in a complex society. If anyone did ever succeed in fully organizing such a society, it would no longer make use of many minds, but would be altogether dependent on one mind; it would certainly not be very complex but extremely primitive——and so would soon be the mind whose knowledge and will determined everything. The facts which could enter into the design of such an order could be only those which were known and digested by this mind; and as only he could decide on action and thus gain experience, there would be none of that interplay of many minds in which alone mind can grow (Hayek 1973: 49).

What Hayek calls the "extended order" of the marketplace, on the other hand, is uniquely capable of accessing, processing, and communicating vast quantities of information.

Much of the particular information which any individual possesses can be used only to the extent to which he himself can use it in his own decisions. Nobody can communicate to another all that he knows, because much of the information he can make use of he himself will elicit only in the process of making plans for action. Such information will be evoked as he works upon the particular task he has undertaken in the conditions in which he finds himself, such as the relative scarcity of various materials to which he has access. Only thus can the individual find out what to look for, and what helps him to do this in the market is the responses others make to what they find in their own environments (Hayek 1988: 77).

Information-gathering institutions such as the market enable us to use such dispersed and unsurveyable knowledge to form super-individual patterns. After institutions and traditions based on such patterns evolved it was no longer necessary for people to strive for agreement on a unitary purpose (as in a small band), for widely dispersed knowledge and skills could now readily be brought into play for diverse ends (Hayek 1988: 15).

The market thus takes on for Hayek what he acknowledges to be a transcendent character, organizing interactions of a scale beyond the capacity of any single mind or organization——beyond even the mind of God.

> There is no ready English or even German word that precisely characterizes an extended order, or how its way of functioning contrasts with the rationalists requirements. The only appropriate word, 'transcendent,' has been so misused that I hesitate to use it. In its literal meaning, however, it does concern that which *far surpasses the reach of our understanding, wishes and purposes, and our sense perceptions*, and that which incorporates and generates knowledge which no individual brain, or any single organization, could possess or invent. This is conspicuously so in its religious meaning, as we see, for example, in the Lord's Prayer, where it is asked that "*thy* will [i.e. not *mine*] be done on earth as it is in heaven ..." But a more purely transcendent ordering, which also happens to be a purely naturalistic ordering (not derived from any supernatural power), as for example in evolution, abandons the animism still present in religion; the idea that a single brain or will (as for example that of an omniscient God) could control and order (Hayek 1988: 72-73).

Barrow and Tipler draw on Hayek's reasoning to argue that in a market system the technological and economic development necessary to support the construction of interstellar von Neumann probes will take place spontaneously. They argue that insofar as

> the economic system is wholly concerned with generating and transferring information ... the government should not interfere with the operation of the economic system ... if it is argued ... that the growth of scientific knowledge is maximized by information generation and flow being unimpeded by government intervention, does it not follow that the growth of economic services would be maximized if unimpeded by government intervention? (Barrow and Tipler 1986: 173)

Indeed, they argue that if the operation of the marketplace is left to run its course, the cost of energy and raw materials relative to wages will decline to the point that humanity will become capable not only of interstellar travel, but ultimately of reorganizing the structure of the cosmos on a macroscale——developments which are both critical for their meliorist physical eschatology.

> ... the price of raw materials and energy have, on the long term average, been decreasing exponentially over the past two centuries ... (Barrow and Tipler 1986: 172).

The sort of interstellar probes which Barrow and Tipler believe are necessary in order to secure the destiny of intelligent life in the cosmos would currently cost between $3x10^{10}$ and $2x10^{14}$, depending on their speed.

> These costs ... seem quite large to us, but there is evidence that they could not seem large to a member of a civilization greatly in advance of ours ... the cost relative to wages of raw materials, including fuel, has been dropping exponentially with a time constant of 50 years for the past 150 years. If we assume this trend continues for the next 400 years ... then to an inhabitant of our own civilization at this future date, the cost of a low velocity probe would be as difficult to raise as 10 million dollars today, and the cost of a high-velocity probe would be as difficult to raise as 70 billion dollars today. The former cost is easily within the ability of ... at least 100,000 Americans ... and the Space Telescope project budget exceeds 10^9. If the cost trend continues for the next 800 years, then the cost of a $3x10^{10}$ probe would be as difficult to raise as $4000 today. An interstellar probe would appear to cost as much then as a home computer does now ... In such a society, *someone* would almost certainly build and launch a probe (Barrow and Tipler 1986: 583).

Tipler's cosmology even has theological implications. Despite his frequent references to Aristotle and Aquinas, and his effort to show the compatibility of his theory with most of the principal religious traditions, these implications tend very clearly towards Calvinist Christianity. This is because of the centrality of what he calls "agent determinism." Realization of the Omega Point is, in one sense, inevitable; it is required by the very existence of the universe itself. But it presupposes the subordination of the interests of individual carbon-based organisms to a larger cosmic plan which involves the displacement of carbon based by machine, and eventually by nonmolecular intelligence. And in so far as this transition is best carried out through the unimpeded operation of rationally inscrutable market forces, it requires the submission of individual carbon based organisms to cosmic imperatives which they cannot understand, with which, at the very least, they cannot fully identify. Eternal life, furthermore, is not something the soul achieves, by becoming actually capable of infinite self-organizing activity, but rather something bestowed on it by the nearly omnipotent and omniscient beings near Omega, simply because it is in *their* self-interest. Tipler makes a game-theoretical argument (1994: 245-259) that these beings will resurrect us, and will bestow eternal life upon us, and that this will be a life of potentially infinite richness and joy——but ultimately the decision is theirs. We have here, in effect, an anthropic cosmological argument not only for Reaganomics but for a peculiar, high tech, Calvinism.

Tipler's cosmology has a number of attractive features. Clearly, Tipler comprehends the cosmos as an evolving system developing towards ever higher degrees of organization, and recognizes human civilization——or rather the social form of matter in general——as a key link in the cosmohistorical evolutionary process. And he makes a strong case that far from being a mere counterpoint to stronger forces of cosmic disintegration, the forces of complex organization will, in fact, triumph in the end. On the scientific level he has attempted, at least, to make his theory testable and thus opened the way towards experimental verification of the claim that life will survive and develop towards Omega, and thus eventually embrace the universe as a whole. And he has made some effort, at least, to draw out the philosophical and political-theological implications of his position.

At the same time, his work has serious limitations. The most fundamental of these has to do the way in which Tipler understands the concept of system or organization. We have already noted that quantum theory calls radically into question the atomistic logic and ontology which has characterized European philosophy since the time of the scientific revolution, and suggests, rather, that matter is relational at the most fundamental levels of its being. Tipler, however, does not seem to have fully comprehended this development.

In order to illustrate this problem, it is useful to distinguish three meanings of the word "system." At the lowest level of integration, a "system" consists of various elements, the *behavior* of which is determined by their relationships with other members of the system. As von Bertalanffy points out, systems of this type can be described by a system of differential equations. If the system is dynamic, i.e. evolves over time, then we must use partial differential equations. And if we wish to make the state of the system depend on its past states, then we must use "integro-differential" equations (von Bertalanffy 1969: 55-56). Despite the difficulty involved in solving such systems of equations, mathematical formalization is, in principle at least, possible.

There are, however, two higher levels of systemic integration. At the second level, the very nature or essence of the elements is determined by their relationships with the other elements, and at the third, their existence is dependent on——or even constituted by——their relationships with the other elements. It is not clear that either of these two meanings of "system" can be formalized mathematically, since most mathematical formalizations of this sort ultimately rely on set theory, which itself presupposes groups of particulars which are related to each other only in an external manner (Harris 1987).

Now Tipler seems to understand systems only at the first level. From the simplest to the most complex levels, his cosmos continues to be constituted by irreducible particles which are externally related to each other, rather than by a system of relationships the nodes of which merely appear to be particular when we abstract them from the system as a whole. That his "elementary particles" are "bits"——units of information rather than of matter——does not really change anything. His understanding of complex organization, life, intelligence, and social evolution is governed by an ultimately atomistic paradigm in which individual particles (bits, organisms, human persons, von Neumann probes) are externally related to each other. If systems are nothing but aggregates of externally related particles, then organization is nothing more than the order which prevails among those particles——the negative entropy or information content of the system. But negentropic and information theoretical approaches to organization and complexity run into serious problems when we attempt to apply them to biological and social systems. IBM scientist Charles Bennett (1988) has recently pointed out that the negative entropy theory has limitations even at the physical level. The human body, for example, is intermediate in *negentropy* between a crystal and a gas, while being more highly *organized* than either. Similarly, organized objects, "because they are partially constrained and determined by the need to encode coherent function or meaning, contain less information than random sequences of the same length, and this information reflects not their organization but their residual randomness." He proposes instead to define organization as logical depth, "the work required to derive a" message "from a hypothetical cause involving no unnecessary ad hoc assumptions," or "the time required to compute this message from" its "minimal description." This definition bears an interesting resemblance to the Marxist labor theory of value, according to which the value of a commodity is equal to the average socially necessary labor time necessary to produce it (Marx 1849/1978: 203-217).

The limitations of the atomistic paradigm become even more apparent when we attempt to explain the evolution of systems towards higher levels of complexity and organization. Random variation, competition, and natural selection——the mainstays of the atomistic paradigm——do not seem adequate to the task of explaining the emergence and development of living——and especially social——systems. Complex systems theorist Ilya Prigogine has shown that the time necessary to produce a comparatively small protein chain of around 100 amino acids by *spontaneous* formation of structure is much longer than the age of the Earth. Hence, spontaneous formation of structure is ruled out ... according to the modern theory of self-organizing systems, classical arguments concerning the "coincidental" realization of a complex living system cannot be employed (Zimmerman 1991).

Evolutionary biologists point out that the theory of natural selection provides no mechanism to explain the origin of progressive, adaptive change. Molecular biologist Barry Hall for example, has found that the bacterium E. coli produces needed mutations at a rate roughly 100 million times greater than would be expected if they came about by chance. Nor can it account for the fact that such changes seem to occur rather suddenly, rather than in gradual increments, as the theory of natural selection would suggest. A retina or a cornea, after all, without the rest of the organ, would have no survival value by itself, and would be unlikely to be preserved in future generations.

Biologists have identified two distinct types of processes which they believe contribute to the development of increasingly complex forms of life. Molecular biologists have found that the genome operates as a complex interrelated totality, with some genes regulating the operation of others. Random mutations in some parts of the genome, where they threaten to undermine well established life processes, are systematically corrected, while others, in areas where experimentation seems promising, are permitted. At the same time, small changes in one part of the genome can trigger fundamental structural changes in the system as a whole. This is why animal breeders, in the process of selecting for certain traits, so often produce undesired side effects. It also helps to explain how whole new structures, (such as the eye), with significant survival value, might have emerged all at once. Variation, furthermore, is constrained by the material out of which organisms are built. Genetic instructions to construct an elephant with legs as thin as those of an ant, Kevin Kelly points out, simply cannot be carried out. Matter, as potential organization, provides life with certain structural options, but systematically excludes others. This is another reason why structural leaps are more common than incremental changes. The genome must undergo considerable internal reorganization before it can produce a new form which is both structurally different and viable. This complex of phenomena has led some biologists to suggest that the genome contains algorithmic search instructions which help it to discover mechanically stable, biologically viable, ecologically progressive, new forms. Life, it appears, is at least incipiently creative and self-reorganizing.

At the same time there is growing evidence that cooperation plays an important role in the evolutionary process. Biologist Lynn Margulis, for example, has argued that nucleated cells, with their specialized organelles devoted to photosynthesis and respiration, and their genetic high command, came about through symbiosis, when membraned cells incorporated bacteria which had already developed these processes and the structures necessary to carry them out.

These new developments suggest that natural selection is only one of many processes which contribute to the emergence of new, increasingly complex, forms of life. And its contribution is largely negative. Natural selection, as biologist Lynn Margulis puts it, is the editor, not the author, of evolution. The creative self-reorganization of matter, and symbiosis, the cooperation between organisms, play the leading role (Waddington 1957, Lenat 1980, Sheldrake 1981 and 1989, Denton 1985, Wesson 1991, Margulis and Fester 1991, Kelly 1992).

If the evolution of biological systems involves more than random variation, competition, and natural selection, this is even more true of social systems. Even if one accepts the "information content theory of organization," it is clear that the marketplace has no access to information regarding the impact of various activities on the qualitative complexity of the ecosystem and the development of human social capacities. On the contrary, all the market "knows" is a quantitative expression of the existing capacities (supply) and current interests (demand) of individuals, as these are expressed in the form of price curves. It has no way to analyze latent capacities or optimum paths of development for either individuals or the system as a whole. It is market forces, after all, which draw people away from preparing themselves to be elementary school teachers, and towards selling crack cocaine.

Indeed, it is not clear that the information content theory of organization can even grasp the concept of the social form of matter. In so far as human beings are constituted and shaped as persons through their interaction with each other, the social form of matter necessarily involves mutual determination of the existence and essence, and not merely the behavior, of the elements in the system. The economic model which Tipler borrows from Hayek fails to understand this, and remains within the horizon of a universe in which human beings are irreducible atoms which interact with each other externally, but do not ever really influence, much less constitute, each other. A really profound understanding of the marketplace, for example, involves not merely writing a system of equations describing the mutual determination of a limited number of state variables, but an in-depth social-psychological analysis of the ways in which people's ideas and desires are shaped, e.g. by the socialization process, which constitutes them as producers, accumulators, consumers, by advertising, etc. But this kind of analysis involves looking not only at the behavior of individual consumers, but also at their psychological make up, including such factors as sexuality, which are manipulated by advertising campaigns. Once we look beyond the marketplace to family relations, the educational process, the organization of the workplace, the building and exercising of political power, the processes of artistic creation, scientific research, or philosophical reflection, of the complex operation of religious institutions, it becomes apparent that we can say almost nothing of interest so long as we confine ourselves to a model of externally related individuals, since all of these processes are in fact centered on the mutual determination of the existence and essence of the elements in the system. The atomism which relativity and quantum theory have made increasingly untenable at the physical level is in fact *prima facie* absurd at the social.

The information content approach to complex organization has, furthermore, some potentially very dangerous policy implications. Working from very different tendencies within the dialectical materialist tradition, Ernest Mandel (1968, 1978) and the dependency and world systems theorists (Emmanuel 1972, Wallerstein 1974, Frank 1978, Amin 1980) have demonstrated that insertion into market relations in fact undermines a country's economic development, measured in value terms——i.e. in terms of the total quantity of socially necessary labor embodied in its products. On the one hand, as a system becomes more technologically developed and thus more capital intensive, the rate of profit declines, and capital is redeployed to low wage, low technology activities on the periphery of the world system, blocking capital formation and holding back technological development. At the same time, differences in productivity and/or the value of labor power lead to unequal exchange between developed and underdeveloped countries, draining the latter of a significant portion of the value they produce, and holding back their development. This, in turn, blocks the formation of demand for high technology goods and high skill services, constituting a further obstacle to social progress. Finally, in order to rectify the resulting tendency towards underconsumption, states attempt to "pump up" their economies through deficit spending on both income-transfer, demand supporting, and military technological programs. The resulting expansion of the public debt further strengthens *rentier* elements, raises interest rates, and leads to overconsumption of luxuries and a crisis in capital formation.

The market system is so destructive of social organization precisely because it treats human beings as individual atoms related to each other in a purely external fashion, and thus undermines the mutual determining relations by which social systems sustain and develop their complexity and that of their constitutive elements. The strategy for social and cosmic evolution proposed by Tipler, far from promoting the development of intelligent life towards an infinitely self-organizing Omega Point, would, in the short run, devastate the world economy by undermining investment in infrastructure, education, research, and development and, in the long run, replace complex, living, intelligent systems with a race of predatory machines which make some of the most frightening artifacts of science fiction, such as Star Trek's Borg (which at least assimilated, rather than completely destroying, other cultures) seem benign by comparison.

It is necessary, finally, to say something about Tipler's vision of eternal life as a kind of computer simulation run on the "hardware" of the Omega Point. Tipler's vision of the Omega point in fact seems to have more in common with a comfortable and even decadent retirement than with any project for the realization of the self-organizing activity of the cosmos in an infinitely creative, powerful, knowing, and loving future. The many minds which Hayek rightly points out are necessary for authentic social development, are here replaced by a single mind which spends eternity amusing itself with simulations of other minds which no longer exist——and which cannot therefore really challenge it to grow and develop in new ways. The neoliberal entrepreneur who boasts of his contributions to human historical——and even cosmic——development here shows his true colors. Behind the entrepreneurial cosmetic lies the decaying corpse of consumerist, *rentier* capitalism.[12]

3. Towards a Synergistic Theory

---

[12]. It should hardly be surprising that theories which seek what amounts to eternal life in a computer simulation should be produced in cities like Santa Fe (home of the Santa Fe Institute and in many ways a suburb of Los Alamos National Laboratory) or New Orleans (where cosmologist Frank Tipler teaches in Tulane's Department of Mathematics). Both cities have, in effect, been transformed into "simulations" of Indo-Hispanic and Afro-Creole culture respectively, by yuppies who don't seem to know the difference between a simulation and the real thing, or who, more likely, prefer the "clean" simulation, free of people who actually work with their hands, and might actually affect them, to the "dirty" reality. Is God just the ultimate yuppie, who, after buying out the universe, spends all of eternity creating simulations of inferior life forms and cultures for his own enjoyment?

It should be apparent by now that the neoliberal interpretation of the new science is fundamentally inadequate. Neoliberal theory, in so far is it adheres to an atomistic view of the world, cannot theorize the high levels of interdependence and holism discovered by such new disciplines as unified field theory, nor can it comprehend the self-organizing dynamic discovered by complex systems theory, postdarwinian evolutionary biology, and dialectical sociology. Clearly the worldview implicit in the new science has much more in common with objective idealism or dialectical materialism than it does with the logical atomism or infotheoretical subjectivism favored by the neoliberals.

But even these problematics are inadequate to the task of theorizing the philosophical implications of the new science. As we have seen, even when objective idealism is adapted to accommodate authentic growth and development (e.g. in the work of Teilhard de Chardin) it made this growth and development the product of an organizing principle outside of matter. But the whole burden of the new science has been to show that matter itself is dynamic and self-organizing. Dialectical materialism, on the other hand, by making complex organization purely and simply the product of material interactions, without recourse to any prior "organizing principle," fails to capture the profound sense in which organized complexity appears to be *necessary* to the very existence of the universe. The result is an atheistic doctrine which fails to provide an adequate ontological ground for organization and development. The highest levels of organization thus come to be regarded as purely and simply the product of human organizing activity. As we have seen this leads inevitably to either an authoritarian disorder, in which the party substitutes itself for the missing ontological ground, or else to postmodern relativism and nihilism, which deny meaning and value altogether. Clearly it is necessary to transcend in a more profound way the form/matter dualism which remains at least as a residue in the objective idealist and dialectical materialist traditions.

In this section, I will sketch out in broad outlines an alternative interpretation of the philosophical implications of the new science. My aim here is not a comprehensive statement, which is clearly impossible within the confines of just a few pages. I am concerned, rather, to provide an adequate logical-ontological, cosmological, and axiological groundwork for the more specifically sociohistorical argument which follows.

The most fundamental philosophical implication of the new science is the radically relational character of the universe. Relativity, quantum mechanics, complex system theory, organismic, developmental, and evolutionary biology, and dialectical sociology all suggest that the entire cosmos is interconnected at the most fundamental levels of its being. As we noted in the previous section, this means not only that the cosmos is a system in the sense that the behavior of its constitutive elements is radically dependent on the behavior of all the other elements in the system, but also that the essential nature, indeed the very existence, of these elements is determined by their interrelationships with each other.

This principle has important logical-ontological implications. It is no longer possible to understand the universe as a composite of immaterial forms and a passive material substrate, or as a set of interacting atoms which sometimes come together to constitute systems. Being is not substance but relation. Indeed, it is first and foremost a *system of relationships*, from which it is possible to abstract certain *nodes* which therefore *appear* particular, but which exist, and can thus be comprehended, only as part of the general system.

A few points of clarification and elaboration are in order. Relationship implies both unity and difference. Being realized as relationship consists neither in simple, undifferentiated unity nor in pure difference. Without difference there is nothing in particular, but only a One which is at the same time Nothing. Without a prior, underlying unity, difference is mere disintegration: the absence of any capacity to connect, to relate, and therefore potentially to act, have properties, etc. Being consists precisely in the capacity to unite things which differ——in the self-differentiating unity which we call "system." The word "system" comes from the Hellenic roots *sys-* and *histanai* meaning "to put together." At the very simplest level, therefore, system refers to the radical interconnectedness of all things, an interconnectedness so profound that the existence of the tiniest subsystem abstracted from the whole implies the system in its entirety. The most minute alteration at any point in the system affects the system as a whole. The fact that I am sitting here at my computer, thinking and writing requires and implies, with iron clad logical necessity, *everything* else in the universe——not only the existence, but the precise disposition of every particular system along every possible world trajectory in the cosmos, from the most intimate thoughts of a young woman on a corner in Bangkok waiting for her lover to the precise disposition of the atoms and molecules in some remote nebula in a galaxy far too distant for its light to ever reach me during my lifetime.

This approach has the merit of clarifying the relationship between appearance and essence. The universe generally, and its various subsystems, *appear* to us as things possessing various properties. The underlying essence or nature of a system or subsystem, however, (what it is), is determined by its internal and external relationships, of which its appearance is merely the expression. *Essence*, in other words, is nothing

other than *structure*, both a system's internal structure and its place in the
larger structure of the cosmos as a whole, which defines both its own
trajectory of development, and its contribution to the development of the
cosmos generally.

Now the structures of various subsystems of the cosmos do not merely
differ from each other. These structures are arranged in a kind of
hierarchy or dialectical scale. We already know from the results of the
special sciences the characteristics of at least several different levels on this
scale. Mechanical systems are merely ordered but do not have the capacity
to combine and form larger wholes which are more than the sum of their
parts. Chemical systems, on the other hand, manifest precisely this kind
of holism. Carbon and oxygen, for example, combine to form a new
whole, carbon dioxide, which has properties which make it quite different
from either of the two elements which compose it. With biological systems
we see the beginning of purposefulness or teleology. Each particular organ
within an organism has a specific function with reference to the whole, the
integrity (and reproduction) of which constitutes its *telos* or goal. Finally,
with the social form of matter, we find the capacity to develop systems
which have new structures, and thus serve new functions, which were not
encoded in the genome of the organisms which created them.

The principle which governs this hierarchy, is nothing other than the
principle of organization itself. By organization we mean the integrating
power which brings systems into being. Chemical valence reflects a higher
degree of this integrating power than mere mechanism, biological organism
a higher degree than mere chemism, and social organization a higher
degree than mere biological organism. The extent to which a particular
subsystem of the cosmos is organized makes it physical, chemical,
biological, or social, while its specific pattern of organization makes it e.g.
carbon rather than oxygen, a sheep rather than a goat, an engineer rather
than a philosopher, etc.

While each one of these types of system expresses to a greater or lesser
degree the organizing dynamic which is embedded in being itself, none of
them express this dynamic perfectly, and, indeed, none of them possess the
characteristics or capacities necessary to account for the fact that there is
something rather than nothing, cosmos rather than chaos. For this it is
necessary to have recourse to the idea of a system whose very essence it
is to organize, and thus to bring into being. This system, "organization
itself," corresponds to the scholastic idea of the *ens realissimum* (Harris
1987: 193) ——that being whose essence it is to be. Such a structure
would contain within itself all other structures, which are, ultimately, just
partial, imperfect expressions of its own divine nature, vanishing moments

in its drive to express itself. Such a form includes the qualities and capacities of all lesser forms (relationality, holism, self-organization, teleology, consciousness, the capacity for love, work, power, creativity, knowledge) as well as others which, located as we are at a more humble place on the dialectical scale, we are unable to even conceive.

This way of theorizing being is superior to even the most advanced forms of objective idealism or dialectical materialism. Objective idealism shows while organization is necessary, but treats it as something external to matter, a view which conflicts with the new science and which is pregnant with authoritarian potential. Dialectical materialist theories of reflection, we have seen, have the merit of highlighting the role of concrete interactions in the process of development, but do not really explain *why* there is organization (and thus being) rather than chaos (and thus nothingness) or why systems evolve towards higher, more complex forms of organization. Our own approach captures the strengths, and avoids the weaknesses, of both doctrines. On the one hand, unlike objective idealism and like dialectical materialism, we make organization internal rather than external to matter. If being *is organization*, then unformed matter simply doesn't exist, while an "organizing principle" is nothing other than organization *in potentia*——what the philosophical tradition has historically understood as the *prima materia*. On the other hand, like objective idealism and unlike dialectical materialism, we show why organization is necessary, and thus supply an adequate ontological ground to the whole process of cosmohistorical evolution. If being is organization, then everything existing participates, to a greater or lesser degree, in the drive towards organization. It becomes necessary, furthermore, to acknowledge the existence of an organizing power sufficient to bring the universe into being in the first place, and which, as organization itself, contains within itself all logically possible forms.

If being is organization, then logic, far from being purely a purely formal science, is, in fact, nothing other than the science of the organizing principles of being itself, the rationally necessary determinations of the very concept of being: system, structure, organization. It is possible, to be sure, to find a place within this system for what has generally been called formal logic, which retains a certain usefulness in theorizing the operation of very simple systems, such as "information systems" and the mechanical or electronic "hardware" on which they run. Specifically, formal logic is the logic of systems in so far as they are treated as systems of only externally related particles. It is also possible to find a place for the kind of "transcendental logic" developed by Kant, Fichte, and Schelling, and further elaborated by the phenomenological tradition and by certain "transcendental" Neothomists such as Bernard Lonergan. Transcendental logic is a reflection on the logical conditions for the existence of the various "essences" or structural forms discovered by the special sciences. Ultimately, however, any complete logic must be dialectical: it must be able to demonstrate the logically necessary interrelationship of all things to each other, so that starting from any one point (even from Nothing) it demonstrates the necessity of everything without making any unproven assumptions (Harris 1987). In demonstrating the necessity of all things, it demonstrates their existence, at least *in potentia*, or what is the same thing, their existence as *matter*.

Our approach also has important implications for the interpretation of such traditional metaphysical categories as "matter" and "spirit." Matter, from our perspective, is simply the potential for organization. We use "potential" here in its authentic sense, to mean a latent, as yet undeveloped, organizing capacity and not merely the ability to receive organization from the outside. Spirit, on the other hand, is actual organization, the developed ability to organize. The "degree of spirituality" of a system is the degree of actualization of its organizing capacity. How does matter become spirit, or what is the same thing, how does "organization as such" act to bring the cosmos into being? Synergism rejects both *taxis* and *katallaxis*——the idea that organization is imposed on matter from the outside, or emerges spontaneously through the interaction of discrete particulars. Rather, we argue, *cosmopoesis* (the emergence of organization) is a complex process which can be understood in three ways. First, there is a unitary, underlying, organizing principle which contains the entire organization of the cosmos *in potentia*, and which of which the cosmohistorical process is simply the gradual, logically necessary unfolding. At the same time, this logical unfolding involves real, material interactions. These are not interactions between discrete particulars (which our theory does not allow), but rather interactions which bring simpler systems into relationship with larger, more complex systems, thus negating their relative particularity and transforming them into something new, more complex, and more highly organized. Finally, all systems are drawn towards the beautiful, the true, and the good, towards the complex synergistic integrity which they already *are*, at least implicitly, and in which alone they can find true joy.

On the basis of this ontology, it is possible to systematize the results of the special sciences and to show that the Universe is in fact a unified, self-organizing, teleological system. Now matter in its simplest form is the metric field, the space-time continuum which constitutes the matrix for the formation of more complex forms of organization. But we have already seen that according to the general theory of relativity the metric field, because of its inherent structure, its "curvature" already necessarily implies such higher forms of organization as gravity. The complete unification of physics will undoubtedly demonstrate that it implies, either directly or indirectly, the other fundamental forces (the strong and weak nuclear forces and electromagnetism) as well.

As one draws out the logically necessary implications of one level of organization, one identifies contradictions which can be resolved only by realizing the existence of yet another layer of organization. Thus at a certain point, in struggling for logical completeness, mathematics becomes physics, physics chemistry, chemistry biology, and biology sociology. By dictating the existence of a certain number of elements with certain definite electromagnetic properties, physics dictates chemistry, but at the same time brings into being a higher level of organization which operates according to laws which are no longer merely physical. Similarly, the most complex chemical compounds, such as DNA, take on the ability to organize other matter into definite structures serving well defined functions, and thus bring into being a new, biological level of organization. Organisms, finally, relate to each other in complex ecosystems and societies. Initially this interaction may be of such a character that it can be comprehended using purely biological concepts. But the concept of the collective action of an animal society on an ecosystem requires, for completeness, a whole range of processes, which define what we have come to think of as the social: technology, social organization, language, culture, etc. Intelligence is necessary to the existence of the universe because it is logically necessary in order to complete all of the lower levels of organization, each of which, in turn, is necessary to complete the level of organization subordinate to it. And merely human intelligence is not sufficient. Logical completeness requires an actual infinity of increasingly intelligent beings, which converge on the infinitely wise, infinitely powerful, infinitely loving being which will exist at Omega. It is not so much that less complex systems require more complex systems in order to design and "run" them (as Barrow and Tipler argue) but rather that because being is organization, and organization implies both further differentiation and higher levels of integration, the lower levels of organization, including the lowest, prime matter or the metric field, contain within themselves the seeds, as it were of all of the higher and more complex levels of organization.

But these relations of logical implication are, at the same time, real material relationships——interactions which gradually bring into being the real world of physical, chemical, biological, and social organization. In their simplest form these interactions are purely physical: they are the four fundamental forces——gravity, the strong and weak nuclear forces, and electromagnetism. The last of these physical forces, however, forms the substratum for more complex chemical interactions. Chemical interactions form the basis of biological relationships, and biological interactions the basis of social relationships. Our theory does not depend on some mystical transition from quantitative to qualitative difference, nor on the idea of reflection——i.e. the capacity of material systems to retain the "imprint" as it were of its earlier interactions with other systems. It is the actual relating which yields progress and development. Any given system, by interacting with other systems, realizes in practice the web of inter-relationships which is implicit in it from the very beginning. To put the matter differently, it become actually connected to, and participates in, the cosmic whole on which it depends for its very existence, and to the existence of which it really and truly contributes. When an electron relates to a proton it becomes part of something larger than and qualitatively different than either: an atom of hydrogen. When that hydrogen atom relates to atoms of carbon, nitrogen, and oxygen, it becomes, under certain circumstances, something larger and qualitatively more complex——an organic molecule. And so on, up the dialectical scale. This underlying relationality is most apparent, of course, in the social form of matter. By interacting with other people we are forced to acknowledge ideas and interests different from our own. While it is possible to attempt to ignore or annihilate the other, it is generally more productive to collaborate. And in the course of collaboration we are changed and expanded. We develop more sophisticated ideas, wider interests, etc. We become more complex people, better able to love, to produce, to build and exercise power, and to organize our experience of the world. Eventually these social interactions will yield a higher, supersocial form of relationship, the nature of which we can only imagine.

Cosmohistorical evolution is, finally a teleological process because every particular system is, at least implicitly, "aware" of the telos of the universe, and is drawn into the very interactions which move the universe towards this *telos*. In the simplest systems, to be sure, this awareness is not in any sense conscious. It is just that the nature, i.e. the structure, of e.g. the electron is such that it is drawn to the proton, and thus drawn into the larger whole which is the proximate manifestation of complex synergistic integrity of the cosmos. And even in complex organic and social systems this awareness of the end may be rudimentary to say the least: a desire for food or for sex or a deeply felt but poorly articulated religious longing. It is only at the highest levels of organization that matter, realizing itself as spirit, grasps the *telos* in anything like a fully adequate manner, and approaches realization of this end.

The evolution of the universe as a whole can thus be understood as the gradual "collapsing" of the cosmic wave function. This function encodes within itself the potential for all logically possible forms. We "collapse" the function through our interactions with each other,[13] gradually realizing these possibilities and thus bringing the cosmos into being, and move it along towards its ultimate goal. In its general form this function grasps the cosmos as a fundamentally relational, but still radically indeterminate system. As the cosmos becomes increasingly organized, this function gradually collapses, as the system realizes an increasingly large portion of its potential, and thus becomes increasingly determinate. At any point on the time space continuum, however, a significant degree of indetermination remains.

As the function collapses, increasingly more complex forms of organization emerge. Physical and chemical systems are ordered, but can *themselves* be said to be organized, i.e. to be structured to serve some purpose, only in the light of the fact that they are structured in such a way as to make more complex systems possible and even inevitable. Only biological systems are actually organized, i.e. structured in such a way as to serve some purpose internal to the system. Social systems take the organizing process one step further. Human societies are not merely organized; they themselves create increasingly complex forms of organization. We should note here that the process by which systems develop towards increasing levels of complexity itself becomes more complex as we move up the organizational scale. Mechanical processes give way to chemical, chemical to biological, and biological to social. But the social form of matter does not exhaust the organizing potential encoded in the wave function. On the contrary, as we have seen, full realization of the potential for organization latent in matter requires an actual infinity of increasingly organized forms which converges on the telos, Omega, for which all things yearn.

Throughout the process as a whole it is organization which collapses the quantum wave function. Human action (including the action of the scientific observer, which is itself a form of organization) does not intervene into the system from the outside, but is, rather, an integral part of the system and thus contributes to the process by which the function collapses, matter becomes more organized, and the cosmos comes into being.

---

[13]. "We" here meaning not only all intelligent being, but all subsystems within the cosmos, including the most primitive.

This way of thinking about the world, which is still difficult for physicists trained in the study of mechanical processes, is not really even counterintuitive for those of us who study human societies. A human person is a complex system with a capacity for a very wide, but still limited range of behaviors. No matter how much information we have regarding a person (relationships, capacities, interests, etc.) we do not know for sure just what that person will do, though we might be able to assess the relative probability of various courses of action. A human life might, therefore, be described by a function which specified the relative probability of each possible course of action. As the person acts, and expends (or realizes) his or her potential, that function collapses, and the person's life takes on greater definition and specificity. Much the same is true of a social system, or an entire civilizational complex.

Indeed, once we have understood the question from this point of view, the discovery that human beings affect the position or velocity of "particle" in an experimental apparatus, which so distressed physicists, seems trivial. For we affect far, far more. Through the whole process of bearing and raising children, through labor, through building and exercising power, through artistic creativity, scientific research, philosophical reflection, and through our spirituality, we add whole new layers of complexity to the organization of the universe. The human civilizational project is an integral dimension of, and indeed, a critical leading force in, the self-organizing activity of the cosmos. Cosmohistorical evolution, therefore is the logically necessary working out of the capacity for organization latent in matter itself. The evolutionary process itself takes the form of real material interactions——physical, chemical, biological, social, etc. But it is also a result of the working out of an axiological (aesthetic, epistemological, ethical, theological) imperative. In diverse ways, appropriate to its various levels of organization, the universe and each of its constitutive relationships and subsystems "wants" to become more loving, more creative, more powerful, beautiful, more true, more good, and more highly integrated. It wants to be One, not in the sense of an undifferentiated unity, but in the sense of being able to generate and relate infinite difference.

It is possible, finally, on the basis of the ontology and cosmology which we have outlined, to derive a new theory of value. In so far as organization is not merely a logical possibility or an actual state of affairs, but also a goal, it is in fact a *value*——it is the real *telos* towards which the universe and everything within it constantly struggles, what Aristotle and the Scholastics called the "final cause."

Particular values are simply the specific forms of organization which we have already encountered at various points along the dialectical scale, themselves realized as ends. These are, at the lower levels of the scale, relationship, holism, self-organization, and teleology. The higher values, however, are those which correspond to the forms of organization which made their appearance only in the social form of matter: the nurturing love which transforms human animals into fully social beings, the creative capacity of labor which reorganizes physical, chemical, biological, and social matter into increasingly complex forms, the power which taps into the diverse talents and interests characteristic of complex societies and channels them into enterprises more complex than any single individual could undertake or even imagine, and, finally the ability to create beauty, grasp the truth, and do good, thereby bringing into being ever more complex forms of organization. These last three values, because they define organization itself (being as such) as a value are called transcendentals.

In so far as the social form of matter is not, by itself, a fully adequate expression of the drive towards holism implicit in being itself, but must ultimately give birth even more complex forms of organization, even the Good, understood as that ordering of social relationships which best promotes the development of social beings and social capacities, is not yet a fully adequate expression of the concept of value. Nor are the specific virtues proper too supersocial forms of matter fully adequate. The supreme value is nothing other than the complex synergistic integrity of the cosmos, the One——God herself——understood as the highest good and ultimate end for which all being, the entire cosmos, strives, and which draws all things to herself.

It should be apparent at this point that the new sciences of organization involve nothing short of a philosophical revolution, a radical break with the atomistic paradigm which has dominated European philosophy for the past three hundred years, and the emergence of a new, synergistic, dialectical philosophy which comprehends the cosmos as a relational, holistic, self-organizing, teleological system. But unlike traditional objective idealism, which regarded the organizing principle of the cosmos as something radically distinct from matter itself, these new sciences regard matter as relationship, and the cosmos as a self-organizing process. And unlike traditional dialectical materialism, which regarded complex organization as simply the result of material interactions, we recognize that matter itself *is* the potential for organization, and therefore develops necessarily from less complex to more complex forms.

We are thus able to affirm with certainty, on rational grounds the ultimate meaningfulness of the universe, and to put to rest once and for all the demons of postmodern nihilism and despair. At the same time, we can also affirm, with certainty and on rational grounds, that we are not mere subjects of a transcendent divine sovereign whose decrees we must obey. Nor is there some realm of forms, separate from matter, in which the music of the spheres plays more beautifully than it does in our own realm, and to which, therefore, we must flee. On the contrary, the social form of matter, and the human civilizational project in particular, play a critical, even leading role in the cosmohistorical evolutionary process. And there is no ontological limit to our participation in the self-organizing activity of the cosmos which has brought us into being——no limit to our ability to compose still more beautiful strains, and to build the instruments on which to make that music play. The purpose which we are realizing is not merely human in character. We are agents of cosmos, impelled by a drive written into the very matter from which we evolved, and drawn on by the incredible beauty of God, to continue this struggle which is already more than fifteen billion years old.

Omega itself is, at least in part, the product of human historical action. Thus the critical importance of understanding as well as we can the basic processes of social organization. We need to understand the social form of matter not simply in order to bring it into harmony with divine law, or to fulfill narrowly human purposes, but rather so as to catalyze its development towards increasingly complex forms of organization, and thus realize the cosmic vocation of the human civilizational project.

CHAPTER THREE:
The Social Form of Matter

We are now in a position to begin our consideration of the principal subject matter of this work: the nature, significance, and general course of development of the human civilizational project. We begin with a discussion of the characteristics of the social form of matter, since it is the social form of matter which generates human civilization. We have already hinted, in at least a preliminary way, at the defining characteristic of the social. Social systems are constantly engaged in generating ever new forms of organization, characterized by ever more complex structures, while subsocial forms of matter at most merely reproduce existing organizational forms. The social form of matter organizes, while biological matter is merely organized.

In the course of our discussion we will make a distinction, which will turn out to be important, between social organization and social structure. By social organization we mean the actual process by which the social form of matter creates new, more complex forms of organization. Social organization includes the rearing of children, the labor process, building and exercising power, communication, artistic creativity, scientific research, philosophical speculation, and human spirituality. By social structure, we mean the system of rules which regulates the way in which these organizing processes are carried out. Social structure includes kinship rules, relations of production (rules for the centralization and allocation of resources), authority relations, and the ideological systems which govern artistic, scientific, philosophical, and spiritual activity. Social organization brings society into being; social structure determines the type of society which emerges. Taken together, social organization and social structure constitute a complex whole, a social formation embracing a structured system of organizations and institutions, which is itself constantly evolving——which is itself a concrete historical process contributing layer after layer to the organization of the cosmic whole. Realized as such it is a conscious project, human civilization, and an active participation in the cosmohistorical evolutionary process.

I. Social Organization

A. Socialization

The transition from the merely biological to the fully social constitutes a fundamental "phase transition" in the organization of matter, an organizational, and thus ontological breach as significant as that entailed in the emergence of life out of complex chemical compounds, or of chemistry out of the physical matrix constituted by the fundamental forces of nature. Sociality cannot, therefore, be regarded as something encoded in the genome of the species *homo sapiens*. It is, rather, itself the product of a definite social process——the process of socialization.[1] This process of socialization consists fundamentally in the development of new organizing capacities which are not encoded in the genome. Broadly speaking there are three dimensions to the socialization process:

a) the process of technical-organizational-intellectual development, i.e. the emergence of the ability to organize physical, chemical, biological and social matter through labor, the building and exercising of power, through artistic creativity, knowledge, etc.,

b) the process of affective-moral development, i.e. the transition of the child from an animal interested only in realizing its own biological drives into a complex human person interested in other human persons, and in the self-organizing activity of the cosmos as a whole, and

c) the process of spiritual development, i.e. the process by which the human person comes to comprehend him/herself as an active participant in the self-organizing activity of the cosmos, and to find his or her realization in that participation.

---

[1]. This is one of the internal contradictions of the neoliberal theory which regards the emergence of complex social organization as a spontaneous process, pointing, among other social institutions, to the family (Hayek 1988: 135-143). What the theory ignores is the fact that the family embodies a social process which, while certainly not the product of a centralized state planning agency, is certainly conscious and, in a broad sense, at least, rational, engaging the best energies of men and (especially) women struggling to raise productive, powerful, creative, intelligent, and loving children.

Clearly understanding the nature of this phase transition is essential to understanding the social form of matter in general.

The importance of the problem notwithstanding, however, there has thus far been only limited progress in understanding the actual character of the socialization process. On the one hand, theories of technical-organizational-intellectual development (e.g. the work of Piaget) have been largely divorced from studies of affective-moral development (e.g. the psychoanalytic tradition broadly understood). On the other hand, theories which have stressed the positive development of human social capacities (Piaget, Kohlberg, Fowler) have said little about how this development actually takes place, while theories which have analyzed the actual process of development have tended to stress the role of repression at the expense of understanding how human beings develop complex new abilities.

It is not possible in this context to remedy the limitations of current social-psychological theory. Rather, we will simply attempt to outline some tentative directions, which are supported by current research, and draw out the implications of these principles for our larger argument.

Initially——i.e. at birth——the complex of biological relations which constitutes the individual human organism expresses itself through various animal capacities and animal drives, which impel it to sustain and reproduce itself. On the one hand, like all of the higher forms of animal life, human children are sentient. They are aware of their surroundings, which, however, appear to them as a kind of undifferentiated unity within which they are able to distinguish neither themselves as subjects nor other things as objects. At the same time, also like the other higher animals, they are driven to sustain and reproduce their own organisms. By themselves these capacities and drives are neither social nor antisocial, but rather simply biological. Specifically, they are the expression at the biological level of the drive towards holism characteristic of all forms of matter.

Let us examine this question in greater detail. All life, or at least all animal life, sustains itself by consuming other living things (plants or animals). This is often regarded as a sign of some underlying aggressiveness structured into animal existence. This is a mistake. Aggressiveness implies some net destructiveness. Most predation, however, involves the assimilation of elements of a lower animal into the body of a higher one. While not creative in the sense of bringing into being a more complex form of organization, predation does, at least, tend to mobilize lower forms of life in the service of higher. Higher animals, furthermore, reproduce themselves through sexual intercourse with other members of the same species. Since all mammals are born from the wombs of their mothers, they have a biologically conditioned bond with other animals of the same species, and a desire to re-establish that lost biological union (For a detailed discussion of various analyses of fundamental drives cf Fromm 1973: 34-54).

Both the drive to consume and the drive for sexual intercourse, are, therefore, fundamentally, good. They represent a certain level of realization of the tendency of matter to develop towards higher levels of organization, and are in fact necessary to that dynamic. At the same time, these drives do not bring into being new forms of organization which were not structured into the genome, and therefore are not, by themselves, fully social in character.

Almost immediately, however, the infant *homo sapiens* comes into contact with other members of his or her species. It is this interaction itself which sets into motion the emergence of authentically social capacities. There are both intellectual and affective-moral dimensions to this process. On the intellectual side, the child begins to be exposed to language, which provides a way in which it can begin to distinguish the various objects in its field of experience, identify their various properties, and begin to understand the relationships between them.[2] The child thus makes the transition from mere sensation (a capacity which it shares with other higher animals) to *perception*. A further process of dialogue (much of which takes place outside the family, in formal educational institutions) begins to make the child aware of both patterns and contradictions in his/her everyday experience. The child learns that there are certain general categories (universals) to which particular things belong, and of which they are simply individual examples, and that the relationships which s/he perceives between these individuals have a law-like character. This is the level of the *understanding*, and represents the highest level ordinarily achieved in our society. In some cases, however, through rigorous philosophical training, human beings may take a further step and begin to recognize that the world is not constituted simply by things with properties, or even by universal laws by which individuals are related to each other in an ordered but still external fashion, but rather by a unitary organizing principle of which particulars are merely the manifestation, related inwardly to each other at the most fundamental levels of their being, and driven by this relationality to develop towards ever more complex and

---

[2]. There is a growing body of evidence that some other animals, including parrots and dolphins as well as apes, are capable of **at least** this level of intellectual development, and must, therefore, be regarded as participants in the social form of matter.

comprehensive levels of organization. This is the level of *reason* proper, or of the *dialectic*.

A similar process takes place at the affective level. The child's immediate awareness of its own unity with the world around it (primary narcissism) is gradually undermined as it recognizes that its relationship to its environment is not one of command and control, but rather of radical dependence. At the same time, in so far as its needs are in fact met, the child gains a sense of self-worth, and of continuing participation in a reality larger than itself. Gradually it becomes aware of the ways in which its desires conflict with those of others (parents, siblings) and becomes engaged in a struggle with them over realization of those desires.

This is a critical moment in the socialization process. Logically, there are three possible solutions to the child's dilemma. It can win its struggle and force others to submit to its desires, without recognizing or attempting to help realize their desires. Aside from being an unlikely outcome, given the power relations between the child and its parents and older sibs, this alternative would mean that the child would fail to develop, because it would never develop a relationship with or interest in any system outside of or larger than itself.

On the other hand, it is certainly possible for the child to be defeated and its desires to be frustrated by the superior power of the other members of its family. This is, of course, the alternative celebrated by Freud, and the orthodox psychoanalytic tradition. This alternative certainly makes it possible for the child to become aware of, and indeed to learn to serve, interests different from and larger than his or her own. At the same time, the child never learns to make those interests his or her own——to develop to the point of actually caring about others, about their joys and sorrows, their fears and aspirations, etc.

Such a level of development is possible only if the child is encouraged to recognize the interests of others as formally equivalent to his or her own, and to recognize those others as human beings, i.e. as other selves. Once the child makes this leap, it can learn to cooperate with others in order to realize its interests along with those of others. It may understand this cooperation in terms fixed by traditional social norms, or else as something fluid, to be negotiated and renegotiated as part of a kind of social contract, or perhaps even as something governed by universal moral norms: social utility, universalizability, etc. This is the highest level of affective development ordinarily attained in our society.

As in the case of intellectual development, however, it is possible for human beings to go further. By developing unusually rich and complex relationships with others they may gradually come to understand both themselves and others not as separate and distinct persons related externally by tradition, contract, or rational moral norms, but rather as participants in a complex interrelated totality (the cosmos) which is constantly developing towards ever greater degrees of complexity and organization, and begin to take an active interest in the ongoing evolution of that totality, of which their own particular interests (and those of others) are mere manifestations (Kohlberg et al 1983: 249ff).

Taken together, the intellectual and moral capacities formed in the process of socialization make the human animal into an authentically social being at once conscious of and interested in the self-organizing activity of the cosmos. Now this ability to at once comprehend and take an active interest in the system as whole is what the religious traditions have historically understood as spirituality. We will have more to say regarding the process of spiritual development at a later point in this chapter. At this point, we will just outline the foundational process of spiritual development in so far as it is part of the ordinary process of socialization.

At the stage of primary narcissism there is no spirituality properly so called. While the child may experience a sense of unity with the cosmos as a whole, it is not aware of the cosmos as a system larger than itself and the element of self-transcendence is thus entirely absent. As the child enters the stage of perception and of the struggle to realize its desires vis-a-vis those of others, it is also exposed to the religious ideology of its society, which mediates knowledge of the cosmos as an infinite, complex, self-organizing system through a symbolic language which, however, the child takes as literal. Its world is inhabited by God(s), angels, demons, etc. in much the same way as it is inhabited by stones, plants, animals, and other humans, each regarded as a thing with properties externally related to other things, each able, albeit in varying degrees, to be used, submitted to, or bargained with in order to realize various particular interests and desires. As the child begins to understand in an elementary way the law-like behavior of the things which inhabit its field of perception, and learns to cooperate with others in order to realize his own interests, he or she likewise begins to ascribe law-like behavior to (the) God(s) and to participate fully in the partially rationalized religious systems which characterize most human social formations. Generally speaking this means regarding the fulfillment of ordinary social duties appropriate to one's station in life as at the same time a fulfillment of one's duties to the cosmos as a whole, so that the person's particular calling is endowed with cosmic significance, however small, and is thus sublated in the self-organizing process of the cosmos as a whole.

For those whose intellectual and moral development proceeds beyond the conventional level the process is more complex. The analytic processes of the understanding soon run up against the multiple internal contradictions of religious language (which generally operates on the level of perception) and begins to discredit religion as "mere myth" or superstition. At the same time, the social norms governing cooperation are rejected as "mere convention" as the increasingly autonomous individual discovers myriad other ways to realize his or her interests in conjunction with others, often through fluid and ever changing arrangements which call radically into question the larger meaning of moral action. In societies characterized by the rapid disintegration of traditional communal structures and by a failure to develop new, more complex forms of social organization, many become stuck at this level and slip into skepticism and nihilism, actually denying the meaningfulness of the human history and of the cosmos as a whole.

At the highest levels of development of the understanding and of rational morality it is possible to recover at least a partial sense of spirituality. There is, after all, an internal logic to human action itself, whether this is regarded from a purely formal standpoint (internal consistency, universalizability) or substantively, as a drive to maximize utility. From this standpoint human social life appears as an island of meaning in a larger chaos, and human life and society are thus endowed with a kind of tragic sacredness which helps to ground science and morality.

The full realization of human spiritual potential, however, presupposes a dialectical level of intellectual and moral development. When we grasp the underlying unity of the cosmos, and its dynamic of self-differentiation and self-organization, we recover at a higher level of rationalization the sense of cosmic unity and meaning which religion mediated to us in symbolic form. When we comprehend ourselves and others as participants in a larger totality we recover at a greater level of emotional depth and maturity the sense of participation in something larger than ourselves which characterized our first encounter with religion. Now, however, we understand that we are called not simply to bring our ideas and interests into harmony with some given set of traditional norms, moral laws, etc., but rather to actively participate in the self-organizing activity of the cosmos, and thus in the creation of the cosmos itself.

We should note here that while the process of development expresses an underlying organizing principle latent in the child, at each and every point in the process, and along all three dimensions of development (intellectual, moral, spiritual) the actual mechanism of development is interaction with other human beings, which makes us aware of the internal contradictions of our thoughts and desires, and propels us towards an ever higher synthesis. Kinship structures which attempt to break the primary bond with the mother, or which attempt to repress (rather than catalyze the differentiation and development) of desire in fact hold back, and can even completely derail, the socialization process.[3]

The result of the socialization process is to transform the individual human organism into a social product and thus part of the web of human social relationships. The person *is* the intersection of this web of relationships with the underlying biological drives. This intersection, however, is not simply the locus of a mechanical interaction of forces. Rather it is itself a new form of organization. The biological drives (partly) determine the aims of the individual, while the socialization process actually creates new and more complex aims. The result is a structure which is relatively autonomous vis-a-vis biological drives. Our drive to eat, to have sexual intercourse, etc., can, at least to some extent, be deferred. The structure is also relatively autonomous vis-a-vis the social forces which are brought to bear upon it. We can, and do, violate social norms, even those which we have internalized in the most powerful ways, making it possible for us to act to reorganize the basic structures of our

---

[3]. There are a number of difficulties with psychoanalytic theory. First, there is a growing body of evidence which suggests that Freud's whole account of infantile sexuality was in fact developed to cover up evidence of widespread child sexual abuse (Miller 1986). Far from being a residue of the family's efforts to enforce the incest taboo, and thus bring over-sexed young human animals into line with the requirements of civilized live, the "superego" and the other structures which Freud regarded as the mark of a healthy personality may in fact be the residues of incest and other forms of sexual and physical abuse.   Second, it is, in fact, precisely the child's bond with the mother (or a surrogate) which introduces it to human society and forms the basis of the child's ability to form bonds with others. There is a growing body of evidence that repression of this bond in fact cripples social development, leading to a whole complex of narcissistic-authoritarian personality disorders. When the child's bond with the mother (or surrogate) is ruptured, it loses its sense of connectedness to other human beings, and thus its sense of self-worth, while at the same time coming to believe that ordered human relationships are possible only on the basis of repression (Lasch 1977, 1979. Chodorow 1978).   Finally, Freud's approach fails on the one hand to grasp the way in which biological organization contributes to the social form of matter, while at the same time deriving social bonds entirely from the biological substrate of which they are merely the sublimated forms, thus negating the radical newness of specifically social forms of organization.

society. The human person therefore, while wholly a product of biological and social forces, and thus radically determined, acts in a way which is itself determined only by the interaction of its own internal structure with the outside world. In this sense the human person is autonomous in and through its overdetermination. It is the *subject* of its own actions.[4]

Not every human subject, however, is characterized by the same degree of autonomy. The process of socialization takes our fundamental biological drives and transforms them into social *interests*. Every human being necessarily acts in accord with his interests. This does not mean, however, that everyone is selfish. The word interest derives from the Latin words *inter esse*: to be between. We are interested in those things to which we have some relationship. Not everyone has the same interests, nor is everyone equally interesting. The infant, before undergoing the process of socialization, is "interested" only in satisfying its immediate physical needs. Gradually it develops more complex interests. Some people never develop interests which go much beyond a partial sublimation of their elementary biological drives. Such persons will always be more or less at the mercy of their bodies, and the immediate social circumstances in which they find themselves. Other human beings develop a real interest in their work, or in complex relations with other human beings not merely as a means of satisfying biological needs, but as ends in themselves. They take pleasure in reorganizing matter into more complex forms, in organizing and directing the activity of others, and in changing other human beings, and being changed by them, through *social* intercourse with equals. A few persons go further, developing an interest in the organization of the cosmos as such, in a way that transcends their own limited participation in that organization, and actually derive pleasure from activity which promotes the development of the cosmos to ever higher levels of organization, even if they themselves derive no immediate benefit from that development.

---

[4]. Jonathan Edwards' distinction between natural and moral necessity is helpful in this regard. Human beings are the subject of their actions, and thus morally responsible for them, so long as they are doing what they want to, even if their desires are the product of biological and social forces external to them (moral necessity). Only external coercion (natural necessity) relieves us from moral responsibility. Where Edwards' analysis falls short, it seems to me, is in his understanding of causality. Edwards was constrained by the physics of his time which had a mechanistic understanding of causality, in the context of which determination by biological and social forces was incompatible with constitution by those forces of an autonomous subject (Edwards 1957a).

The wider the scope of a person's interests the greater his or her capacity to exercise power, and thus the greater his or her autonomy. A real interest in others permits us to grasp their concerns and talents and thus to create networks of collaboration which permit us to realize those interests which we share, but which lie beyond the scope of our individual capabilities. The wider the scope of our interests, furthermore, the less constrained we are by either our own biological drives, fulfillment of which is less interesting to us than our complex relationships with others, or the internalized norms which make up our superego, which we recognize as merely the crystallization of the interests of a limited social system. A broad understanding of our own self interest, in this sense, is the foundation of creativity, power, knowledge, and spirituality.

The human psyche realized as a subject is thus a complex system articulated over a field which includes many other systems: physical, chemical, and biological systems, but also and most importantly other psyches, and indeed organizations and institutions and their products which "interest" the subject. In this sense, it is impossible to speak rigorously of a "psychic" level of organization distinct from the social form of matter. The psyche is itself a social organization, and it is impossible to form a coherent concept of the psyche apart from an understanding of the social form of matter in general.

In order to be fully autonomous, of course, a system would have to be determined only by its internal organization, which was, so to speak, self-caused and unaffected by any external relationships. This is possible, however, only for the general system——i.e. for the cosmos as a whole——and only when that system is fully developed, and thus capable of consciously and rationally organizing all of its parts. Human persons participate actively in the self-organizing activity of this system precisely to the extent to which they develop their capacities for productivity, power, knowledge, and love. In this sense human spiritual development is just an extension of the socialization process. At the same time, in so far as we always and ever fall short of actually *becoming* God, we remain subject to an element (a very large element in fact) of determinism and heteronomy.

B. The Dimensions of Social Activity

1. Labor

The process of socialization brings into being a subject capable not only of reproducing biologically and socially but also of performing labor.

By labor we mean the process by which human beings reorganize physical, chemical, organic, or social matter, and raise it to a new level of complexity. In this sense the biological process of eating is not labor. The matter consumed is simply decomposed and assimilated to the already existing structure of the body. From the moment, however, that human beings begin to hunt and gather food which they store away so that they can devote time to activities other than hunting and gathering——e.g. various forms of religious life——the activity serves the development of a human social world and is thus properly regarded as labor.[5]

Labor is the most fundamental social process not because it is the most distinctively human characteristic, or the highest level of development of the human capacities, but rather, on the contrary, because it is the simplest determination of the concept of sociality and the key to understanding the more complex determinations of human social activity. *All human activity is, in the final analysis, labor, in the sense that all human activity reorganizes existing matter (physical, chemical, biological, social) in accord with some human purpose.*

In this sense the notion of "unproductive labor" is a contradiction in terms. All activity which increases the level of organization of matter is productive. All activity which does not increase the level of organization of matter is not labor. Political and cultural activity which authentically develops human capacities is productive. Political and cultural activity which merely reproduces existing relations of production, power relations, religious mystification, etc. is unproductive. Labor which does not actually increase the level of organization of matter, but which creates the social conditions for such productive activity, in a society in which such productivity would otherwise be impossible, we call *indirectly productive labor.* Such, for example, is the labor of the entrepreneur in a capitalist society, who brings together the capital and the labor necessary for

---

[5]. We should note here that we reject Hannah Arendt's distinction between labor and work. By labor Arendt means activity which serves to support the life process. By work she means something which leaves behind an artifact of some kind which becomes part of the human social world. Labor to support basic life processes, however, is the essential precondition for work which creates artifacts, or indeed any other human activity. Even the simplest forms of labor create the free time in which to engage in other activities: it creates the social space in which civilization unfolds and develops, even if it does not itself make an enduring contribution to the civilizational process (Arendt 1958).

production. The necessity of such indirect labor is a mark of contradictions between the process of social organization and the structure of the society in question, since such indirect laborers generally appropriate for themselves a share of surplus which might otherwise be invested in expanded production (on productive and unproductive labor cf Marx 1857-1858/1978: 221ff).

Human productive capacities are such that even at their lowest level of development human beings can produce more than they need for subsistence. We can thus make a distinction between necessary and surplus labor. *Necessary labor* is the labor which is needed to reproduce the existing system of social relations at its present level of organization. *Surplus labor* is labor performed above and beyond this point which is available to raise the overall level of social development. Labor is thus a dynamic process which tends, over a period of time, to develop towards ever higher degrees of complexity and productivity.

It is possible to identify several discrete levels of development of the human capacity for labor. We will consider each of these levels later in more detail as constituent elements of the complex social totalities of which they formed a part. For now we are interested only in the very different degrees to which they manifest the essential characteristics of the human labor process (On the various stages of human technological development cf Lenski 1982).

Hunting and gathering is labor only in a very rudimentary sense. The activities of hunting and gathering differ from activities performed by various animals principally in the use of tools to kill animals, and in the creation of various containers in which to store plants which have been collected for food purposes. This makes it possible to accumulate a sufficient surplus of food to permit some time to be devoted to other activities: making tools, developing new social relationships and deepening old ones, ritual practices, etc., which may contribute to the overall development of human productive capacities.

Gradually human beings learn how to tap into the self-organizing dynamic of matter itself, in order to find ways to satisfy their subsistence needs and generate the surplus product necessary for development ——though they do not, at this stage, actually understand how this self-organizing dynamic works. This is the stage of horticulture[6]

---

[6]. Horticulture comes from the Latin root *hortare* which means to encourage, and thus captures well the unique quality of this mode of production, which is centered on tapping into and encouraging the self-organizing processes of biological matter, rather than on breaking down the existing structure of matter to release energy, and then using

————cultivation of plants without the aid of the plow. On the one hand human beings have become involved in an actual reorganization of the natural world: clearing, planting and harvesting fields, etc. On the other hand, this labor produces a larger surplus which permits the development of a world which is entirely of our own making: the world of the village with its permanent dwellings and its lodges and ritual centers. Some advanced horticultural societies have even been able to support the development of cities. This surplus also made possible the emergence of a leadership of priests and protoscientists who studied the movements of the stars and the cycle of the seasons, and who taught their people the arts of cultivating plants and raising animals. We will see that the period during which horticulture was the dominant mode of production was one of tremendous human development, during which humanity made most of the important technological innovations and scientific discoveries necessary for the development of human civilization. It is little wonder that philosophers in later advanced agrarian societies looked back on this as a golden age, or that we find surviving horticultural societies on our own planet (e.g. the Hopi) to be such an extraordinary source of wisdom.

From this point of view the emergence of agriculture is an ambiguous development. Agriculture makes possible the creation of a huge social surplus which can support the city with its temples, palaces, garrisons, workshops, marketplaces, etc., and eventually its public assemblies and universities. At the same time, agrarian production is characterized by a growing reliance on mechanical processes (plowing, irrigation, etc.) which often disrupt the complex integrity of ecosystems and lead to deforestation and desertification. Whereas horticulture requires careful attention to be paid to each and every plant and animal, and taps into the unique interests and abilities of each of the individual members of still relatively small societies, agriculture treats plant, animal, and human simply as a source of energy or labor power, without regard to its specific characteristics or its optimum use within the system. In so far as it reduces the vast majority of human beings to mere beasts of burden or instruments of labor, agriculture introduces into human society a fundamental dualism between the organizer and the organized, which, as we have seen, is the social basis for the form/matter dualism, and related ideological disorders.

The industrial revolution extends this contradictory process of development and disintegration still further. Industrial production is characterized by the decomposition of existing physical, chemical, biological, and social structures in order to harness the energy they contain,

---

this energy to impose a new structure.

and produce new structures characterized by an even higher level of organization. Thus we burn chemical and nuclear fuels, decomposing complex compounds and nuclear structures in order to produce energy and drive machinery which creates products designed to serve some human social purpose. We break down the existing structures of family, clan, village community etc., and organize human beings in new ways, under the discipline of the factory, to harness their labor for purposes which they would not spontaneously have embraced. On the one hand industry makes every member of society in some sense a participant in the civilizational project, in that everyone is involved in producing artifacts with some enduring quality which become a permanent or semipermanent part of the social world. Furthermore, industry permits the accumulation of a surplus sufficient to permit between one third and one half of the population to become involved in the political and cultural activities which were formerly the province of a very small ruling class. At the same time, industry extends even further than agriculture the tendency to ignore the rich variety of social interests and abilities, and to exploit both nature and humanity in the drive for increased production.

In this sense both agriculture and industry, while in themselves more productive than horticulture, do not necessarily increase the overall complexity of the biosphere. They may in fact simply displace organization from one location (a fragile ecosystem) to another (a complex of machine tool factories).

The full development of human productive capacities requires a *synergistic* mode of production which is based on a scientific understanding of the self-organizing character of the material world——something which is only now becoming possible. Such a mode of production would draw on renewable nonpolluting energy sources and raw materials. Production would be organized in such a way as to qualitatively increase both the overall complexity of the ecosystem and the level of development of human social capacities. At their outer limit human productive capacities develop to the point where routine labor is abolished, or becomes largely the province of machines, with human beings devoting themselves either to the design, monitoring, and reorganization of the productive process, or to political and cultural activities. We will have occasion to discuss the conditions for the development of such a mode of production in subsequent chapters.

It should be clear from this discussion that human labor is, in the final analysis, the motive force of human development and human history. It is by reorganizing matter that we create a human social world and make it possible for ever larger sections of the population to spend an ever larger portion of their time in such higher order activities as politics, art, science, religion, and philosophy. This statement requires, however, certain qualifications. First of all, human labor is itself determined by the complex social totality of which it forms an integral dimension. The level of development of the labor process includes the organization of the work force, which is a political reality, and such cultural factors as the artistic creativity of the craftsman, the level of development of our scientific understanding, and the religious meaning attached to work within a particular social formation. Second, the role of the labor process in determining political, and cultural development must not be understood in positivistic fashion, as if the ecological, technological, and economic facts of the labor process determined the political and cultural facts of the superstructure. The development of a higher level of material interdependence at the level of the productive forces, characteristic of advanced industrial society, for example, does not automatically produce at the level of consciousness a rational grasp of this interdependence and thus the basis for global solidarity. Rather, political and cultural processes must themselves be regarded as labor processes with the difference that the object of labor is now the organization of society itself. These activities will be constrained, in ways we will indicate below, by the overall level of development of the productive forces. The interdependence of economic, political, and cultural processes is, furthermore, overdetermined by the system of relations across which those processes are articulated.

2. Power

Labor, realized as the capacity to transform social relations as well as physical, chemical, and organic matter, is power. In so far as all labor is social, and thus involves social organization of some kind, power is present, at least in a rudimentary form in even the simplest social formations. To persuade one member of the community to gather fruits and berries while I hunt, to develop a division of labor in the course of the task of hunting itself, so that several members of the party herd the animal into a narrow ravine, where I am waiting to make the kill——even these very simple tasks involve the organization of human beings for a particular task and thus involve power.[7]

---

[7]. In this sense the Thomistic tradition was correct in arguing against Augustine, that political life is a constitutive dimension of human social existence, and not simply a result of social contradictions.

The elements of power are, first of all, the interests of the various persons involved in the organization. Human interests, in so far as they are themselves relationships imply both connectedness and difference. Between any two or more people, there will be both shared interests and conflicts of interest. Realizing my interests may help you to realize yours; or else it may undermine your effort to realize those interests——e.g. by consuming scarce food. What mediates between the interdependent but contradictory interests within any group is a plan, which specifies just how a coordinated pattern of actions by various members of the group will realize at least some of the interests of some of the members. The more interests a plan can accommodate, the more individuals it will be able to organize and mobilize, and the more power the group will be able to exercise.

It should be apparent from what has been said so far that power is not something which can be possessed, but is rather something which is exercised between two or more people who have intersecting interests, and a common plan for realizing those interests.[8]

The human capacity to build and exercise power develops historically along with the capacity for labor, with which its development is radically interdependent. In hunter gather societies there is relatively little differentiation between the interests of the different members of the society, all of which move within a rather narrow orbit. On the one hand this makes it very easy to achieve unanimity. On the other hand it means that the total power of the society will remain very limited.

In horticultural and agrarian societies, the existence of a larger social surplus permits a segment of the society to develop wider interests——i.e. in the study of the natural world for example, of the rotation of the seasons, the movement of the stars, the patterns of flood and drought. This knowledge permits those who possess it to introduce innovations in the productive process which benefit the society as a whole——i.e. irrigation systems, more precise timetables for planting, etc., and eventually to organize and direct the process of production as a whole, in return for which they are able to extract a share of the social surplus

---

[8]. Bogdanov gives an interesting illustration of this point. "Experience from the French colonial wars has shown that with equal armaments the average Arab soldier in a one to one encounter is no worse than the average French soldier; but a detachment of 200 French soldiers is already stronger than an Arab detachment of 300-400 men; and a force of 10 thousand Frenchmen is able to demolish the army of natives numbering 30-40 thousand men. European tactics give a much more perfect summation of military forces, and mathematical calculation is in fact refuted (Bogdanov 1980: 45)."

product. At the same time, even if their control over a society is absolute, the limited differentiation of the population limits their power. They can persuade the population to grow food, to turn over a share of the surplus, to work on public works projects, to participate in a centralized cult, and perhaps even to fight in wars of defense or conquest——projects which are in the real or imagined interest of the peasantry. There is very little possibility, on the other hand, that they could transform the peasantry into a class of philosopher-cultivators, or persuade them to carry out a systematic rational reorganization of society——a lesson which philosophers and prophets and would-be messiahs have had to learn again and again throughout the course of history. This is because society has not yet developed to the point to allow the cultivation of a widespread interest in such pursuits.

The most critical transition in this regard is the democratic revolution. The differentiation of interests engendered by industrial forces of production in fact requires conscious political action either to directly allocate products, or, more often, to establish legal norms which will govern the spontaneous exchange of products through the medium of the marketplace. In so far as everyone is involved in producing goods which they themselves cannot consume, but must exchange for other goods, everyone has an interest in the terms which govern this exchange. As interests become more and more differentiated, it becomes increasingly difficult to create unanimity within the population, or even to form organizations within one sector of the population around a broad range of issues. At the same time, this differentiation of interests makes it possible to mobilize thousands and eventually millions of smaller cross cutting groups to carry out a tremendous array of different projects.

In industrial societies most collaboration is motivated less by an interest in power as such than in the realization of interests which lie outside of the political process. Generally speaking these interests have to do with the production or consumption of wealth. As human productive capacities develop, however, human labor is less and less absorbed in direct material production and more and more devoted to the organization of the production process and thus to the reorganization of *social* relationships, or else to artistic, scientific, and philosophical creativity. Those who expend their labor in organizing and directing the work of others have a direct interest in building power. Those engaged in cultural activities directed at comprehending the self-organizing activity of the cosmos have an indirect interest in building power, as a means of realizing the vision which emerges from their creative activities. It thus becomes possible to organize directly around the task of increasing the total power of the society——around the all-sided development of human social capacities, and thus to realize the democratic promise to which industrial society could give only a partial fulfillment.

This is the basis for the emergence of a synergistic form of power. At this stage of development the particular interests of each and every person is, immediately and directly, the development of the cosmos as a whole. It thus becomes possible to tap the whole productive capacity of every individual, so that all of their energy is deployed in such a manner as to best promote the self-organizing activity of the cosmos, without any need to make concessions to limited visions or imperfectly focused desires, and without any infringement on the dignity or autonomy of persons, because each in fact, in giving everything, does precisely what he or she wants to do. Power thus becomes perfect, within the limitations imposed by the finite nature of the social form of matter, so that the whole capacity of the social formation is organized optimally to serve the process of cosmohistorical evolution.

3. Culture

Our ability to organize nature and society depends on our ability to communicate——to represent to ourselves the various elements, operations, and relations of the systems in which we participate, and to exchange these representations with other human beings. This communication process must organize our experience of the cosmos——or, what is the same thing, it must make explicit the organization which is already present implicitly in the physical, chemical, biological, and social forms of matter.

When we communicate we engage in a form of labor which (like all labor) consists in the reorganization of matter. The matter we reorganize however, is not the physical, chemical or biological matter of the production process, nor is it the directly social matter——the fabric of social relations——which is the object of political activity. It is, rather, the system of signs which mediates those social relationships, and by which we represent those relationships as a system. We call this process *semiosis*. Semiosis appears first of all as the relationship between signifier and signified. This relationship may take many forms. It may be iconic: the signifier may in some way resemble the signified. It may be symbolic: the signifier may simply be an arbitrary representation of the signified. Or it may be indexical: the signifier may point to the signified. T h e s e different forms of semiosis are not simply indifferent alternatives, but, rather constitute an ascending dialectical scale. Iconic signification corresponds to the level of intellectual development which we have called perception, which treats the world as a system of interrelated things, each with its own properties. Symbolic signification——the kind of signification used in ordinary spoken and most forms of written language——corresponds to the level of intellectual development we have called the understanding, which treats the world as a system of atomic particulars only and externally related by formal rules. Paradigmatic rules define the ways in which signifiers relate to signified. This system of rules defines the semantic or meaning complexes of the language. Syntagmatic rules, on the other hand, define the ways in which signs relate to each other. Indexical semiosis corresponds to a fully dialectical form of thought in which universe is realized as a pure system of relationships within which "things" or "particulars" are simply nodes of intersecting relations. Communication is realized as a form of relationship building, in which the representation of systems is simply a moment in a larger organizing process.

Semiotic, like economic and political, systems thus develop over time. As social formations become more complex iconic and symbolic forms of semiosis become increasingly inadequate, and gradually give way to the indexical forms characteristic of dialectical thought. Just what a purely indexical system of communications, completely free of representation and symbolism, would look like, however, is not yet clear, and may ultimately presuppose a transition to supersocial levels of organization.

Semiosis, therefore, *is* consciousness——the cosmos' own consciousness of itself as an organized system, developing towards ever higher levels of complexity. The recognition of language, the material form of thought, as a social product, in no sense implies that the world represented or comprehended in thought is anything other than the "objective" world existing outside of consciousness. On the contrary, in so far as language emerges from the physical, chemical, biological, and social organization of matter (which is what we actually mean when we speak of the "objective" world), it includes and encodes within itself these less complex forms of organization to which, as it evolves, it gives an ever more adequate expression.

Human semiosis takes place first of all through the medium of natural languages, semiotic systems which have emerged spontaneously and apparently independently in different regions of the planet. Gradually, however, human beings have begun to reflect on the process of semiosis itself, and have created second order semiotic systems within the framework of natural languages——artistic, scientific, and philosophical discourse——which organize our experience in a more complex way.

In this sense, semiosis becomes a creative process which brings into being a totality of relationships. This creative process is, first of all purely imaginary. Human societies reorganize the system of signs (and the meaning complexes these signs convey, which have been endowed by the socialization process with profound affective significance) in such a way as to represent to themselves the whole of the cosmos, and thereby reflect on their position within that cosmos. This representation does not initially grasp the real structure of the cosmos. It is *only* a representation, an image, but at the same time one endowed with profound affective significance. It is *art*, in all its various forms (visual, musical, literary), and in all its various stages of development, from archaic myth and ritual, which take themselves to be a representation of the actually existing cosmos, through the "fictional" forms dominant in industrial capitalist society, to the as yet undeveloped synergistic forms of artistic creativity.

At a certain point, however, language begins to represent the representational process itself. At this point, humanity begins to grasp the concept of organization and its various determinations. Thus the emergence, in many advanced horticultural and agrarian societies, of various arithmetic and geometric systems, in which order, pattern, etc., themselves become objects of experience and representation. This is the critical epistemological rupture which stands between the realm of representational thinking and on the one hand, and scientific research on the other hand.

Once humanity has grasped the concept of organization, it can then use this concept to analyze the various specific forms of organization which characterize the physical, chemical, biological, and social forms of matter. In the process of doing this, however, it necessarily becomes lost in the rich diversity of the cosmic manifold, isolating and identifying the relationships between an ever increasing number of variables. The result is a atomism which is written into the scientific method, and which represents the world as a system of only externally related particulars.

Gradually, however, this analytic process begins to run up against what increasingly appears to be an infinite regress. Behind every presumed "elementary" particle lies a complex system of still more fundamental particles, and so on, until science gradually becomes conscious of the fact that it is actually theorizing a system of relationships within which "particles" characterized by a definite position and velocity, mass and energy, etc. appear increasingly to be merely theoretical constructs or experimental artifacts. Realization of the radically relational character of the cosmos leads necessarily to a grasp of the self-organizing dynamic at work in the universe, and eventually to recognition of the cosmos as an organized, meaningful, totality.

Even before the emergence of such a theory, however, science begins to reflect on itself, and on the whole body of evidence it has accumulated, and to become aware of the "community and interrelationship of all things" of which art was aware but could only capture in representational form. Such theories attempt an approximation of the organizing principle of the cosmos, and use this principle to explain the movement of matter towards ever more complex levels of organization, of which science itself is one of the most developed. In so doing, it uncovers the *beauty* of the cosmos, its *truth*, and its *goodness*, which consist precisely in this development towards increasingly complex forms of organization, in its gradual realization of its own complex synergistic integrity. Science, therefore, which has understood not merely the underlying structure of a particular form of matter, but which has grasped the totality and teleology of matter, passes over necessarily into *philosophy* or the love of wisdom. Wisdom itself *is* the organizing principle of the cosmos, which is, as it were, the very mind of God, written into the fabric of the world, and reproduced in human language.

Philosophy, in this sense, is at once knowledge and love, and recovers on the basis of scientific truth the affective dimension which artistic symbolism captured at a lower level of development. In philosophy we come to know and love the cosmos, realized as a system of relations, or what is the same thing, to know and love God. We thus realize the end towards which matter itself has been evolving for over 15 billion years.

C. Spirituality

The cosmos which philosophy comprehends as an organized totality is the same cosmos in which we participate actively by bearing and raising children, reorganizing physical, chemical, and biological matter, by building and exercising power, etc. This active participation in the self-organizing activity of the cosmos, become conscious of itself *as* participation, is *spirituality*. Spirituality is thus not merely an intellectual attitude (belief in God) nor is it merely disposition or a feeling (love of

God). Rather it is the whole act of being, in so far as this act is conscious of its real cosmohistorical significance.[9]

We have already noted that our ability to participate in the self-organizing activity of the cosmos is a function of the level of our development as human persons. Partly, of course, this means our level of technical skill, our capacity to build and exercise power, our artistic, scientific, and philosophical creativity. The development and above all the deployment of these capacities is, however, determined in part by the breadth and depth of our interests. The wider our interests, the more we will be able to produce, organize, know, etc. Both the level of development of our individual capacities, and the breadth of our individual interests is, furthermore, radically constrained by the overall level of social development. We cannot be interested in, much less love, something which we do not know.

This means that human spirituality develops historically in much the same way as human productive, political, and scientific capacities. In relatively simple societies, in which human organizing capacities are as yet insignificant, the organization of the cosmos appears as at once static and mysterious. When the impact of the social form of matter is as yet relatively insignificant, it is the cyclical patterns of physical, chemical, and biological matter which predominate: the harmonious movements of the heavenly bodies, the cycle of the seasons, of menstruation, birth, death, and rebirth. When human science has grasped only the fact that organization exists in the cosmos, or at most the concept of organization, the way in which that organization operates, and perhaps even its ultimate purpose, remains a mystery. Human beings have little hope of really participating in the self-organizing activity of the cosmos (of participating in the life of God) and thus little hope of contributing anything of significance to the development of the cosmos——unless God himself somehow infuses into us divine capacities which transcend our own human nature. At this stage we understand our spiritual task as largely one of conforming to a pre-existing cosmic order. The highest calling remains that of the contemplative.

The advent of industrial capitalism marks a qualitative leap forward for human organizing capacities. Industry remakes the natural world in the image of humanity. Democratic revolutions reorganize the structure of

---

[9]. In this sense the Thomistic theory of virtue, which stresses the actual development of human capacities, is superior to the Edwardsean theory, which regards virtue essentially as a disposition (to love Being in General) (Aquinas, Summa Theologiae, I-II: 90, 91, 93, 94, 95, 99; Edwards 1957c).

society, and the scientific revolution begins to penetrate the hitherto mysterious processes which govern the natural world. Human beings come for the first time to see themselves as participants in the self-organizing activity of the cosmos. At the same time, in so far as industry breaks down the existing organization of the natural world, and reorganizes nature in accord with specifically human purposes, rather than in accord with the requirements of cosmic evolution per se, (which remain mysterious), there is a tendency for the spirituality of industrial societies to become centered on a struggle for mastery over rather than participation in nature and history. This can lead either to the dangerous illusion of human sovereignty over a process which in fact transcends human knowledge and control, or to the conviction that human activity is merely a locally meaningful island in an ultimately hostile and entropic universe.

It is only when human labor, the exercise of power, and the production of knowledge are realized as participation in the self-organizing activity of the cosmos, that a fully developed human spirituality can emerge. On the one hand, such a spirituality recognizes that human labor makes a real contribution to the development of the cosmos. Our task is not merely to conform to a pre-existing natural order, but rather to participate in creating the complex organization which brings the cosmos into being. We provide to this process not merely muscle, but authentic creativity. Humanity makes a difference in the development of the universe. At the same time, human organizing activity has an end outside of merely human history. Our task is not to remake the universe in our own image, nor to create an island of human meaning in a hostile universe, but to participate in the creation of complex new forms of organization which will ultimately surpass a merely human level of development. Because of this, human creativity, power, knowledge, and love can find their fulfillment only in something which radically transcends merely human capacities. This is the Omega point, the infinitely creative, powerful, knowing, and loving end which we help to bring into being, and the desire for which sustains us on our difficult journey through a segment of space time which, for all its beauty, can never fully satisfy our desires.

II. Social Structure

Now if the social form of matter is, like being in general, first and foremost an organizing process which constantly develops towards increased complexity and integration, then (also like being in general) any given society will be characterized by a specific structure which determines just what kind of society it is. The basic organizing processes of society are thus regulated by structures which determine how the resources necessary for social development are centralized and allocated. More specifically, social structure determines the centralization and allocation of the intellectual and affective resources necessary to socialize children, and of the raw materials, tools, and labor needed for production, as well as the way in which power is exercised and art, science, and philosophy produced. In short, it determines the form of human participation in the self-organizing activity of the cosmos, and thus the shape of human spirituality. The specific structures governing these various dimensions of social life taken together form an integrated whole, which we call social structure.

Different social structures are not merely equal alternatives, "different ways of doing things." Rather, some structures are better able than others to facilitate the continued development of human social capacities, and thus to further the process of cosmohistorical evolution, realizing ever more fully the potential latent in the social form of matter. The various types of social structure are, therefore, arranged in an ascending dialectical scale. In so far as the social form of matter is finite, however, there is no one optimum structure which makes possible the "perfection" of human society. The perfection of any finite system necessarily lies in a new level of organization (in this case a supersocial level of organization).

We will discuss the specific structures regulating the various dimensions of social organization, exploring the evolution of each in turn.

A. Kinship and the Regulation of Socialization

The earliest human societies——bands of roving hominid omnivores——were probably somewhat amorphous in structure, with sexual relations taking a rather free form character as individual organisms followed the shifting patterns of their personal preferences, and with the whole band sharing responsibility for rearing the young.

Eventually it became advantageous to form stable alliances with other bands. Intermarriage provided a ready way in which to accomplish this. Stable marriage, however, involved restriction of the free form sexual expression which had hitherto prevailed. Kinship systems emerged which arbitrarily divided human societies into definite groups and then required that marriages take place outside of those groups, leading to the formation of a complex network of intergroup alliances (Levi Strauss 1969a). This also had the effect of setting in motion a differentiation of the relationships within each band. The establishment of permanent or semipermanent marriage bonds permitted the gradual separating out of the whole manifold of predominantly nonsexual relationships: the relations between parent and child, between siblings, cousins, and other kin, etc., and thus laid the ground work for the development of greater affective complexity.

At this point, however, the network of alliances between bands is purely external in character, and the distinction between them purely negative. The fundamental social bond is with someone whose defining social characteristic is that they are "not my kin," while the directive to marry outside one's kin group takes the purely repressive form of the incest taboo.

This kind of structure, which we call a *tribal* form of social organization, opens up the possibility of serious social deformations. Gayle Rubin has argued the gift relations between groups which emerge as a result of kin networks are fundamentally relations between groups of fathers (or, in matrilineal societies, between brothers) who exchange wives. Such kinship systems thus transform women into property (Rubin 1975). Even if we do not accept her verdict that all tribal societies are patriarchal, it is clear that many actually existing tribal societies are, and Rubin's analysis of the precise nature of the tribal patriarchy is quite perceptive.

Furthermore, the role of kinship rules in regulating the socialization process, as we noted earlier, is fundamentally negative. By organizing socialization around the internalization of the incest taboo, and thus a rupture of the bond with the mother, kinship structures undermine the basis on which the child actually develops the capacity to understand and organize the world around it, and on which it develops more complex social relationships and interests. Submission to law, enforced by fear (initially fear of rape, castration, etc.) becomes the basis of a social order characterized by a high degree of social control (i.e. the ability to prevent certain actions) but a relatively low level of power (i.e. the ability to organize and mobilize human interests and energy to get something done).

For this reason, societies regulated primarily by their kinship systems tend to be relatively undeveloped, even if they have managed to acquire horticultural or even agrarian technologies.

A note of caution is, however, in order. We should approach with some suspicion the notion that the all human societies pass through a "tribal" stage, at least in the sense in which we have defined it. There is some evidence that contemporary hunter gatherer tribal societies are in fact degenerate forms of more complex societies which had already developed horticulture, the village community, etc. (Levi-Strauss 1963). We simply do not know how archaic paleolithic societies were structured. It is quite possible that what we have defined as the tribal stage was simply an transitional form of social organization, intermediate between the band and the village community, and that in contemporary tribal societies these transitional structures became ossified and blocked all further development.

In any case, it is only with the emergence of the village community, which brings together members of several different bands or clans into a single settled community, bound together by a shared vision of the cosmic whole which they serve, that makes possible the further development of human social capacities. With the emergence of the village community the narrowness of vision and of drive which characterizes undeveloped human nature is constantly challenged by the demands of other members of the community so that, without ever being forced to repress his/her own ideas or desires, the child is gradually initiated into the larger world of the village and the cosmos. This is the significance of the African proverb which says, quite correctly, that "the whole village raises the child." It is also the significance of the African initiation ritual for infants, during which the child is raised up to face the starry sky as the initiator pronounces the following words:

Behold, the only thing greater than thyself.

B. Economic, Political, and Ideological Structure

1. Relations of Production

We have already established that human labor (unlike the feeding of animals) raises matter to a higher level of organization and is thus able to produce a surplus above and beyond what the laborers themselves require for their sustenance. It is precisely the production of a surplus which leaves behind a more complex structure than existed before the labor was performed. The performance of surplus labor is thus the basis of the antientropic vocation of the social form of matter and the foundation of the human civilizational project.

The possibility of producing a surplus raises several difficult questions. First of all, it is necessary for a society to develop some means of insuring that a surplus product is actually produced. Second, it must determine how this surplus will be allocated——something which in turn will exercise a powerful influence over what is produced during the next cycle of production, and thus on the overall pattern of economic development. How much will be allocated to maintaining the integrity of the ecosystem and the social fabric? How much to new development? What resources, if any, will be siphoned off for luxury consumption?

It is possible to identify several distinct ways of extracting and disposing of the social surplus product. (For an excellent account of the debate around the various "modes of production," cf Amin 1980). Societies which do not produce a surplus, either in the form of some tangible product, or in the form of complex kinship and mythic-ritual systems, we call *band* societies. In nearly all cases, however, human beings are able to produce some surplus, even if it is only in the form of "free time" when the members of the community, having produced sufficient food, shelter, clothing etc. devote themselves to informal and formal (i.e. ritual) social interaction——that is they perform surplus labor producing the society itself by creating the web of concrete and mythical relationships which holds the society together. Societies of this kind have traditionally been called *tribal.*

Gradual development of the labor process, and more specifically the development of horticultural and agricultural forces of production, creates a sufficient surplus to permit a small minority of the population to be liberated from direct production altogether, to devote themselves to the political and cultural work of centralizing surplus, building cities, and engaging in artistic, protoscientific and religious activities. The vast majority of the population, the peasantry, supports the activities of these leaders by means of rents, taxes, labor service, temple sacrifices, tithes, etc. The system of extracting this surplus may be centralized or decentralized, and may rely on persuasion, coercion, or a mixture of the two. To the extent that the surplus is extracted primarily on the basis of the prestige of leaders whose knowledge and organizing ability contributes to the productivity of the society as a whole, and to the extent that the surplus is invested primarily in supporting public works, the development of a democratic public arena, and on protoscientific and artistic activities, we may speak of *communitarian relations of production.* To the extent that the surplus is extracted through coercion or religious mystifications of authority, and is used primarily to support military conquest, such societies are characterized by *tributary relations of production.*

Development of industrial forces of production creates a complex division of labor and a high level of material interdependence which can no longer be organized by means of the centralized redistributional mechanisms of tributary relations of production. Increasingly production is not of use values but rather of commodities, the exchange of which is mediated by the marketplace. The marketplace operates by transforming the complex relations of interdependence characteristic of industrial society into things (commodities) and more specifically into one "general equivalent" (money) which can be exchanged for any other thing. The operation of the marketplace permits the extraction of surplus by indirect economic means. In a market economy, the value of a commodity is determined by the average socially necessary labor time consumed in its production. But labor power is itself a commodity, with a value equal to the average socially necessary labor time consumed in producing it——or, what is the same thing, in producing the commodities necessary to its reproduction. But when the capitalist purchases the labor power of the worker, he purchases the entire product of that worker's labor, and not merely the necessary labor for which the worker is being reimbursed. He thus extracts surplus value from the worker, which, if he can sell his products at or above their value, he will realize as profit. A society in which extraction and disposition of surplus is organized by the marketplace is characterized by *capitalist relations of production* (On capitalism cf Marx 1867/1978, Mandel 1968, Amin 1980).

The marketplace has no access to information regarding the impact of particular activities on the complexity of the ecosystem and on the development of human social capacities. Indeed, there is considerable reason to believe that as the technological level of an economy increases there will be a tendency for capital to be redeployed to low wage, low technology activities on the periphery of the system, holding back technological development. Exchange between low wage and high wage firms, regions, etc. will, furthermore, result in a transfer of surplus from the former to the latter, blocking capital formation, and thus also blocking development. Resources in a capitalist society are thus allocated in such a manner as to maximize the profits, and ultimately the individual consumer interests of those who control capital, and not the qualitative complexity of the social system as a whole. And market relations make all activity simply a means of realizing individual consumer interest, undermining human organizational, artistic, scientific, spiritual, and political development. As a result, movements emerge to gain social control over capital and to allocate resources in accord with a central plan. Thus the advent of *socialist relations of production* (cf Amin 1982).

Socialism represents a reassertion of the redistributional structures which governed communitarian societies, with the single and very important difference that these structures are now articulated across a field of productive organizations driven by industrial, rather than horticultural or agrarian forces of production. Generally speaking, socialist systems are able to restrict or even abolish the market in capital, but leave the markets in labor power and consumer goods intact. Resources are centralized and allocated systematically in accord with the planners' understanding of what will best promote the development of human social capacities, understood as technological, economic, organizational, artistic, scientific, and philosophical progress. Such rationalized redistributional systems are generally more effective than markets at centralizing resources necessary for development (Therborn 1992). In this sense they have played a progressive role. In so far as they do not, however, abolish the markets in labor power and consumer goods, they fail to remove the social basis of the alienation engendered by the marketplace (Chiang Chung-chiao 1975). And centralized planning mechanisms have proven themselves unable to mobilize the interests and creativity of the complex working class of advanced industrial societies (Gorbochov 1987). The result is a tendency towards the restoration of capitalism as the penetration of market relations breaks down the social fabric, engendering consumerism and undermining the social basis of the communist parties.

Complete transcendence of the marketplace presupposes the emergence of synergistic technologies which comprehend and tap into the latent potential form organization which is present in matter generally, and the social form of matter in particular. At this stage of development, the performance of labor becomes a direct expression of the antientropic character of the social form of matter, which has now achieved maturity, and is organized according to the rational imperatives immanent in the labor process itself. Each contributes to the labor process in accord with his abilities, and draws as much from the common pool as will maximize the complexity of the ecosystem, of human society, and of the cosmos as a whole.

Synergistic social organization requires new planning mechanisms which are able to grasp the complex relationships of interdependence characteristic of advanced societies *better than the marketplace*. Synergistic planning must, like the marketplace, begin from the actual interests and capacities of the members of the society——and not from abstract requirements of a Plan. It must be able, however, to find ways to link up these interests in qualitatively more productive ways. Rather than allocating the skills of an engineer, for example, to the production of luxury automobiles, it might allocate these skills to the development of automobiles which are more fuel efficient, or to development of new types of transportation altogether. It must accomplish this by creating, on the part of the engineer, a knowledge of the objective social needs at stake, and an actual interest in fulfilling them, without recourse to material incentives which make work simply a means of realizing individual consumer interests.

In a synergistic society, therefore, the leaders must have a highly developed ability to analyze social interests and abilities, and to bring together people and organizations which are willing, and able, to carry out tasks which promote the all-sided development of human society——and of the ecological system in which that society is embedded. The people, for their part, must understand what this process of development requires, and have an active interest in serving that it.

Clearly there remain important unresolved questions regarding just what social mechanisms are necessary in order to centralize and allocate surplus and regulate production in such a way as to increase, rather than merely displacing, the complexity of the ecosystem, and to promote the most rapid, balanced, all-sided development of human social capacities.

2. Authority

It should already be apparent that securing the performance of surplus labor, and centralizing and allocating the resulting surplus product, necessarily involves the exercise of power. Until all human beings have developed to the point that they comprehend fully the imperatives of cosmohistorical development——something which will happen only at Omega——there is a continual danger that, enclosed within a narrow vision of the possible, and thus a narrow understanding of their self-interest, human beings will work only so much as is necessary to sustain themselves, and those with whom they have intimate relations——family, clan, village, etc., and that they will allocate any surplus which they do produce to luxury consumption rather than to investment in the further development of their own social capacities. If matter is to escape entropy, it is necessary either to force or to persuade the vast majority of humanity to perform surplus labor from which they will not derive a *direct* benefit. This means exercising power.

This power may be either linear or relational in character[10]. In exercising linear power a leader asks "What do I want? What stands in my way?" It is thus essentially coercive in character. This is the kind of power which is exercised by a raider over the people of the village he is raiding, or by a master over his slaves. The difficulty with linear power is that it is unable to engage the creativity of the people it coerces. Continuous, high levels of expenditure are necessary to maintain an adequate level of coercion. I may be able to get a group of villagers to part with their accumulated wealth, but I will not get them to produce any more for me unless I leave posted guards in the village. I may be able to get a slave to work until he dies, but I will not be able to get him to exercise his creative capacities in making innovations in the productive

---

[10]. On relational power cf. Mansueto, Maggie. "Relational Management: Building Dynamic Complexity," Mansueto, M. 1993.

process. Relational power, on the other hand, is based on a real or perceived unity of interests. The leader asks "What is my interest and who else is involved?" Relational power makes it possible for a leader to reduce the costs of repression and to tap into the creative capacities of the community he or she is leading.

In band, tribal, and communitarian societies, where there is little internal differentiation of interests or abilities, the exercise of relational power presents few problems. For the most part, authority is traditional in character. The leaders simply ask people to do what they already believe they should. Occasionally an especially creative person emerges who develops innovations which benefit the community as a whole, and who exercises authority based on his/her ability to make specific contributions to the common welfare.

With the emergence of warfare, exploitation, and class struggle this changes. The perception of common interests is based not on a real, scientific grasp of human interdependence, but rather on the mere representation or even the mystification——the alienated expression——of the common good in religious symbols (a matter which we shall discuss shortly). The leader is regarded as either himself divine, or else as standing in a special relationship with the divine, either in virtue of his personal qualities (charisma) or in virtue of his office (institutionalized charisma). This is the form of legitimation of power which is most common in societies characterized by tributary relations of production (Gottwald 1979).

In industrial societies, on the other hand, the real diversity of interests determined by the division of labor is recognized, and the elaboration of a common plan is legitimated by reference to a system of formal rules (themselves a reflex of the marketplace) which is supposed to permit all interests to be represented, addressed, compromised, etc. Authority is exercised by persons who occupy a particular office, who must act in accord with certain fixed rules. While the bureaucratic leader may issue particular decisions which are contrary to the interest of particular individuals, members of the society have an interest in maintaining the larger system of rules which, they believe, allows them to struggle to realize their own particular interests with a minimum of constraint (Weber 1921/1968).

In synergistic societies, power is based on a rational grasp of human interdependence, recognized not merely as a means of satisfying individual interests, but as a value in itself, which is realized in the very process of creating and exercising power. No longer is it a question of the leader having a "right" to lead, based either on divine status or election, or on his

legitimate appointment to an office———nor even, for that matter, on the
basis on his superior ability[11]. Society as a whole, rather, has developed
to the point that the vast majority of people have a real, active interest in
promoting the overall development of society, and the ability to grasp and
evaluate rational arguments. Furthermore, in so far as an individual is
really interested in serving the common good, the leader never asks
anything which is contrary to that person's interest. Those who lead are
simply those who are more advanced———who grasp the general line of
social development more clearly and more quickly than others, who
understand the strategic imperatives imposed by that line of development
and the various specific forms of activity required by it, who have the
ability to identify, train, and organize others with leadership ability, and
who, in general, can best deploy the resources of the society in the
interests of its own development and that of the cosmos generally. The
formal rationality of bureaucratic authority thus gives way to the
substantive rationality of a fully realized relational exercise of power.

3. Ideology

Power realized as a common plan of action based on a recognition of
the interdependent character of human society presupposes a representation
of that interdependence———a representation which, furthermore is endowed
with sufficient affective power to engage the interest of the individual
members of society. Just as the socialization process is governed by
kinship rules, production by relations of production, and power by
authority relations, human communication, artistic, scientific, and
philosophical production, and human spirituality, are governed by definite
ideological problematics. An ideological problematic is not a content,
which can be supplied only by artistic, scientific, or philosophical labor.
It is, rather, a way of regulating that labor, governed by the level of
development of the labor process and by the other dimensions of the social
structure itself.

In band societies engaged in hunting and gathering, and which as yet
have very little internal differentiation thought of necessity remains very
simple. The experience of the emerging synergy of the group provides the

---

[11]. This sort of meritocratic criterion should be regarded as a transitional form of
authority, associated, for example, with the socialist societies of the twentieth century,
in which the content of the rules governing selection of bureaucratic cadres, as well as
the criterion governing the exercise of authority by them, is increasingly rationalized, but
the **form** of bureaucratic authority nonetheless remains.

basis in the experience for a diffuse concept of a universal which operates in and through particular individuals, organizing and directing their action. This universal is the soul, and the resulting ideology we call *animism*.

In tribal societies, where the labor process is dominated by more complex hunting and gathering activities which make use of tools, storage containers, etc., where the social bond consists exclusively of the sexuality sublimated by the socialization process, where each group is formally identical with every other group, human sociality is represented immediately by the things which the group finds ready at hand and which are necessary for its survival——i.e. plants and animals, rocks, etc. Each clan is represented by a particular thing, which becomes its totem. This totemic scheme, furthermore, becomes the matrix for organizing the entire cosmos. The social bond is reproduced by means of rituals in which the totemic object is consumed, thereby establishing a direct material bond between individuals who are otherwise relatively independent (Durkheim 1911/1965).

In horticultural, communitarian societies human productive capacities, and thus human participation in the self-organizing activity of the cosmos, take on an increasingly important role. This labor is characteristic especially, however, of the minority of the population which has begun to understand the motions of the heavenly bodies, the cycle of the seasons, and the techniques for cultivating plants and domesticating animals. The self-organizing activity of the cosmos is thus represented partly in the natural elements (sun, earth, water, vegetation, animals) which continue to exercise a preponderant regulating influence on the population, and partly by various cultural heroes and instructor-deities (katchinas, etc.) who lead the development of human social capacities (cf., e.g. Waters 1963). In its most complete form, communitarian society gives birth to the idea of the cosmos——i.e. of the universe as an organized totality, of which the village community itself serves as the type. This idea is represented symbolically in the cult of the Great Mother, who is both goddess of the earth and of fertility, but also the heavenly queen, divine wisdom, etc.

In tributary social formations a double contradiction appears between men and women on the one hand, and between the ruling classes——the priests, the warriors, and to a lesser extent the artisans——who are entrusted with the universal labor of organizing and regulating the life of the city, and the peasants who participate only indirectly by consuming themselves in the production of food, on the other. Human participation in the organizing activity of the cosmos seems to be primarily an activity of these ruling classes, while the peasantry appears to be simply raw material for their creative action. This is the social basis for idealist ideologies which regard matter as inert, chaotic, and even dirty, and which teach that order only and always comes to matter from the outside. The activity of the ruling classes, furthermore, consists increasingly in ritual sacrifice (in the case of the priests) and warfare. The creative, self-organizing activity of the cosmos is, therefore, represented symbolically as the sacrificial (even self-sacrificial) warlike activity of the gods. Humans participate in this process, and thus sustain the cosmos, through the performance of ritual sacrifices, and, secondarily, through the military conquests which secure the surplus product necessary for the sacrifices (cf. e.g. Brundage 1985).

The emergence of industrial forces of production involves everyone either directly or indirectly in labor which builds up civilization, which creates some lasting social product. The growing division of labor creates a profound level of material interdependence (Durkheim 1893/1964). But the organization of the labor process and the mediation of this interdependence through the medium of the marketplace obscures our interdependence and, furthermore, makes the destiny of each individual dependent on economic forces beyond his control (Marx 1844/1978). On the one hand the marketplace, by serving as a catalyst for economic growth and development, promotes the common good and the development of society as a whole. But it does so only abstractly, and in a way that is susceptible only to quantitative, statistical analysis. It may well reward me for my labor, but it might also cast me down into poverty and uselessness, my level of skill and my hard work notwithstanding. The marketplace, furthermore, makes all human activity simply a means of realizing individual consumer interests, and transforms what is objectively a highly collaborative economic process into a war of all against all. We thus experience human beings as creative, powerful, knowing, and loving in a way we never have before, but also, at the same time, as radically selfish, or else as simply a quantity, as the exchange value of their creative, powerful, knowing and even loving activity.

This is the basis of the multiple contradiction, characteristic of bourgeois ideology, between art and science, on the one hand, and within the sciences, between humanistic and positivistic ideology and between the social and the natural sciences. It is also the basis of bourgeois religion. In bourgeois art the aesthetic product is a totality created by human labor and is thus a reflection of the cosmos as a human (i.e. meaningful) totality. But this totality is merely fictional. Unlike the mythic cycles of precapitalist societies, bourgeois art does not claim to represent the actually existing cosmos, but only an imaginary one which is merely the product of the artist's mind. In bourgeois science, on the other hand, this totality is sundered. According to the positivistic problematic dominant in the natural sciences the universe consists of particulars (atoms) which are related to each other in a purely external, and predominantly linear fashion. We know only facts and events and the constant patterns which we observe in the play of facts and events. At this level, then, the human world appears as a limited sphere of meaningful activity within an otherwise meaningless universe. Similarly, within the context of bourgeois science, the dominant positivistic problematic, on the one hand, regards society, like nature, as simply a play of facts and events, while the competing humanistic problematic regards the study of human society as primarily the analysis of the meanings individuals attach to their actions. In either case we are dealing with the ideological reflex of the market which "knows" the totality of human labor only as consumer preference on the one side and the play of supply and demand and various derivative quantities on the other hand (Lukacs 1971). This contradiction reaches its most extreme form in bourgeois philosophy, which treats the universe as a system of only externally related particulars. While the dominant positivistic trend is content to confine itself to an analysis of the formal rules governing the relations between these particulars, the subaltern "existentialist" trend searches for some way in which finite humans can endow what they acknowledge to be the purely mechanical system of the universe with meaning.

This contradiction is "resolved" in bourgeois religion. Human interdependence, and indeed the relationality of the cosmos as a whole, appears only as an abstract totality (reflex of the marketplace) which is utterly beyond human comprehension, a mysterious "other" which judges me not according to any human criteria but according to inscrutable norms of its own. Since I cannot know this other, I cannot love it. It is not accessible through any work I might perform, but only by the negation of my work, its sublation in the marketplace, and my own radical surrender to the judgement of that other, my willingness to accept its verdict on my value. This is the basis of the Calvinist theology which demands radical surrender to a transcendent divine sovereign whose plan for the cosmos we can never hope to understand (Fromm 1941).

In late capitalist society, where confidence that the marketplace serves the common good begins to break down, or where the concept of the common good has eroded entirely, human interdependence and the cosmic totality disappear entirely from consciousness and the pietistic religion of the early bourgeoisie turns into its abstract negation which is atheism or secularism. To put the matter differently, as society becomes increasingly complex, the order which initially seemed to transcend any but a divine mind, gradually comes to seem too complex to be the product of any single mind, even that of God (Hayek 1988: 72-73).

Only when the marketplace has been transcended can we recognize the cosmos as cosmos, i.e. as at once the principle and the product of the various organizing processes which bring it into being. On the one hand this rules out ideologies which attribute organizing to one or a few ordering minds. At the same time, it also rules out the view that the organization of the cosmos is radically transcendent, i.e. beyond human comprehension. Rather, we grasp the cosmos as a system which is at once constitutive of, and constituted by, its particular elements, which derive their being and essence from the whole which they nonetheless help to create. At this point the contradiction between art and science is authentically transcended. The underlying unity which artistic creativity intuits, and portrays in representational form, science understands in and through its explication of the diverse organizing processes which constitute the cosmic manifold, and philosophy comprehends as a unified, rational system. Human spirituality sheds dualistic, self-sacrificial character which it acquired with the emergence of the warlord state. Participation in the self-organizing activity of the cosmos (the life of God) is directly accessible through the ordinary human activities of working, exercising power, producing knowledge, loving, raising children, etc. The place of these activities in the self-organizing activity of the cosmos, has, furthermore, been rendered more or less transparent, so that far from being confined by the visible horizon of our own segment of the space time continuum, human knowledge and love are able to reach beyond themselves, towards Omega, to the cosmic whole of which they are a constitutive part, and in which alone they find their realization.

C. The Structure of Human Society as a Whole

The ensemble of particular structures regulating specific social processes together constitute a general structure. Actually existing societies in fact reflect complex combinations of organizing processes and regulating mechanisms, all at different levels of development. Notwithstanding this internal diversity, however, any given society, taken as whole system, is governed by definite structural rules which determine the way in which these various structures are related to each other, and in general the way in which the society in question carries out its general vocation——i.e. the work of developing complex organization in the universe. Actually existing capitalist societies, for example, may incorporate bands, tribes, village communities, warlord states, etc., which continue to function internally according to their own structural rules, but the system as a whole will be regulated not by some amalgam of these forms with the market system, but by the marketplace, which subordinates each of these forms to its own imperatives, destroying some and reinforcing others, according to its own distinctive logic.

In the chapters which follow we will have an opportunity to trace the development of social structure in detail. At this point, however, it will be useful to present the principal structural forms in an outline format, for purposes of definition, and so that the logical relations among these various forms is clear from the beginning.

1. In preindustrial societies,

    a) prior to the emergence of the warlord state, we find:

        i) band societies, which are just emerging from the animal state and as yet lack sufficient internal differentiation to yield a definite structure,

        ii) tribal societies, organized by kinship systems which divide the social whole into arbitrary groupings and then establish marriage alliances between those groupings, and

        iii) communitarian societies, which have developed complex mechanisms such as the village community, the urban center, or the state, for centralizing, redistributing, and investing surplus in such a way as to promote social development.

    b) The warlord or tributary state uses coercion and religious mystification to extract rents, taxes, or forced labor from dependent peasant communities in order to support warfare and luxury consumption.

    c) The emergence of salvation religious forces a partial rationalization of the tributary state, so that part of the surplus extracted is invested in human development, once again unleashing, if only partially, the development of human social capacities.

2. In capitalist societies the marketplace regulates the centralization and allocation of resources for investment, ultimately shaping the way even

nonmarket institutions socialize, produce, organize, and engage in artistic, scientific, philosophical, and religious activities. Since the market has no access to information regarding the impact of various activities on the development of human social capacities, market societies tend to hold back the long-term development of human social capacities.

3. The limitations of market societies give rise to a variety of movements which attempt to unleash human productive capacities.

a) Socialist movements of both idealist and materialist inspiration represent a reassertion of the centralized redistributional dynamics of communitarian societies.

b) In the absence of an adequate grasp of matter as self-organization these movements are unable either to wholly replace market forces which continue to erode the social fabric, or to tap into the latent potential of the working classes.

c) Development of a synergistic social structure presupposes the development of the whole population to the point that they comprehend and are devoted to the cosmic vocation of the social form of matter and thus work freely, taking from the common store what they require to further develop their creative capacities. Synergism also presupposes the development of complex organizing mechanisms which identify, develop, organize, and deploy ecosystem, technological, and human resources in such a way as to best promote the all sided development of complex organization in the universe.

These different structural forms constitute a dialectical scale ranging from relatively simple, undifferentiated unity, through increasingly complex and differentiated forms. Synergism, which does not yet exist, is defined as a kind of optimum "limit case" without ruling out the emergence of still more intermediate forms——or of still more adequate formulations of the optimum.

III. The Human Civilizational Project

We need now to consider systematically the relationship between the various dimensions of social organization themselves, and between the process of social organization generally and the social structures which regulate that process.

The basic processes of socialization, labor, power, communication, and cultural production take place only through the definite system of relations which constitutes an actual society. Indeed, the distinction between forces and relations of production, between power and authority, between communication and the rules of signification, etc. is in the final analysis only an analytic distinction. The social form of matter actually exists as a complex totality embracing both the basic social organizing processes and the structures which regulate those processes.

This social totality is first of all a particular *organization*: i.e. a particular system of human relationships structured in order to achieve a particular purpose. Organizations in turn are organized into *institutions*. By an institution we mean a particular cluster of organizing activities carried out in accord with definite social norms——a system of organizations governed by a single, common organizing principle. Each institution has a definite function within the social totality, but each involves economic, political and cultural processes. The corporation for example, is an economic institution, and it has the function of organizing the production process. But corporations themselves are in fact a structured ensemble embracing not only the material production process and the relations of production by which production is regulated, but also power and definite political relations, communication, ideology, etc. Religious institutions have for their principal function the organization of human spirituality (human knowledge and love of the cosmic totality). Religious institutions must, however, secure revenue, and organize their members as well as producing and reproducing religious ideology. Taken together, these institutions constitute a *social formation*——a system of institutions which, like each particular institution, is itself governed by a single organizing principle.

Now, in so far as each social formation is an expression of the underlying organizing principle of the cosmos as a whole, pregnant with creative potential, it gives rise to forms of activity which cannot be accommodated within the context of the existing social structure. Social development is not only quantitative, but also qualitative. Far from being static systems, social formations undergo constant self-reorganization. They are thus *historical* systems.

The historical process is first of all the *organizing process* itself ——socialization, labor, the exercise of power, the production of knowledge, etc. But the structure of human societies becomes, at certain stages of development, an obstacle to the further reorganization of matter. There is thus a contradiction between the process of social organization and the structure of any given society. Resolution of this contradiction involves the transformation of the underlying structure of the society (relations of production, authority structure, ideology), which in turn unleashes new developments in the field of production, power, knowledge, spirituality, etc. which had been held back by the old regime.

At the most fundamental level, it is the contradiction between social organization and social structure which constitutes the driving force in the historical process. Concretely, however, this contradiction appears as a struggle between definite social groups which are defined by their position in the forces and relations of production. On the one hand certain sectors of the laboring population begin to engage in new, highly creative activities which, if permitted to flourish, could vastly increase human productivity, power, knowledge, etc. To some extent, even the ruling class benefits from these innovations. On the other hand, the patterns of surplus extraction and centralization, the legitimation of authority by means of religious mystification or formal rules, and the mystification of the production process itself hold back development of these new activities, while at the same time benefiting the old ruling class, which therefore defends the existing order. Historical change thus takes on the form of a *struggle between social classes*. Often this struggle takes the form of strikes, uprisings, etc., and other violent clashes. At other times it is matter of the gradual reorganization of institutions and the contest of ideological problematics. In this sense "all history is the history of class struggle (Marx 1848/1978)."

Two points are, however, in order here. First, the class struggle is not simply a struggle over the division of the surplus, nor even a question of resistance to oppression. It is, rather, first and foremost an expression of the creativity of the laboring classes. Second, it is important to remember that the class struggle can produce only those historical changes for which an adequate social basis exists in the technological, economic, political, scientific, and spiritual development of the society. The social basis of revolutionary change is always the superior capacity of the rising class to promote the reorganization of matter at a higher level of complexity, interdependence, etc. While a rising class may gain control of key institutions, including the dominant political institutions of the society, simply on the basis of superior organization, strategy, tactics, etc., the old order will not be overthrown until the new order has reached a sufficient level of maturity.

Now, in so far as the class struggle is not simply a struggle between conflicting social interests, but rather a struggle to unleash human creativity, it is a rational, conscious process undertaken by organized groups of human beings with a more or less developed understanding of the "conditions, line of march, and ultimate general result" of their struggle. It thus gives birth to *social movements*. By a social movement we mean a complex network of social individuals, which may or may not have achieved the level of formal organization, struggling for some common goal which, however they represent it to themselves, can be adequately comprehended only as an effort to raise human society, and thus the cosmos as a whole, to a higher level of organization.

Now the way in which these movements understand themselves will be largely determined by the overall level of development of human organizing capacities. Initially, in horticultural and agrarian societies, these movements understand themselves simply as instruments of a pre-existing cosmic order or divine will. They will attempt to bring a society which they perceive as disordered into harmony with this pre-existing will or order, either through force or through persuasion, without comprehending their own role as creators of order. This is the origin of the great salvation religions.

The industrial, democratic, and scientific revolutions, together with the emergence of a market economy have an ambiguous effect on the development of human social movements. On the one hand the tremendous increase in human organizing power makes the social movements of the industrial-capitalist era conscious of their own creative power, as authentic participants in the creation of human history. At the same time, the social disintegration and alienation engendered by the penetration of market relations undermines their ability to comprehend their place in the larger process of cosmic evolution. They may regard themselves as the agents of a transcendent deity accessible only through faith (Calvinism). They may regard themselves as serving a radically incomprehensible process of social progress in and through their struggle to solve particular human problems (the entrepreneurial movement of capitalism). Or they may see themselves as struggling for mastery over nature and history, in order to put these processes at the service of humanity (socialism).

It is only on the basis of the whole history of human social movements that we come to comprehend these movements as the most advanced, conscious expression of the reorganizing dynamic which is implicit at all levels of the social form of matter——and indeed of matter itself——a dynamic which is at once an integral dimension of and an authentic contributor to the cosmic evolutionary process. As we analyze the history of social movements in the next three chapters we will see that they all have as their ultimate, if always only implicit aim, nothing less than Omega itself. Participation in the struggle to reorganize a society comprehensively on the basis of new ways of raising children, new modes of production and of building and exercising power, new artistic styles and scientific paradigms, and new levels of spiritual development, represents the most advanced form of human participation in the self-organizing activity of the cosmos.

From this vantage point, human history takes on a new dimension. Far from being simply a "struggle" between particular social groups over control of surplus, it is in fact a rational process of matter organizing itself at higher and higher levels. The purposefulness of history is only implicit in the ordinary expressions of the social form of matter: socialization, labor, power, artistic creativity, scientific research, philosophical speculation, etc. though each of these activities itself reflects a high degree of self-conscious, teleological, holism. It is only at the level of transformative social movements that the purposefulness of human history becomes fully apparent. And even here, it is only as social movements become conscious of *both* their larger cosmic function, *and* their active, creative, role that humanity can become conscious of its history as a fully meaningful process: i.e. as something that makes a difference which extends beyond history to the cosmos itself.

Thus realized as a purposeful process directed towards a larger end, human history becomes a self-conscious endeavor, what we will call the *human civilizational project*. This project embraces the whole of human organizing activity, now comprehended, however, as a meaningful whole directed at the creation of dynamic complexity, and an integral, even constitutive dimension of the self-organizing activity of the cosmos. This project is teleological by its very nature, though it is only through a long process of historical development that humanity comes to understand the real character of the end towards which it is struggling.

In the chapters which follow we will trace the gradual unfolding of this project.

CHAPTER FOUR:
The Mandate of Heaven

I. The Emergence of Human Society

The social form of matter emerges when intelligent organisms begin to create new forms of complex organization rather than merely reproducing the biological structures encoded in their genome. Specifically, this means the development of tools, the creation of organized social groups, and the development of semiotic capacities sufficiently complex to create within the human mind, and within the intersubjective reality of human culture, an image, however inadequate, of the organized totality we call *kosmos*. Which of these developments came first, or whether, as we suspect, they came more or less all at once as part of a unified phase transition, we cannot say with certainty.

The first stages of this "phase transition" are already visible to us in the behavior of various animal communities. There is significant evidence that dolphins and possibly even parrots have the capacity for symbolic communication. Canines and other carnivores have a highly developed band organization which permits them to cooperate effectively in hunting animals far larger than themselves. And subhuman primates, especially the chimpanzees, have demonstrated the capacity for organization at all three levels: the use of tools, the formation of organized bands, and use of symbolic communication.

The earliest human societies——the societies of the periods archaeologists call the Lower and Middle Paleolithic——probably represented little more than an intensification of these same dynamics to the point that tool-making, band-organization, and symbolic communication, rather than innovation at the level of the genome, became the driving force of human evolutionary development. These societies procured their means of subsistence primarily through hunting and gathering. The activities of hunting and gathering differ from activities performed by other higher animals principally in the use of tools to kill animals, and in the creation of various containers in which to store plants which have been collected for food purposes. This makes it possible to reduce the time spent on finding food, and even to accumulate a surplus food supply. More and more time can be devoted to developing new and better tools, to ritual practices which build and maintain social relationships, and to exploring the world of nature and society——all things which contribute to the overall development of human productive capacities (Service 1966).

The earliest form of social organization was the *band*, a relatively undifferentiated unity the internal structure of which was determined largely by the imperatives of reproduction and childrearing, of hunting for game and gathering wild plants, and possibly of small scale sparring with other bands over hunting rights, etc.   At this level there is no division of labor beyond that established by mammalian biology: women bear and nurse children, men do not.   There appears to be little reason, however, to assume that men hunted and women gathered, or that there was any other significant secondary sexual division of labor.   Where else would humanity have acquired its images of great hunter-goddesses, if not from the memory of real women hunters?

Of the religious forms of this stage in human historical development we know little.   We can infer from surviving hunter-gatherer societies the belief that not only human beings, but everything in the universe, is infused with and governed by something like what more complex societies call the soul——an early attempt to grasp the organizing principle of the cosmos. Our ancestors left among their artistic records images not only of animals and plants, but also of the human form, and not only of humans alone, but of humans engaged in distinctly social action: dancing.   There is some evidence of goddess worship far back into the paleolithic period, but the details remain obscure, and it is not clear if this cult extends back into the earliest periods (Stone 1976: 13).

For some reason, the nature of which remains obscure, sometime around 35,000 B.C.E.   the rate of technological innovation began to increase rapidly.   There were only five major innovations between 4,000,000 and 100,000 B.C.E.   and only three between 100,000 and 35,000 B.C.E.   But there were 16.5 such innovations between 35,000 and 10,000 B.C.E., and 15.5 between 10,000 and 7,000 B.C.E. (Lenski 1982: 110).   Gradually, as human tool-making capacities improved and population density increased, the undifferentiated unity of the band gave way to increasingly complex tribal structures organized on the basis of kinship relations (Service 1966).   Kinship rules divided the society into various groups, which anthropologists generally call clans.   Marriage rules prohibited marriage within one's own clan, and often prescribed rather narrowly from which clan one had to take a spouse (Levi Strauss 1969a).

By thus arbitrarily dividing the society into groups and then establishing marriage relationships between groups, the kinship system created a society out of the undifferentiated unity of the band a complex network of specific relationships.   Internalization of the incest taboo, during the course of the socialization process, furthermore, required the individuals to subordinate their spontaneous individual desires to larger social imperatives, creating within the personality a structure (the superego) which could later internalize other moral norms (Rubin 1975).

Members of these tribal societies understood their relationships with each other and with the world around them through the medium of a totemic religious system. Each clan had its own peculiar totem, generally a plant or an animal, which served as its collective representation. Often consumption of the totemic species was forbidden throughout the year, except at the time of the annual feast, when members of the clan would gather for a kind of communion feast, at which the sacrifice and consumption of the totemic species served to rekindle the bonds of unity among the otherwise relatively autonomous members of the group (Durkheim 1911/1965).

The system of totems served, furthermore, as a kind of general classificatory scheme. Not only the members of the society, but everything in their cosmos belonged, in some sense, to a particular clan. This totemic system permitted human beings to begin to think in general, if not yet authentically abstract terms. The existence of various taboos, furthermore, which were external to the individual and exercised power over him, provided the basis in experience for the emergence of ideas of force and attraction, and ultimately of causality (Durkheim 1911/1965). Totemism thus represented an immediate grasp of the cosmos as an ordered totality, a system, of which the tribe was a kind of microcosm.

The social structure of tribal societies, however, and the totemic ideological problematic, set real limits to the development of human social capacities. First of all, tribalism marks the first stage in the millennia long oppression of women. We have already noted Gayle Rubin's (1975) claim that organizing society around the system of marriage relationships between clans creates what amounts to a "traffic in women." Socialization organized around the internalization of the incest taboo is first and foremost socialization organized around internalization of respect for the father's, or the mother's brother's, property in "his" women. It is also socialization organized around the repression of biological drives, rather than the development of properly social interests. As such it is fundamentally negative in character. The residues of this kinship system in our own society form the basis for the emergence of the authoritarian personality which has become so prominent in our own time. Even if archaic tribal societies, unlike those which survive into the present, had not yet succumbed to patriarchy, it is clear that a social structure based on kinship, more so than advanced communitarian structures centered on the village community, open up the road to harshly patriarchal trajectories of social evolution.

Second, societies organized exclusively on the basis of kinship have no way to centralize and accumulate surplus for investment in infrastructure, organizing activities, or artistic, scientific, and religious pursuits. This holds back the development of human social capacities and thus retards the process of cosmohistorical evolution.

Finally, totemic religious ideologies, while permitting humanity to grasp the concepts of order and power, do not really permit them to understand the leading role of human labor in the cosmos, and in particular the concept of organization. The clans are represented not as organized groups of laboring humans, but rather as animals, as though the cosmic order was realized more perfectly in the harmonious relationship between animal species than it is in the organized activity of human beings. This was undoubtedly a mark of the fact that humanity had not yet fully distinguished itself from these less complex forms of matter. Having not yet emerged as fully social being, humanity could not yet grasp itself as social, nor could it understand the cosmos as an organized, humanly meaningful totality.

Gradually human beings learned to cultivate plants and——in those areas of the planet where potential domesticates were available——to domesticate animals. This process of domestication was not so much a triumph of human power over various plants and animals, which humans bent to their will, as it was a kind of symbiosis, which tapped into and extended the evolutionary dynamic which had led to the emergence of human beings in the first place. Recent research regarding both human evolution and the process of domestication has centered around the phenomenon of neoteny. Neoteny refers to the retention by a sexually mature organism of juvenile and even infantile physical and behavioral characteristics. Adult humans look like nothing so much as infant apes, just as domestic dogs resemble young wolves and domestic cattle the calves of their wild bovine ancestors. Neoteny apparently permits the extraordinarily rapid development of new, evolutionarily progressive, characteristics, in particular a loss of fear of new situations, and a tendency to abandon specialized ecological niches for generalist behavior, which it would otherwise take thousands of generations of intentional breeding or millions of generations of random selection to achieve. Many ecologists now believe that the existence of human settlements, with their tendency to clear away wild plants, and to accumulate piles of scrap food, in effect created a niche into which opportunistic members of various plant and animal species settled, and which created powerful forces which selected in favor of rapid neotenization.

> Plants that humans collected in the wild and scattered the remains of about a campsite would get a foot in the door; it would take relatively little human action to encourage their growth.   The natural favoritism man would apply in gathering wild plants——for example favoring those that had ripened simultaneously, those that had fewer natural defenses such as bitter toxins or thorns, or those having large seeds with thinner coats and seed heads less likely to shatter——would automatically steer the evolution of these plants in the direction of what we, in hindsight, term domestication, since many of the characteristics that make a plant worth gathering make it worth cultivating.   The substitution of human cultivation for more natural methods of propagation would have, in the long run provided a strategy for survival of evolutionary significance from the plant's point of view ... The deliberate practice of saving seed and clearing land and sowing could have come much later——indeed after the essential steps in the process of domestication were complete (Budiansky 1992: 85-86).

Much the same is true of the domestication of animals.

> ... systematic neoteny, selected as a way to adapt to a changing world, would have laid an even more solid foundation for the interactions of humans and other animals.   The curiosity, the lack of a highly species specific sense of recognition, and the retention into adulthood of juvenile care soliciting behavior (such as begging for food) of neotenates would all have been powerful factors in inducing wolves, sheep, cattle, horses and many other occupants of the Asian and European grasslands of the late glacial era to approach human encampments and to allow humans to approach them.   The neoteny that is part of our own evolutionary heritage may have likewise made us more wiling to enter into relationships with animals other than the highly specialized one of predator to prey (Budiansky 1992: 80).

The domestication of plants and animals, and the emergence of the horticultural mode of production, was not, therefore, so much a triumph of human control over nature as it was a development within nature itself of a more complex mode of organization, and one that advanced the survival interests of all of the species involved.   Horticulture was part of the self-organizing movement of the cosmos from less complex to more complex forms.   It represents the emergence within the cosmos of organisms capable of grasping the latent potential of matter, and working matter in such a way as to release that latent potential.

Humans now established permanent settlements, characterized by an emerging division of labor. Growing surplus permitted an increasingly large part of the population to be relieved from the responsibility to produce food. Specialists in various crafts emerged, as well as a priestly stratum, which organized more complex productive activities, and engaged in protoscientific study of the motion of the sun and stars, the rising and falling of the tides and of rivers, and other aspects of the natural environment which bore on their people's survival. This was, in many ways, a remarkably productive period in human history.

Before the agrarian revolution comparatively poor and illiterate communities had made an impressive series of contributions to man's progress. The two millennia immediately preceding 3,000 B.C. had witnessed discoveries in applied science that directly or indirectly affected the prosperity of millions of people and demonstrably furthered the biological welfare of our species by facilitating its multiplication.

Among the discoveries of this period we should note

artificial irrigation using canals and ditches; the plow, the harnessing of animal motive-power; the sail boat, wheeled vehicles; orchard husbandry; fermentation; the production and use of copper; bricks; the arch; glazing; the seal; and, in the earliest stages of the revolution——a solar calendar, writing, numerical notion and bronze (Childe 1951).

This rapid development of human social capacities was facilitated by the emergence of new socioeconomic forms, the most important of which was the village community, which gradually displaced the clan and the tribe as the basic units of social organization. The domestication of plants and animals made possible the establishment of permanent settlements which usually, though not always, included people drawn from different clans. This in turn facilitated the development of inter-clan relationships based on something other than marriage. Sometimes these relationships grew out of collaboration in the clearing of fields or the construction of simple public works such as the irrigation systems or ritual centers. At other times it was the need to mobilize the able bodied adults of the village for warfare which catalyzed the emergence of inter-clan ties. In still other cases it was the religious cult itself which united members of different clans, as in the case of the various religious societies of the Hopi, Zuni,

Keres, Tewa and Towa. Because of the role of the village community in organizing production, we call these societies *communitarian*.

The village community was the effective owner of the land, which in most societies was parcelled out to individual families for cultivation, and periodically redistributed in order to insure a rough equality. Collective labor was not uncommon, and woodlands and grazing land were generally used, as well as held, in common. Mandel has collected extensive documentation of this phenomenon.

> The cooperative organization of labor implies, on the one hand the carrying out in common of certain economic activities——building the huts, hunting the larger animals, making paths, felling trees, breaking up new land——and, on the other, mutual aid between different families in daily life. The American anthropologist John H. Province has described such a work-system in the Diang Dyak tribe, who live in Borneo. All members of the tribe, including the witch doctor, work alternately on their own paddy field and on that of another family ...
>
> Margaret Mead describes a similar system prevailing among the Arapesh, a mountain people of New Guinea. The cooperative organization of labor in its pure form means that no adult holds back from participation in labor ... The work is planned by the community in accord with custom and with ancient rites based a deep knowledge of the natural environment ... The chief, if there is one, is merely the embodiment of these rites and customs, the correct fulfillment of which he ensures ...
>
> ... in Dahomey ... in ancient times... there were no chiefs and the dokpwega [directing the communal work] was in command of the village (Mandel 1968:32).

Closely related to collective labor was collective appropriation of the surplus product.

> It is not possible to list all the sources that confirm the existence of .. common ownership of the land among all civilized peoples, at a certain phase of their agricultural evolution. The Japanese village community called the *mura* is described by Yoshitomi ... In Indonesia "the village community represents the original community," writes Dr. J.H. Boeke. Wittfogel has analyzed the *tsien-tien* system of dividing the fields of Chinese villages into nine squares, and discovered there the village community which has descended from the collective appropriation of the soil. The work of Professor Dykmans on the ancient empire of the Egyptian Pharaohs states explicitly that there the land was originally clan

property with periodical redistribution of the holdings ... M. Jacques Weulersse, describing the agricultural system of the Arab people called the Alaouites, has found among them even today traces of collective ownership ...

In respect of Central and East Africa, the semi-official African Survey states that:

> " It is true to say that through that part of Africa with which we are concerned, there is a prevailing conception of the land as the collective possession of the tribe or group. "

Speaking of the Polynesian economy of Tikopia, Raymond Firth notes "the traditional ownership of orchards and garden plots by kinship groups. "

Historical research confirms the existence of collective ownership of the land in Homeric Greece, in the German Mark, in the ancient Aztec village, in the ancient Indian village of the time of the Buddhist writings, in the Inca village where the plowed fields were called *Sanslpacha*, that is the land (*pacha*) which belongs to everyone, in Egypt, Thrace, Asia Minor and the Balkans, before the Slav colonization, in ancient Russia ... among the South Slavs, the Poles, the Hungarians, etc. In a study undertaken for the F.A.O. Sir Gerald Clausen confirms, furthermore that everywhere, in the beginning, agriculture was carried on within the framework of an agrarian system based on communal ownership, with periodical redistribution of land (Mandel 1968: 35-36).

In addition to organizing cultivation, the village community was responsible for adjudicating conflicts among the villagers, and for organizing the religious life of the community. Religious societies developed, often cutting across clan lines, which planned various rituals, and which passed on the traditions of the people (Waters 1963: 125-247).

Much has been made of the fact that these village communities were not strictly egalitarian. Two points are in order here. First, there is good reason to believe that most horticultural communitarian societies were matriarchal, or were at least characterized by relative equality between the sexes. Matrilineality is far more common among horticultural than among hunting or pastoral societies (Lenski 1982: 155). While it is true that most of the matrilineal horticultural societies which have survived into the present period are in fact patriarchal, power being exercised by the community of brothers rather than by a monarchic *paterfamilias*, there is reason to believe this may not always have been the case.   Literary evidence for the Neolithic period itself is scare, but the literature of ancient Mediterranean societies is full of stories of ancient or remote communities in which "mother-right" survived and women ruled.   Diodorus Siculus is typical in this regard.   Writing of Egypt he says that

> ... it was ordained that the queen should have greater power than the king and that among private persons the wife should enjoy authority over the husband, husbands agreeing in the marriage contract that they will be obedient in all things to their wives (Diodorus Siculus in Stone 1976: 36).

Similarly in Ethiopia

> All authority was vested in the women, who discharged every kind of public duty.   The men looked after domestic affairs just as the women do among ourselves and did as they were told by their wives ... (Diodorus in Stone 1976: 35).

H.W.F. Saggs writes that "the status of women was certainly much higher in the early Sumerian city state than it subsequently became (Saggs in 1976: 39)." Heraclides Ponticus writes of the Lycians that "from of old they have been ruled by women (in Stone 1976: 46).   Indeed, this seems to have been the pattern throughout the Mediterranean basin.

> Among the Mediterraneans as a general rule society was built around the woman, even on the highest levels, where descent was in the female line.   A man became king or chieftain only by formal marriage and his daughter, not his son succeeded so that the next chieftain was the youth who married his daughter ...   Until the northerners arrived, religion and custom were dominated by the female principle (Seltman in Stone 1976: 47).

The Amazons of ancient lore may well have been communities of women who banded together to resist the new patriarchal order (Stone 1976: 46).

Analyzing impact of the European conquest on the Pueblo peoples of Southwestern North America, Ramon Guttieriez notes that it was those communities which retained a matrilineal kinship system which were better able to resist assimilation and Christianization.   This suggests that the original structure of these societies may have been more globally matriarchal, and that the drift towards patriarchy and patrilineal organization already underway before the conquest undermined the integrity of the Indian communities, creating a social structure which was more like that of the European *conquistadores*, and which, therefore, provided fertile ground for Christianity.

Second, it is true that the village community facilitated surplus extraction, and eventually paved the way for the development of state taxing mechanisms. Thus Mandel points out that

> ... the custom of carrying out tasks in common which is found very late in class-divided societies is doubtless the origin of *corvee*, that is of unpaid extra work which is carried out on behalf of the State, the Temple, or the Lord. In the case of China, the evolution from the one to the other is perfectly clear (Mandel 1968: 32).

But it is this, precisely, which made the village community superior to the kinship system as a means of organizing the labor process. It was, in fact, in the interest of the community as a whole to allocate a certain portion of the surplus, which might otherwise never be produced, or simply be consumed in religious festivals, to support the activities of chiefs, who developed interclan and eventually intervillage relationships, which proved vitally important when the tribe came under attack from outsiders. Communitarian societies which failed to develop strong political structures, such as the Hopi, found themselves unusually vulnerable to domination by even technologically less developed invaders, such as the Na-Dine (Waters 1963: 251-338). Even more important were the activities of the priests, who studied the motions of heavenly bodies and the progress of the seasons, and who thus knew when to plant and when to harvest. The priests, in virtue of their superior understanding of the self-organizing activity of the cosmos, were able to develop more effective methods of production, and, because of their ability to help the community realize its interest in survival, they were able to exercise authentic, relational power.

A society without any means of surplus extraction would be a static society, one that failed to fulfill the antientropic vocation of the social form of matter. It is not the mere existence of surplus extraction, but rather the way in which that extraction in carried out (by relational or coercive means) and most importantly the way in which the surplus is used (to promote the development of human social capacities or for warfare and luxury consumption) which sets later warlord states and tributary empires apart from communitarian societies.

In this sense the *polis* or city-state must be regarded as first and foremost simply an extension of the village community, a mechanism for extracting the surplus necessary to undertake projects too large for a single village: complex irrigation systems, roads, and other types of physical infrastructure, the construction of observatories and cultic centers, with their attached schools and libraries, etc. The fact that we do not have firm archeological or literary evidence for clearly communitarian cities does not mean that such cities did not or could not exist, and that the city is an organically exploitative entity which only and always preys on the countryside, but only that in most or even all cases communitarian societies were brought under the sway of warlord states before they could develop *polis* level structures.

Participation in the village community provided a basis in experience for grasping the universe as an organized totality. Day to day collaboration in the work of bearing and raising children, growing food and producing tools, building the complex social relationships necessary to carry out these tasks, and investigating the latent potential of the various forms of the material universe made the village quite literally a microcosm of the organizing process which constitutes the universe itself. Each of the tasks of daily life provided not only a window on, but an actual means of participating in, the work of the cosmos.

Humanity's emerging vision of the universe as an organized totality was symbolized in countless different ways. Horticultural communitarian societies developed complex calendrical systems based on sophisticated observations of the heavenly bodies and complex mathematical computations. The horticulturalists of the Americas, for example, were able to calculate the movements of Venus far more accurately than their "advanced agrarian" European contemporaries, and developed the mathematical concept of "zero" long before Europeans. Horticulturalists also tend to develop an intimate relationship with the plants and animals, the growth and development of which they have learned to partially comprehend, and to catalyze, so that the natural world begins to shed something of its opaqueness. Thus the emergence of archaic cosmological systems, which attempted to grasp the universe in all its complexity.

At the same time, the worship of plants and animals loses some of its importance. It is increasingly the human rather than the plant or animal which provides the symbolism for representation of the divine. Ancestor worship becomes increasingly central, reflecting the enduring importance of kinship ties, but also a marked advanced in understanding kinship as a system of relationships among human beings.

It is no doubt also from this period that the memory of "culture heros" derives——gods, demigods, or rulers who taught humanity the art of civilization. Consider the following account from China:

> The ancient people ate the meat of animals and birds. At the time of Shen-nung there were so many people that the animals and birds became

inadequate for people's wants and therefore Shen-nung taught the people to cultivate ... During the Age of Shen-nung people rested at ease and acted with vigor. They cared for their mothers, but not for their fathers. They lived among deer. They ate what they cultivated and wore what they wove. They did not think of harming one another (Chang 1963)

The *kachinas* of the Hopi might be regarded in much the same light (Waters 1963), as might the Greek Prometheus and the other Titans.

An individual who obeys the law of laws and conforms to the pure and perfect pattern laid down by the Creator becomes a *kachina* when he dies and goes immediately to the next universe without having to plod through all the intermediate worlds or stages of existence. From there, traveling through the vast wastes of interstellar space, he comes back periodically with *kachinas* of other forms of life to help mankind continue its evolutionary journey.

Hence *kachinas* are not properly deities ... they are respected spirits: spirits of the dead, spirits of mineral, plant, bird, animal and human entities, of clouds, other planets, stars that have not yet appeared in our sky; spirits of all the invisible forces of life. (Waters 1963: 165-167).

And yet it is precisely around the *kachinas*——around those who have helped humanity along on its evolutionary journey——and not around the gods properly so called, that the Hopi cult revolves. This is clearly a reflection of the symbiotic character of horticultural technology and the relational and communitarian power structure of Hopi society.

It is interesting to note that while most communitarian societies have some concept of a high god, their operative religious structure is radically polytheistic, with high gods, demiurges, gods of various natural phenomena, and culture heros all participating in important ways in the mythic cycles, and all sharing in the community's worship. As Durkheim suggests (1911/1965) the emerging concept of a high god probably reflects some sort of dawning awareness of other peoples, and thus of a social reality which transcends the tribe. The actual activity of organizing society, however, remains the work of the community as a whole. There is no warlord who alone organizes the society.

To the extent that communitarian societies did worship a high deity, this deity was more likely to be a goddess than a god. The religious systems of communitarian societies seem to have been predominantly gynocentric. This should not surprise us given the matrilineal and matriarchal character of the societies in question. Indeed, it would not be too much to speak of an almost universal cult of the *Magna Mater*, to use the Latin form of the Goddess' name. Archeological evidence supports this claim, the most common religious artifacts of the Neolithic period (7000 B.C.E. - 3000 B.C.E.) being small statues of the Goddess (Stone 1976: 14-18, 24), usually as a pregnant woman. More important however is the evidence of the great mythic cycles of Bronze and Iron Age societies, which almost all conceal an archaic gynocentric layer which probably reflects a survival from the religion of earlier times (Stone 1976: 9-29).

The goddess was, of course not infrequently represented in terms of the symbolism of agrarian or human fertility——the fertile earth or the grain which it produces, the cow, or the pregnant woman. Thus Demeter, Kore, and Persephone; thus the Nahuatl Tonantzi. But she was also understood as a heavenly ruler and as source of all wisdom. The Mesopotamian Inanna and the Egyptian Au Set (which means "exceeding queen"), for example, seem to have functioned as authentic high gods, bearing titles such as "Great Goddess ... Queen of Heaven, Lady of the High Place, Celestial Ruler, Lady of the Universe, Sovereign of the Heavens ... (Stone 1976: 22). As Ishtar, Inanna was the Morning Star, and more generally the Goddess of the whole night sky. Au Set was the cosmic writer, whose pen directs the course of the world, librarian of the gods, a great physician and healer, and the one who established justice in the land (Daly 1984: 118, Stone 1976: 32). Similarly, Spider Woman, the great goddess of Southwestern North America, is known in Keres as Sussistinako, or "thinking woman," "who thought outward into space and what she thought became reality (Tyler 1964: 89)." In Hopi she is known as Huruing Wuhti, or "hard beings woman," and is credited, in some accounts at least, both with creating the universe and with instructing humanity in the arts of civilization (Tyler 1964: 82ff). In short, the goddess represents the creative self-organizing power latent within matter, a power which manifests itself in the beautiful harmony of the heavenly bodies, in the tremendous drive of life to reproduce itself, and in the complex products of human civilization.

In communitarian societies, in other words, the universe is understood as in some sense the expression of an underlying, organizing principle, often symbolized as a divine creatrix. But it is also a collective process incorporating the diverse efforts of a variety of different kinds of beings. Organization is immanent to matter itself, just as the capacity to give birth to new life is immanent in the Great Mother who gives life to all, and does not require the intervention of an ordering power from on high. The hierarchization of the various deities under the mantle of the Great Mother is less a mark of their subordination to her than of their integration into an organized cosmic totality.

It is interesting to speculate on the course humanity might have taken had the earliest communitarian societies been permitted to continue their development. Indeed, a really complete theory of the social form of matter would include a specification of all logically possible paths of social development. Here, since we are concerned with the actual course of development of human society, we offer only a few, very tentative indications.

One imagines a much more rapid, but very different, process of technological development. While the surplus centralized might well have been smaller than in tributary societies, it would most likely have been invested almost entirely in promoting the development of human social capacities——in inventing new productive technology, in developing and maintaining social bonds, and in deepening humanity's understanding of and participation in the self-organizing activity of the cosmos. The technology developed would undoubtedly have been symbiotic in character, tapping into and developing the potential for organization latent in matter, rather than breaking down existing forms of organization to release energy and drive simple machines. Rather than enslaving other humans, humanity might have concentrated its efforts on cultivating in domesticated animal species the capacity for labor, challenging those species to develop, while freeing themselves for more complex tasks. Having understood from the beginning that labor is first and foremost a participation in the self-organizing activity of the cosmos, and not the subordination of an inert nature to human control, humanity might have focused its attention more on tapping into biological processes rather than breaking down existing forms of the organization of matter (e.g. coal, oil) to produce energy and then recombine material elements mechanically or chemically. Communitarian forms of social organization, such as the village community, might have become the cells of even more complex social networks, maintaining intact a rich social fabric of nonmarket interpersonal relationships even while widening the scope of human social organization. The tension, which today is so profound, between seeking union with the organizing principle of the cosmos (mysticism) and transforming the world in order to advance that self-organizing activity (asceticism) might never have emerged. It is little wonder that human utopias, whether expressed in religious form, or in contemporary works of science fiction (see for example, Doris Lessing's vision of Canopean society in *Shikasta* and *The Sirian Experiments* 1979, 1980) have always reflected an yearning for a fully developed, symbiotic, communitarian social formation.

II. The Warlord State

Such, however, was not the actual path of development on our planet. At a certain point the rapid development of human social capacities which characterized horticultural communitarian societies ground to a halt. It is possible to specify this point, with respect to the civilizations of Asia, Africa and Europe, with some precision. It came within a few hundred years of the invention of the plow, and of bronze technology.

The plow, which first appeared in Mesopotamia around 3000 B.C.E. permitted farmers to control the growth of weeds and to maintain the fertility of the soil by bringing nutrients to the surface, and to cultivate increasingly large areas by attaching the plow to domesticated cattle or other animals. This vastly increased the surplus available to support specialization in crafts, and continued artistic, scientific and philosophical development. At the same time, the shift from horticulture to agriculture did mark a fundamental change in the nature of the productive process itself. Horticulture is by definition intensive cultivation. The development of each and every plant and each and every animal is carefully observed. This means that the production process itself is likely to be the site of new technological and scientific innovations, and also that human welfare is closely bound up with the welfare of the domesticated plants and animals. The emergence of agriculture, on the other hand, means the development of more extensive means of cultivation. Seeds sown in a plowed field grow largely on their own, without the individualized attention of the cultivator. Where horticulture encourages the latent productive capacity of plants and animals, agriculture simply exploits this capacity.

Bronze appeared just a little bit later, during the early part of the third millennium B.C.E. Harder than copper, bronze was well suited to the production of weapons. One scholar writes of the situation in China:

> In the course of a few centuries the villages of the plain fell under the domination of walled cities on whose rulers the possession of bronze weapons,chariots and slaves conferred a measure of superiority to which no community could aspire (Watson 1961).

The story was much the same throughout India, in the Mediterranean basin, and in Africa. As Gerhard Lenski puts it

> For the first time in ... history, people found the conquest of other people a profitable alternative to the conquest of nature ... One might say that bronze was to the conquest of people what plant cultivation was to the conquest of nature (1982: 145).

Victorious warlords put the villages which they conquered under tribute, forcing the villagers to perform unpaid labor on their fields or to build temples, palaces, and fortifications. They imposed taxes, or distributed village lands to their retainers, who imposed rents. Warlords, meanwhile, continued to fight each other, with the strongest eventually emerging as monarchs and emperors.

It would, however, be a mistake to attribute the emergence of the warlord state to bronze technology alone. Much the same pattern developed in the Americas where metal technologies were far less developed. And there do seem to have been people, such as the Hopi and the "goddess cultures" of the Mediterranean basin, which did not opt for warfare and the warlord state. Why, then, did some (but only some) peoples take this path of development?

The most likely answer is that for some societies, which occupied relatively unfavorable ecological niches, warfare represented the optimum, indeed possibly the only, route to further development. The drive for wealth and power is not, after all, intrinsically evil. On the contrary, it is the centralization of resources which makes possible the development of art, science, philosophy, and religion and thus the progress of human civilization generally. It is only when this drive is not ordered to a higher good——the development of human social capacities and of complex organization in general——that contradictions begin to emerge. And in the absence of a larger intertribal or intercommunity structure this higher good could be pursued only at the level of the village or tribe, which had to conquer in order to progress. Those societies which had some alternative path of development open to them generally chose that path, at least until they were drawn forcibly into the downward spiral of warfare and destruction by the appearance of conquering hordes on their horizon. And those societies which developed along the warlord road did so, in a certain sense, in conformity to the same cosmic law——the law of self-organization——which governed the village communities they conquered. This is why, as the internal contradictions of the warlord road became manifest, movements emerged within the warlord states to transcend these contradictions and once again unleash development. All paths of cosmic development ultimately converge on the same infinitely self-organizing Omega.

The emergence of warfare radically altered the internal structure of human societies. The first change was the emergence of patriarchy. In horticultural and early agrarian societies the land was held and worked by the women. The men worked their wives' land, or contributed to subsistence through hunting, trade, or other activities. Their economic significance, if not marginal, was rather limited. This, in turn, curtailed their power. With the advent of warfare, on the other hand, men were able to offer their communities captured booty, and eventually the service of slaves and forced labor levies, or income from rents and taxes imposed on dependent peasant communities. In return, however, they demanded the right to pass their property and position down to their sons, who would join them in their military campaigns. This, as Engels has suggested, required the ability to establish paternity, and thus implied the establishment of monogamous families and patrilineages. They also brought back slave women and concubines, which further increased their bargaining power vis-a-vis the women of their home communities (Engels 1891/1972).

This pattern seems already to have been well-established among the conquering peoples, such as the Indo-Europeans, who make their first appearance around 3000 B.C.E. The change appears to have been more gradual however, among the advanced communitarian societies which they conquered. Many ancient texts from Egypt and Mesopotamia speak of a strange form of kingship, in which the monarch attained his office only by marrying the high priestess of the goddess, and was, in some places, ritually sacrificed after serving for only a year (Stone 1976: 129-153). This institution may well have been a mechanism developed by matriarchal communitarian societies to raise a standing army, the commander of which was the king, while preventing the emergence of a permanent male dominated monarchic system. Gradually, however, the king gained the upper hand, and the priestess was reduced to a figure of the second rank. This is shift is reflected in the myth of Gilgamesh, who refuses to marry Ishtar on the grounds that all of her previous husbands have met with an unhappy fate. The goddess is insulted and her sacred bull slaughtered. Enkidu, a "wild man of the woods" is sacrificed as a substitute for Gilgamesh, who makes his escape. By the time of Sargon, on the other hand, ritual regicide is clearly a thing of the past, though royal authority was still legitimated through a ritual marriage to Ishtar (Stone 1976: 142). Diodorus Siculus tells of a Nubian king who resisted ritual regicide, slaughtered the clergy, and proclaimed a permanent kingship for himself. Eventually of course, women lost control of even the priesthood. Thus the Aztec tlaoni or snakewoman, probably the lineal descendant of archaic Nahuatl "inside" or ritual chiefs, is male, as are the inside chiefs of surviving Hopi, Zuni, Keres, Tewa, and Towa communities.

The second change wrought by the advent of the warlord state is, of course, the emergence of mechanisms for the systematic exploitation of the peasantry. Both communitarian societies and the warlord state had mechanisms for the extraction and centralization of surplus. The difference between the structures has to do with the way in which surplus is extracted, and the way in which it is used. We have seen that in communitarian societies surplus extraction is noncoercive, and generally based on the perception on the part of the community as a whole that it will benefit by supporting full time scientists, teachers, ritual specialists, healers, etc. And the surplus is used exclusively for activities which benefit the community as a whole. In the warlord state, on the other hand, surplus is extracted either through force or through the operation of mystifying religious cults and is largely allocated to warfare or luxury consumption. Because the extraction of surplus has the character of tribute paid to a conquering lord, we call these societies *tributary social formations*.

The emergence of the warlord state was reflected in the degeneration of religion into an increasingly bizarre system of patriarchal, sacrificial sacral monarchic cults. There are two dimensions to this process. First of all, as warlords extended their control over the villages, and displaced the progressive organizing activity of the village chiefs and priests, organizing activity generally came to seem more and more the province of one person alone. Thus the gradual tendency for communitarian polytheism to be replaced by a cult of divine monarchs presiding over a pantheon of heavenly retainers. Increasingly the productivity of the land, on which the community depended for its survival, was attributed to the activity of the monarch and his divine counterpart. We call this the land/lord/god complex. On the one hand the leading members of the emerging ruling class——the local warlords or monarchs——are either deified or regarded as having a special relationship with the deity. On the other hand the already divinized forces of nature (sun, rain, soil, or cultivated plant) are identified with these new warrior gods, creating a complex nexus linking the fertility of the soil with service to, or payment of rents and taxes to the ruling class. Thus the Canaanite god *ba'al*, whose name means lord, master, owner of land, and husband. The term was used not only for the deity, but for the local warlords who were identified with him. The result was an extraordinarily effective system of social control. Rebellion threatened to bring down not only the sword of the local ruling elite, but also the wrath of the gods. Fail to pay your taxes or perform forced labor and the sun itself will cease to shine, the rains will not come, the soil will lose its fertility. This pattern is nearly universal in the religion of tributary social formations. In some cases, as in Egypt the ruler was regarded as divine in his own right, the son of the Sun God. The Chinese Emperors took the only slightly more modest title of *Tian Tzi*, or "son of Heaven." Mesopotamian rulers, on the other hand, were regarded as mere tenants of the gods, to whom they conveyed the tribute extracted from the peasant communities.

The emergence of the warlord gods was accompanied by a relative decline in the importance of the goddess. This change is often reflected in the mythic cycles themselves. The Egyptian cycle is especially interesting in this regard. Merlin Stone describes the incorporation of the Indo-European god Horus into the myths surrounding Isis.

> Hor, (later known as Horus to the Greeks) was described in various texts as fighting a ritual combat with another male deity known as Set ... generally identified as the uncle of brother of Hor. The fight symbolized the conquest of Hor over Set ...
>
> In Sanskrit the word *sat* means to destroy by hewing to pieces ... It was Set who killed Osiris and cut his body into fourteen pieces. But it may be significant that the word Set is also defined as "queen" or "princess" in Egyptian. Au Set, known as Isis by the Greeks, is defined as "exceeding queen." In the myth of the combat Set tries to mate sexually with Horus; this is usually interpreted as being an insult. But the most primitive identity of the figure Set, who is also closely related to the serpent of darkness known as Zet, and often referred to by classical Greek writers as Typhon, the serpent of the Goddess Gaia, may once have been female ... perhaps related to Ua Zit, the Great Serpent, the Cobra Goddess of Neolithic times (Stone 1976: 89-90).

The final form of the myth of Isis and Osiris, in other words, actually encodes the defeat of the original form of Isis (Au Set or Ua Zit) by Horus, the sun god of the patriarchal Indo-European invaders who introduced kingship into Egypt.

Sometimes there is an explicit attempt to rationalize the displacement of the goddess in terms which might be acceptable to the peasantry. Inanna, for example, was said to have lost her royal scepter to Enki, who discovered how to build irrigation canals (Stone 1976: 83). Spider-woman, according to Hopi legend, is punished by Sotuknang, the nephew of the high god Taiowa, for encouraging the people to use their power to melt the mountains of ice which closed up the Bering Strait, the "back door" to the North American continent, making it possible for the Athabascan enemies of the Hopi people to cross over the Bering straights and into the Americas (Waters 1963: 40, 255). The implication is that Spider-woman lacked military acumen and was thus unfit to rule.

At the philosophical level the displacement of the goddess by the warlord gods was reflected in the emergence of "idealist" philosophies which regard organization not as a potential latent in matter itself, but rather as something imposed on matter from the outside, whether by divine fiat or by the action of some prime mover.

The second dimension of this process was the emergence in most tributary societies of a more or less developed sacrificial cult. Sacrifice was, to be sure, a well established practice in both tribal and communitarian societies. In tribal hunter gather societies, however, the central element of sacrifice was the shared meal, which reconstituted the social bond among the members of the otherwise scattered clan. In predominantly communitarian horticultural societies sacrifice retained this character, with food offerings to departed ancestors or to the gods themselves regarded as a way of including the deity in the communal meal. The creative or instructing activity of gods and ancestors, however, was itself never conceived as a self-sacrificial act.

In tributary social formations, on the other hand, we witness the emergence of religions in which the sustenance——and in some cases the very creation——of the cosmos is itself regarded as a sacrificial process. Thus in ancient Mesopotamia, the gods were believed to have created human beings in order to serve them by providing food and drink (Kramer 1963). According to the *Rig Veda* (X.90) humanity was created out of an original cosmic human *Purusha*, who was sacrificed by the gods.

> When they divided *Purusha* how many portions did they make?
> The *Brahman* was his mouth, of both his arms was the *Rajanya* made.
> His thighs became the *Vaisya*, from his feet the *Sudra* was produced.
> The moon was generated from his mind, and from his eye the sun had birth ...
> Fourth from his navel came mid-air; the sky was fashioned from his head;
> Earth from his feet ...

On the one hand, this text reflects and enduring recognition of humanity as a microcosm of the universe as a whole. At the same time, as the creative dynamic of the cosmos is retheorized in sacrificial terms, the cosmic totality is simultaneously retheorized as a kind of hierarchy. The system of *varnas* into which ancient Hindu society is divided is encoded in the very structure of the cosmos itself.

Similarly the *Satapatha-Brahmana*, says that

> ... the dawn is the head of the sacrificial horse, the sun its eye, the wind its breath, the fire its open mouth. The year is the body of the

sacrificial horse, the sky its back, the air its belly, the earth the under part of its belly.

A similar motif is probably reflected in the myth of Isis and Osiris, though here, as is typical of Egypt, where the gynocentric religious dynamic was particularly resilient, it is the wisdom of Isis in piecing Osiris back together, rather than the dismemberment itself, which is regarded as creative and salvific.

The sacrificial cult reached its highest level of development in the religion of the Aztecs. The Aztec empire was a tributary social formation characterized by an advanced hydraulic horticulture. Land was held collectively by the *calpulli*, or village communities, or else by the temples and various state institutions. The *macehualtin* or peasants were required to perform forced labor for the *teuctli* and *pilli*——the ruling classes (van Zantwijk 1985).

But the unique characteristics of Aztec society can be understood only by analyzing Aztec cosmology. According to Aztec tradition, there are to be five epochs in the history of the cosmos, each illuminated by a different sun. Each of these epochs begins when one of the gods sacrifices himself by means of self-immolation, and is reborn as the sun of that epoch. Each epoch ends in a cosmic catastrophe.

These suns or epochs are not part of a continuous cycle. Each, rather, is a stage in an ultimately doomed cosmic experiment in creation through self-sacrifice which will come to an end when the present sun, the fifth and last, is extinguished, annihilating not only humanity, but the entire cosmos, and with it the gods as well (Brundage 1985: 27).

In the context of the cosmic economy, the motive force of which is self-sacrifice, human beings are, in effect, food for the gods (Brundage 1985: 36). Just as the cosmos is sustained by the sacrifice of the gods, so the gods themselves are sustained by the sacrifice of human beings. This sacrifice might take the form of bleeding or self-mutilation, but heart sacrifice, in which captured warriors were relieved of their still beating hearts, was by far the most important aspect of the cosmic economy.

Presiding over this process was the god Tezcatlipoca, the "smoking mirror." The name suggests that originally Tezcatlipoca was the god of the night sky, and more specifically of the Milky Way. But as the Aztecs transformed themselves into a militaristic empire this cosmological deity was himself transformed into the most warlike of all gods. Tezcatlipoca was the first god to sacrifice himself, and became the first sun. He is known by the names "we are his slaves," and "master of the lords of the earth." He ruled all the cities of the Aztecs through his *ixtpla*, a captured warrior who was offered symbolic obedience by the high priest and ruler throughout the year, and was then sacrificed and devoured. As One Death (Aztec Gods were often known by their date-names) he was called Yaotl, "the enemy of both sides," who fomented war in order to insure a steady supply of captives for human sacrifice, and thus an adequate supply of food for the gods. So central was the practice of human sacrifice to the Aztec cosmic economy that when there was no *ratio ad bellum* cities closely allied with each other would engage in flower wars: ——mock battles which had no other purpose than the exchange of captives who were then held for sacrifice. And as the god One Jaguar, Tezcatlipoca was a warrior painted in black, ensconced in the interior of the earth, wielding sacrificial knives, lying in wait to steal the fifth sun and thus to destroy the cosmos once and for all (Brundage 1985: 84-99).

Tezcatlipoca was worshiped universally throughout all of the Aztec cities. Huizilopochtli, on the other hand, was the tribal god of the Mexica, the dominant Aztec grouping.   According to Aztec tradition, Huizilopochtli, whose name means "hummingbird on the left" was the son of the earth goddess, Coatlicue. He slaughtered the stars, or the southern warriors, the Huizilin, and murdered his sister, Coyolxuaqui, and then led the Mexica from their homeland in the north, Azatlan, to the valley of Mexico, where he founded the city of Tenochtitlan.   Each night Huizilopochtli re-enacts this drama, doing battle with the Huizlin (the stars). Each day he drives the sun across the sky. And to sustain him in this work he requires, like all of the other Aztec deities, a steady diet of beating human hearts (Brundage 1985: 128ff).

The new tributary structure proved to be a nearly insuperable obstacle to the development of human social capacities. While there is some evidence for the use of bronze to produce plowshares, the supply was always strictly limited. Bronze production depended on the mining of tin, a rare element, the trade in which always remained a ruling class monopoly. Fearful that the peasants would produce weapons which could be used against them, or reasoning that their wealth could be increased more easily through warfare than through investment in agricultural production, the ruling class blocked the extensive development of bronze agricultural technology. This is why the promise that they might someday "beat swords into ploughshares" evoked such powerful emotions in the peasants of Israel and Judah.

Indeed, the emerging tributary mode of production put a real brake on technological development in general. The emerging ruling class was able to extract almost the entire surplus product from the direct producers. This had the effect of removing any incentive for innovation on the part of the peasantry, which knew that it would not benefit from its own increased productivity. In many cases of course, the peasants were ground down below the subsistence level, and had neither the time nor the energy to innovate. The ruling class on the other hand, invested its energies in extensive accumulation through conquest——extending the area subject to its taxing power——rather than intensive accumulation based on improved agricultural techniques. Warriors and priests increasingly held all forms of manual labor in contempt. The two thousand years after the advent of the warlord state

> say from 2,600 to 600 B.C. produced few contributions of ... importance to human progress ... They are the "decimal notation" of Babylonia (about 2,000 B.C.); an economic method for the smelting of iron on an industrial scale (1,400 B.C.); a truly alphabetic script (1,300 B.C.) and aqueducts for supplying water to cities (700 B.C.) (Childe 1951).

The structural impediments to human progress created by the tributary social formations were not confined to the technological and economic realms. An authority structure centered on coercion and mystification, in which the people as a whole had very little interest in building the power of the monarchy, was ultimately very fragile. Tributary states were subject to constant revolt, and to invasions assisted by oppressed peasants hoping that their new overlords would be better than the last. Religious ideology became an obstacle to scientific research, which constantly threatened to rediscover the real laws of nature, and to expose the mystifications of the court priests for what they were. Perhaps most destructive, however, was the impact of tributary religious structures on human spirituality. Under the hegemony of sacrificial sacral monarchic cults, human beings came increasingly to regard service to the common good, surplus labor which promoted the development of human society, as a sacrifice, something which negated their own particular interests. To participate in the life of God meant first and foremost to give up one's own life, whether through the payment of rents, taxes and forced labor, martial heroics, or participation in ritual self-sacrifice. As we will see, even the most progressive religious movements all bore the marks of this deformation to a greater or lesser degree, radically undermining their capacity to lead humanity out of its tributary dark ages.

III. Restoring the Mandate of Heaven

As human society began to seem increasingly out of harmony with the underlying order of the cosmos, the people resisted, and began to struggle for the wholeness and order which they had lost. The social basis for this resistance was located in the peasant communities. Peasant revolts are attested for most of the principal civilizational complexes on the planet, including China (Wolf 1969, Naquin 1981), Israel (Gottwald 1979), India (Gough 1973, 1981, 1989), the Mediterranean basin (de Ste. Croix 1981), and the Americas (Waters 1963: 118, Stein 1978). It was the partial success of peasant revolts in Greece which led to the emergence of Greek democracy, and the defeat of the Italian peasantry which permitted the emergence of the Roman empire. Much later, in Europe, partially successful peasant revolts in England and Northern Italy laid the groundwork for the industrial and democratic revolutions, while the defeat of the peasantry throughout Southern and Eastern Europe set the stage for centuries of stagnation and authoritarianism (Anderson 1974).

In other areas, where large expanses of arable land made it possible, peasants simply withdrew from areas under the hegemony of the warlord state and migrated to less populated areas, where they attempted to build new communities. This seems to have been the pattern in the Americas, where the Pueblo communities, for example, withdrew to remote areas of the wilderness so that they could preserve the "pure pattern of creation (Waters 1963: 118)."

The resistance of the peasantry was rooted not in the fact of their oppression, as if, having been forced to work for others rather than merely for themselves, they were somehow magically transformed from particular into universal humanity, but rather in the specific institutional forms of the village community, which, as we have seen, formed a kind of microcosm of the universal order and sustained, even under the worst conditions, a basis in experience for comprehending the underlying organizing principle of the cosmos. This gave the peasants not only a basis on which to formulate and/or comprehend critiques of the warlord state, but also a basis on which to envision a new and brilliant future.

Eric Wolf describes the general course of development of peasant revolts in China. The pattern which he discovered in there holds for most of the major civilizational complexes.

> Superficially static, Chinese society was in actuality subject to both repeated rebellions and to periods of disintegration followed by new cycles of consolidation and integration ... These conformed to a patterned sequence ... During the first stage of such an uprising, a number of peasants, driven from house and home for any number of reasons, would seek sanctuary in the wilderness. Turned bandits, they would raid travelers or rich landlords ...

During the second stage the band would extend its radius of action, thus encroaching on the zone of operations of other bands. The resulting conflict would lead to the elimination of the less viable units, and establish the dominance of the strongest and best organized band. When this happened, rivals could no longer threaten the village base of the bandit band ....

During the third stage, the band began to encounter resistance from landowners forced to pay additional amounts of tribute. Attempting to resist, the landowners called on the government of the nearest town. The bandits therefore attacked the town, and attempted to cut off this source of assistance ...

During the fourth stage, the victorious band extended its sway over additional towns, and prepared to defend its booty against government troops. To achieve further success, they had to enter into ever closer alliances with the scholar gentry of the region, since these held the monopoly of bureaucratic and social skills required for efficient administration. The bandits first adapted themselves to the norms of the gentry; later they adopted them as their own. Thus the victorious bandit leader became a general, a duke, or an emperor ... (1969: 112-114).

The first Han emperor, and the founder of the great Han dynasty (202 B.C.E.- A.D. 221) had himself been a bandit who became, in the course of events, emperor of China and bearer of the Mandate of Heaven (Wolf 1969: 108).

Peasant revolts led more often to the establishment of new dynasties than to the restoration of communitarian social relations, and where communitarian social relations were restored, the resulting societies were often unable to resist foreign conquest. This is because effective defense requires the maintenance of a standing army and a unified military command, and this in itself tends to encourage the re-emergence of patriarchal social organization and a centralization of coercive authority. The resulting diversion of psychic energy and economic resources into the military undermined the creative dynamic of the communitarian order and made the re-emergence of a ruling class of warriors almost inevitable.

It would be a mistake, however, to regard this process of peasant resistance leading to the formation of a new dynasty as simply a vicious cycle of a stagnation. On the contrary, while they never realized their millenarian visions, these peasant revolts played an important role in periodically rationalizing the tributary state, and unleashed, for a time at least, the development of the productive forces and the rationalization of political and religious relations. What Eric Wolf says of China is true for many of these new states.

> Movements which began as peasant rebellions frequently became, if successful, the means for a renewed concentration of power at the helm of the state, permitting ... society to reintegrate and consolidate itself. The new ruler would favor the gentry in his own following with appointments to official positions, while depriving opposition gentry of offices and landholding. Frequently such a period of overturn was accompanied by widespread distribution of land taken from the enemies of the regime——distributions calculated to win the support of wide segments of the peasantry and local gentry for the new ruler. With renewed centralization of the governmental bureaucracy and greater efficiency in taxation, it also became possible to consolidate and expand the great hydraulic system ... increasing both the quantity and the productivity of irrigable land (Wolf 1969: 115).

Closely associated with the peasant movements is the emergence of a prophetic-philosophical intelligentsia.[1] The warlord states had displaced the priesthood from the leading role which it had played in communitarian societies. Many priests submitted to the yoke of the warlords, and devoted themselves to elaborating the sacrificial sacral monarchic ideologies which we analyzed earlier in this chapter. Others, however, resisted. Some found themselves at the head of peasant movements, elaborating revolutionary religious ideologies. Others stood to one side, using the room created by successful or partially successful peasant movements to recover their independence and resume their primary role as students of the cosmos. In either case, the result was a recovery and a deepening of the lost wisdom of the horticultural-communitarian period, and a renewed

---

[1]. Far too much is made of the difference between the various types of revolutionary intellectuals, and in particular of the difference between the "prophet" and the "philosopher." Both the prophet and the philosopher are deeply formed in the traditions of their people, both comprehend the contradictions within which their societies have become entangled, and both have the capacity to envision a higher synthesis which resolves those contradictions and once again unleashes blocked potential. The difference is simply that the prophet stops short at the intuition of a higher synthesis, while the philosopher is able to go further, demonstrating that the synthesis which he or she envisions, is rationally necessary and not **merely** a moral or religious imperative. At the same time, of course, it is possible for a prophet to intuit a synthesis more complex than anything s/he, or any living philosopher, can rationally demonstrate or comprehend.

commitment to use that wisdom as a guide to the organization of human society, and of humanity's participation in the cosmos.

Together the peasantry and the intelligentsia charted the next steps in the human civilizational project. Peasant movements created space for the emerging intelligentsia to exercise leadership, and as the intelligentsia organized and directed the peasant movements and systematized the peasantry's immediate and inarticulate grasp of the underlying order of the cosmos, and the ways in which the development of human society had fallen out of harmony with that order. The result was the emergence of the salvation religions. The word *salvare* originally meant to make whole, and the salvation religions represent humanity's aspiration to recover, at a higher level of complexity, the lost holism and integrity of the communitarian social order.

The salvation religions took three principal forms. Some, such as such as Taoism and Confucianism, Hopi religion, or the cult of the *Magna Mater*, were mystical in character, and sought salvation or wholeness in harmony with the indwelling order of the cosmos, or the pattern of life written in creation itself. Others, such as Hinduism and especially Buddhism, grasped the underlying unity of the cosmos only in its radical negativity, as "Brahman," or more radically still as "dependent origination," so that the manifold diversity of the cosmos was regarded as mere illusion. In still others, such as Hellenic philosophy or Judaism, we witness the emerging recognition that the cosmos, realized as an organized totality transparent to human reason, represents a *positive* expression of the underlying principle of order or a creative act on the part of a sovereign God and that salvation or wholeness thus requires not withdrawal from the cosmos, as in Hinduism and Buddhism, but rather active participation in the process of cosmohistorical evolution accord with the principles of divine law. In so far as human society has become disordered, it is necessary to reorganize it in accord with divine law.

Let us now consider each of these movements in turn.

A. Cosmic Mysticism

Chinese civilization emerged a bit later than the civilizations of Mesopotamia and Egypt, and the memory of the "golden age" of communitarian society was still very much alive when the Chinese warlord states entered a period of crisis. China's first great tributary empire, that of the Chou, collapsed in 722 B.C.E., setting in motion a period of social contradiction and disintegration. The peasantry gradually emancipated itself from feudal constraints, and gained possession of a significant portion of the land——not however, under communitarian tenure, but rather as private property, which could be bought and sold. This in turn led to a differentiation within the ranks of the peasantry, as some accumulated large estates, joining the ranks of the aristocracy, and others sank into destitution. A growing agricultural surplus made possible the emergence of a money economy and of simple commodity production. In short, China took on the characteristics of an advanced agrarian society, which, like ancient Greece, had significant commercial features. In this new, more fluid social order, it was possible for the sons and daughters of the peasantry to rise, through scholarship, into the ranks of the scholar gentry and the bureaucracy. At the same time, this scholar gentry found itself increasingly proletarianized, and forced to sell its services to a plethora of competing warlords.

It was out of the ranks of this partially dispossessed and proletarianized scholar gentry that the great philosophers of China emerged. Historically, Chinese philosophy has followed two broad trends: Confucianism and Taoism.

The background to the first or Confucian trend can be found in the

Chou-li, an ancient document purported to have been written by the Duke of Chou, the prime minister of Wu Wang, who vanquished the Shang dynasty in the twelfth century B.C.E. The Duke of Chou envisaged a tightly organized feudal state in which

various grades of vassals would all owe allegiance to

the Son of Heaven ... At the bottom of this pyramid was the peasantry, organized around agricultural units of nine farms, one of which was to be public, while the surrounding eight farms were to be private. The peasants were to till the public farm as well as their own private lots.

The rise of the warlord states had undermined this archaic order. Confucius sought nothing less than to restore it.

Restoration of social order, however, involved more than political-economic reorganization. It required a correct understanding of the various social relationships on which that order was founded, and a cultivation of the virtues specific to those relationships. Confucius identified five "great relationships" and specified the terms of each.

Kindness in the father, filial piety in the son
Gentility in the eldest brother, humility and respect in the younger.
Righteous behavior in the husband, obedience in the wife.

Human consideration in elders, deference in juniors
Benevolence in rulers, loyalty in ministers and subjects.

Correct relationships required the cultivation of the virtues, of which Confucius also identified five: *jen*, or benevolence, *yi*, or justice, *xin* or fidelity, *zhi*, or wisdom, and *li*, or religion.

The purpose of government was to maintain correct relationships, and thus the right order of human society, by cultivating virtue in its subjects. The *jen* and *yi* of the ruler together with the *xin* of the subjects ensured social harmony. *Zhi*, or wisdom, and *li*, or "religion," played an important part this process. Confucius does not seem to have been concerned with speculative theology. Wisdom was primarily the practical wisdom of the court scholar, who advised the ruler regarding the right ordering of his kingdom. And his principal interest in religion had to do with ritual. By participating in the ancient rituals, the principal relationships which defined the social order were rectified and maintained, and thus brought into harmony with the larger order of the cosmos.

> This *li* is the principle by which the ancient kings embodied the laws of heaven and regulated the expressions of human nature. Therefore he who has attained *li* lives, and he who has lost it dies. ... *Li* is based on heaven, patterned on earth, deals with the worship of the spirits, and is extended to ... rites and ceremonies ... Therefore the Sage shows the people this principle of a rationalized social order, and through it everything becomes right in the family, the state, and the world (*Li Ki* XXVII, in Lin Yu-tang 1938).

In effect, Confucius wants to restrict as much as possible the role of the warrior and to rationalize that of the priest, to the extent that both are subordinated to the scholar or philosopher, who understands the function of each in strictly rational terms. Limited extraction of surplus from the peasant population will support a program of public works designed to promote the continued development of the economy, and to support the studies of the scholar gentry who, in all but name, have become the ruling class.

The second, Taoist, trend was deeply rooted in the residual communitarian traditions of China. The word Tao means the way——the eternal way of the cosmos. The *Tao* itself cannot be known discursively, yet everything that exists is a manifestation of its unrealized power. It is the source of both the form or organization and the active capacity or *Te* in everything existing.

The mightiest manifestations of active force flow solely from the *Tao*. The *Tao* itself is vague impalpable. Yet within it there is form. How vague, how and impalpable! yet within it there is substance. How profound, yet how obscure! Yet within it there is a Vital Principal (*The Sayings of Lao Tzu XXI*, in Giles 1905).

The *Tao* is a kind of natural necessity from which human beings and human society can deviate only temporarily, and only at their own peril. It inevitably reasserts itself over the long haul.

Nature is not benevolent; with ruthless indifference she makes all things serve their purposes, like the straw dogs we use at sacrifices (*The Sayings of Lao Tzu V*, in Giles 1905).

What is contrary to the *Tao* soon perishes (*The Sayings of Lao Tzu V*, in Giles 1905).

Taoism had definite political implications. Taoist texts reflect a profound grasp of the essentially relational character of power. If we hope to effect change, we must know the way of the cosmos, and yield to it. As we yield so will the world around us.

Yield and overcome
Bend and be straight;
Empty and be full;
Wear out and be new
Have little and gain
Have much and be confused ... (Tao Te Ching: Twenty Two)

Whenever you advise a ruler in the way of Tao
Counsel him not to use force to conquer the universe
For this would only cause resistance.
Thorn bushes spring up wherever the army has passed
Lean years follow in the wake of a great war.
Just do what needs to be done.
Never take advantage of power (Tao Te Ching: Thirty)

Not surprisingly, we find in the Taoist tradition, an understanding of the exploitative and deformed character of the tributary social formation which had grown up in China.

Why are the people starving:
Because the rulers eat up the money in taxes.

Therefore the people are starving.

Why are the people rebellious
Because the rulers interfere too much.
Therefore they are rebellious (*Tao Te Ching: Seventy Five*).

The emergence of a warrior aristocracy, which invested the surplus product
in military conquests rather than in raising agricultural productivity marks
the departure of Chinese society from its natural course of development,
its deviation from the Tao.

When the Tao is present in the universe
The horses haul manure.
When the Tao is absent from the universe
Warhorses are bred outside the city (*Tao Te Ching: Forty-Six*).

Taoist monasteries cultivated the martial arts as well as scholarship and
meditation, and at certain points in Chinese history became centers of
political military resistance to the Empire (Deng Ming-Dao 1990: 13).
The Taoist masters of Huainan were actively engaged in the process of
reconstruction which followed the end of the period of Warring States
(Cleary 1990: vii). The Taoist tradition in effect counsels a return to the
communitarian norms of pre-tributary China which, they believed,
represented a natural and healthy pattern of social development.

The basic task of government is to make the populace secure. The
security of the populace is based on meeting needs. The basis of
meeting needs is in not depriving people of their time. The basis of not
depriving people of their time is in minimizing government exactions and
expenditures. The basis of minimizing government exactions and
expenditures is moderation of desire. The basis of moderating desire is
in returning to essential nature (The Masters of Huainan in Cleary 1990
3-4).

Taoists did not reject the use of armed struggle to restore human society
to harmony with the Tao.

When greedy and gluttonous people plundered the world, the people
were in turmoil and could not be secure in their homes. There were
sages who rose up, struck down the forceful and violent, settled the
chaos of the age, leveled the unevenness, removed the pollution, clarified

the turbulence, and secured the imperiled. Therefore humanity was able to survive (Masters of Huainan in Cleary 1990: 49).

What the Taoists rejected was warfare for the sake of conquest.

The martial Lord of Wei asked one of his ministers what made a nation perish. The minister replied, "Numerous victories in numerous wars." The lord said "A nation is fortunate to win numerous victories in numerous wars——why should it perish thereby?"
The minister said, "When there are repeated wars, the people are weakened, when they score repeated victories, rulers become haughty. Let haughty rulers command weakened people and rare is the nation that will not perish as a result (Masters of Huainan in Cleary 1990: 12)."

Wars of liberation, however, were sometimes necessary.

The military operations of effective leaders are considered philosophically, planned strategically, and supported justly. They are not intended to destroy what exists but to preserve what is perishing. Therefore when they hear that a neighboring nation oppresses its people, they raise armies and go the border, accusing that nation of injustice and excess.

When the armies reach the suburbs, the commanders say to their troops "do not cut down trees, do not disturb graveyards, do not burn crops or destroy stores, do not take common people captive, and do not steal domestic animals."

Then the announcement is made. "The ruler of such and such a country shows contempt for heaven and the spirits, imprisoning and executing the innocent. This is a criminal before heaven, an enemy to the people."

The coming of the armies is to oust the unjust and restore the virtuous. Those who lead plunderers of the people, in defiance of nature, die themselves ...

The conquering of the nation does not extend to its people ...

The peasants await such armies with open doors, preparing food to supply them, only worried that they won't come (Masters of Huainan 1990: 50).

The Taoist masters argued for an organization of society based on relational power.

> ... those with common interests will die together; those with common feelings will strive together; those with common aversions will help each other. If you move in accord with the Way, the world will respond to you; think of the interests of the people and the world will fight for you (Masters of Huainan in Cleary 1990: 52)

Cosmic mysticism is not by any means an exclusively Asian phenomenon. In the Mediterranean Basin, the cult of the Magna Mater, ancient religion of the matriarchal communitarian societies which inhabited the region before the Indo-European conquests, was gradually transformed into a religion of salvation and social transformation. This transformation took place along two distinct routes.

One current within the tradition flows out of the old story of the rape of Persephone, which is said to have taken place beside Lake Pergusa, in the Province of Enna, in the parched grain growing uplands of the Sicilian interior.

This myth is usually told as a story explaining the origin of the seasons. Demeter, distraught by the abduction of her daughter by Hades, deprives the earth of her graces, and it becomes dry and barren. The people, threatened with famine, beseech Zeus for assistance. After the intervention of Mercury, sent by Zeus as an emissary, Hades agrees to release his prisoner, provided she has eaten nothing during her stay with him. Persephone, however, has swallowed six pomegranate seeds, and is thus condemned to pass six months of every year with Hades in the gloomy nether regions of Dis. It is during these six months that Demeter, ever distraught at the absence of her daughter, is mourning, and that the earth becomes dry and barren.

But this reading of the myth, like every naturalistic interpretation of the sacred, conceals another, more profound, and intensely political meaning. Demeter is the grain mother, and Parthenia/Persephone the new and the ripe, harvested grain. Hades, the god of the underworld, is so called because he is the god of the great underground storehouses where the grain was held. Thus the identification with Pluto, the God of wealth. The half year that Persephone spends with Hades is the half of the grain——the half-year's surplus labor——which customarily belonged to the landlord. The pomegranate seeds represent the loan of seed grain, the interest on which indebted the peasant to the landlord, a debt which was often used to justify the extraction of rents, taxes, or forced labor.

This story should be read as a kind of "myth of the fall." For the peasant the rents, taxes and the corvee to which he is subject are somehow integrally bound up with the barrenness of the land. There is an important truth embodied in this "mythological" notion. Historically, the destruction of mixed subsistence agriculture and the ensuing transformation of the uplands into pasture and wheat fields contributed in no small way to the deforestation and soil depletion which have been eroding the fertility of the Sicilian soil for millennia. The peasants of Western Sicily hoped as late as the end of the last century that the expected revolution would bring about a transformation of the climate and a restoration of the soil as well as a new social order (Hobsbawm 1959: 60, 183).

The myth speaks to us as well of the integral relationship between economic exploitation and the oppression of women. At one level the story can be read as an account of an event all too familiar to the women of the countryside: rape and abduction by a member of the ruling classes, who held steadfastly to the "right of the first night" well into the nineteenth century (Birnbaum 1980:136) and who had few scruples about usurping the "right" to any other night they might please. At a more profound level, the myth points towards the identification of sexuality and fertility with power and domination, which is the psychological basis of women's oppression (Rubin 1975), and which, in its turn, makes economic exploitation seem to be simply an integral, in fact a necessary, part of the natural order. Rape becomes identified with sexual prowess and fertilization; the forcible seizure of surplus, sanctioned by the cult of Pluto/Hades, is likewise the precondition for the fertility of the soil. The myth exposes this identification and tells us that confiscation of the surplus is, in fact, nothing more or less than the rape of the peasant and the rape of the land. The emergence of the warlord state and the development of commercial agriculture, far from being progressive forces, in fact hold back the development of human social capacities and even undermine the integrity of the ecosystem and the social fabric. "Therefore the whole creation is groaning in travail ..." (Romans 8:32).

In Christian times the myth of Pluto and Persephone was absorbed into the cult of Mary, of which it forms a kind of irreducible ideological substratum. In Enna, in the Cathedral, there stood until this century an old pre-Christian statue of a woman, and her female child. It was this statue which formed the center of the local Marian cult (Moss and Campannari 1982). The roots of Sicilian Marian devotion in the myth of Pluto and Persephone are, furthermore, apparent in the cult of *Maria Adollorata*, the "Mother of Sorrows." Each Good Friday in Trapani and other Sicilian towns the *Madonna Adollorata* is carried from church to church in search for her lost son, just as Demeter searched for her lost Persephone.

> She is the symbol of the suffering of the Sicilian people, but above all the suffering of the women who are subject to rape and harassment by the ruling class, whose men have been forced to migrate in search of work (Birnbaum 1980: 132).

One old man, when shown a picture of the *Adollorata* said

> I know her, she is very sad. The police they kill her son. I am not sure, but I think his name was Jesus. So sad she is, she is the one we worship. (oral testimony)

Within the context of this tradition matter itself is sacred, for it is the complex, self-organizing matrix, the great womb, out of which life and human society emerge. The disorder introduced into the cosmos by the emergence of the warlord state is real and destructive: women are raped, their sons are slaughtered, and the land itself is rendered barren. But this is not the last word. Since matter is sacred, it possesses its own unique healing power. Life which was cut down in its youth springs up anew. The peasant communities persevere through the centuries, confident that in the end their wisdom will triumph over the harsh gods of their overlords.

The cult of the *Magna Mater* was not, however, an exclusively peasant phenomenon, nor was it exclusively a fertility cult. On the contrary, certain forms of the cult seem to have exercised a powerful attraction for elements of the Mediterranean intelligentsia, and may constitute a kind of prehistory, or even an ongoing cultic context for, the emergence Hellenic philosophical tradition. We have already noted that Au Set, or Isis as she was called by the Greeks, was first and foremost the cosmic librarian, and the cosmic writer whose pen directed the course of human history. In most forms of the myth of Isis and Osiris, it is not the death and dismemberment of Osiris which is regarded as salvific, but rather his reintegration. And this reintegration is the work of Isis, in her capacity as goddess of wisdom.

> What then is the task of Isis? Plutarch tells us "(Typhon or Set) tears to pieces and scatters to the winds the sacred writings, which the Goddess collects and puts together and gives into the keeping of these that are initiated into the sacred rites." Isis is, then the re-assembler of lost knowledge (Matthews 1991:74).

It is interesting to note that the biographies of many of the early Greek statesmen, mathematicians, and philosophers involve a journey to Egypt. And some of Plato's most important dialogues, including the *Republic* and the *Timaeus* take place during or immediately after one or another feast of the Goddess. Indeed, the *Timaeus* tells of an older Athens whose institutions were ordered by the Goddess, which resembled quite closely the ideal polis which Plato proposes in his *Republic*. Philosophia, the love of wisdom, clearly represents a rationalized form of the ancient cult of the *Magna Mater* understood not as fertility goddess, but rather as goddess of wisdom.

Here it is not so much matter itself, but the philosopher's grasp of the organizing principle immanent within matter which is the means of salvation. Knowledge of this principle makes it possible to restore the archaic harmonies, to repair the ruptured fabric of the universe, to restore the integrity of the cosmos.

Over the course of history, these two forms of the cult of the goddess have often tended to diverge. Sicilian peasants carrying a crowned statue of the virgin through the streets of their village seem to be involved in a very different activity than the philosopher struggling to piece together our fragmentary knowledge of the universe. And it is certainly true that the popular religion of the peasantry has often lost the revolutionary dynamism encoded in the ancient story of Pluto and Persephone, while the philosopher has forgotten that the order s/he seeks is not something imposed on the universe from the outside, but rather a potential latent in matter, which evolves through time towards wholeness and completion.

Ultimately, however, the two forms of the cult are one. The Latin *mater* (mother) and *materia* (matter) share a common root. This linguistic relationship suggests the existence of a profound link between the cult of the virgin mother and the conviction that matter itself, independent of any "form" acquired from the outside, is inherently self-organizing, and thus capable of being fertile, creative, powerful, knowing, loving, etc. This conviction finds symbolic expression both in the story of Demeter and Persephone, who organize what amounts to a cosmic agrarian strike in order to restore the archaic harmony, and in the story of Isis who uses her wisdom——her grasp of the organization latent in matter——to restore the lost integrity of the universe. Peasant and philosopher march along different paths towards a common goal. In referring the complex and multiform cult of the goddess, we shall, in the future, us the term *mater/ia* or *mater/ia*lism, because of its suggestion of a profound link with later doctrines which recognized the potential for self-organization latent in matter itself.

Yet another form of cosmic mysticism developed in the Americas. The Hopi myths suggest that far from being a "pristine" communitarian society, the Hopi people emerged out of a movement of resistance to corrupt warlord states.

According to Hopi cosmology there have been three previous worlds, each of which was destroyed because the people began to forget the plan of creation.

In the First World they had lived simply with the animals. In the Second World they had developed handcrafts, homes, and villages. Now in the Third World they multiplied in such numbers and advanced so rapidly that they created big cities, countries, a whole civilization. This made it difficult for them to conform to the plan of Creation and to sing praises to Taiowa (the Creator) and Sotuknang. More and more of them became wholly occupied with their own earthly plans ...

Under the leadership of the Bow Clan they began to use their creative power in another evil and destructive way ... Some of them made a *patuwvota* [shield made of hide] and with the creative power made it fly through the air. On this many of the people flew to a big city, attacked it, and returned so fast no one knew where they came from. Soon the people of many cities and countries were making and flying on them to attack one another (Waters 1963: 17-18).

Because of this the Third World as well was destroyed. The Fourth World, the present one, has been designed to be especially harsh and barren, so that the people would not fall once again into evil (Waters 1963: 19). On arriving in the Fourth World, the people had, furthermore, to undertake a series of migrations. During the course of one of these migrations, certain of the clans journeyed far into the tropical south, where they built Palatkwapi, the Red House, also known as the Mysterious Red City of the South. The city had three sections——one for living quarters, one for storage room, and one for ceremonial purposes.

On the first or ground floor the *kachina* people taught initiates the history and meaning of the three previous worlds and the purpose of this Fourth World to which man had emerged. On the second floor they taught the structure and functions of the human body and that the highest function of the mind was to understand how the one great spirit worked within man ...

On the third story initiates were taught the workings of nature and the uses of all kinds of plant life ...

The fourth story was smaller than the three below, making the ceremonial building resemble a pyramid. To this top level were admitted only initiates of great conscience who had acquired a deep knowledge of the laws of nature. Here they were taught the workings of the planetary system, how the stars affected the climate, the crops, and man himself. Here too they learned about the open door on top of their heads, how to keep it open and so converse with their Creator (Waters 1963: 67-68).

Eventually, however, evil entered this city as well "because the people found life too easy and did not resume their migrations (Waters 1963: 70)."

This cycle of myths suggests that the Hopi have vivid memories of having once been a part of a complex tributary civilization that destroyed itself, largely due to the growing power of the warlords, and that they see themselves as a kind of tiny remnant of humanity which has retained knowledge of the plan of creation, and which has chosen to live under especially harsh circumstances in order to prevent the reemergence of tributary civilization and thus to preserve knowledge of the plan for humanity as a whole. According to Frank Waters

The Hopis believe that the early Mayas, Toltecs, and Aztecs were aberrant Hopi clans who failed to complete their fourfold migrations, remaining in Middle America to build mighty cities which perished because they failed to perpetuate their ordained religious pattern. This may well be a case of the tail wagging the dog. It is more likely that the people who later called themselves Hopis were a small minority, perhaps a religious cult, who migrated north to the Four Corners area of our own Southwest about 700 A.D. (Waters 1963: 118).

Unlike other many other peasant movements, the Hopi successfully reestablished an essentially communitarian social order. The people are organized into totemically named matrilineal, exogamous clans. Land is held by the clans and is worked collectively. The importance of a clan is determined exclusively by its religious status: whether or not it completed its migrations, and the value of the ceremonies it contributed. The chiefs of the four most important clans, the Bear, Parrot, Eagle, and Badger are known as the Naloonongmomgwit, and represent the four principal directions. There is no formal interclan political structure, and interclan unity is established primarily at the religious level, through participation in the ritual cycle.

For the Hopi the entire purpose of their emergence into this Fourth World, and of their long migrations, was the

> ideal of unifying at their permanent homeland, the Center of the Universe, where they are to consolidate the universal pattern of creation. To this end all clan rivalries and disputes simply served as a weeding-out process to test the people's adherence to this traditional ideal (Waters 1963: 122).

Concretely, maintaining the universal pattern of creation meant first and foremost the annual reenactment of a cycle of ceremonies which expressed in symbolic form the process of the creation of the universe, humanity's emergence through the various worlds, and its migrations in this world.

> Man's total existence is ... circumscribed by the pure pattern of Creation and is led in both stages of one great cycle. In this unbroken continuity, he lives in the upper world from birth to death in one stage of existence and from death to birth in the lower world ...
>
> ... there may be two deviations from this slow and ceaseless course of mankind on its evolutionary Road of Life. If a man adheres rigidly to the ritually pure pattern in this stage of existence, he is released at death from plodding through the remaining three worlds of this universe and goes directly to the next multiworld universe as a *kachina*. On the other hand, if he is evil and becomes a witch or sorcerer, his return to the underworld is painfully slow (Waters 1963: 191).

The focus of Hopi ritual, however, is not on personal salvation, but rather on helping humanity through its gradual evolution towards higher levels of participation in the self-organizing activity of the cosmos. There is no formal priesthood, and every Hopi participates in at least one ceremony each year (Waters 1963: 192).

Ultimately, the purpose of this ritualism was to preserve knowledge of the plan of creation for the time when the Hopi would lead the rest of humanity (or what remained of it) into the Fifth World.

> World War III will be started by those peoples who first received the light [the divine wisdom or intelligence] in the other old countries [India, China, Egypt, Palestine, Africa].

The United states will be destroyed, land and people, by atomic bombs and radioactivity. Only the Hopis and their homeland will be preserved as an oasis to which refugees will flee ...

The war will be "a spiritual conflict with material matters. Material matters will be destroyed by spiritual beings who will remain to create one world and one nation under one power, that of the Creator."

That time is not far of. It will come when the *Saquasohuh* [Blue Star] Kachina dances in the plaza. He represents a blue star, far off and yet invisible, which will make its appearance soon ...

The Emergence into the future Fifth World has begun. It is being made by the humble people of the little nations, tribes and racial minorities. "You can read this in the earth itself. Plant forms from previous worlds are beginning to spring up as seeds. This could start a new study of botany if people were wise enough to read them. The same kinds of seeds are being planted in our hearts. All these are the same depending how you look at them. That is what makes the Emergence to the next, Fifth World (Waters 1963: 334).

***

It would, perhaps, be useful to say something at this point about both the power and the limitations of cosmic mysticism as a religious current. On the one hand, the movements we have been analyzing reflect in many ways the purest expression of humanity's drive to restore human society to harmony with the underlying order of the universe. At the same time, it was not cosmic mysticism, but the more ascetic trend represent by Hellenic philosophy and Judaism, which eventually overpowered the warlord states and once again unleashed the development of human social capacities. The reason for this is not hard to understand. Victory over the warlord states could only be a political-military victory, and mobilization of military power presupposes patriarchy, the establishment of a standing army, etc.: all things which cosmic mysticism by its very nature resists. For this reason, the future——or at least the short-term future——belonged to other traditions. At the same time, we should not assume that the potential of cosmic mysticism has been entirely spent. On the contrary, the great revolutionary philosophy of the nineteenth and twentieth centuries, dialectical materialism, itself represents a partial recovery of the ancient *mater/ia*list insight into the self-organizing power of matter. This is an insight which synergism will take even further. Nor should it come as any surprise that the present period, which has rediscovered the self-organizing dynamic latent in matter, should witness a revival of interest in Taoism, the goddess religions, Hopi spirituality, and similar movements. It may well be that when those of us who have taken the ascetic road finally arrive at Omega, we will be greeted by an assembly of Taoist masters, Hopi elders, and Mediterranean peasants, devotees of Demeter and Au Set, who arrived there, by their own mysterious paths, long ago.

B. Acosmic Mysticism

Having grasped the underlying unity and order of the cosmos, and having recognized that human society has fallen out of harmony with this order, there is a possibility at least that the unity and order of the cosmos will come to seem something quite distinct from the material universe which human beings in habit——that this latter will come increasingly to seem like a realm of conflict and illusion to be fled, rather than simply an imperfect or deformed manifestation of the cosmic ground. It was precisely this possibility which was realized in the tradition of Indian philosophy which gave birth to Hinduism. Buddhism, and their progeny. Here the "pure pattern of creation," or the "community and interrelationship of all things" is comprehended, but only in purely negative form, as *Brahman* or as the principle of dependent origination, and the complex manifold of the existing, multiform, evolving universe as rejected as mere illusion. We call this tradition acosmic mysticism because it locates the principle or ground of unity and order outside the material universe of nature and history, and seeks wholeness or salvation in withdrawal from, rather than engagement with, the cosmohistorical evolutionary process.

Why Indian religion developed differently than Chinese, Mediterranean, or American is difficult to determine. Clearly elements of acosmic mysticism may be found in all of the great civilizational traditions, but it was only in India and to a lesser extent Indochina that this strain became dominant. Louis Dumont (Dumont 1980) suggests that the hegemony of the Brahmin priests prevented the emergence of a unified state apparatus in India. Because of this interdependence was mediated not by the single organizing principle of the state but rather by a complex web of mutual obligations between the various castes. The whole thus appears not as a positive organized totality, but rather as a mere "vanishing moment" which is implicit in the complex network of interdependence, but which never achieves concrete institutional form. Salvation or wholeness is possible only by means of withdrawal from the world, into the hidden unity of

Brahman, or beyond it, into the nothingness of *Nirvana*. This analysis has considerable merit, for we shall see that in Indochina, where the monarchic principal was stronger, acosmic mysticism gradually shaded over into an incipient cosmic asceticism. But it remains to determine why the priesthood was so strong in India, and the monarchy so weak.

There are three principal stages in the development of acosmic mysticism. The first stage, associated with the production of the Upanishads, was essentially an attempt to spiritualize the sacrificial ideology of the earlier Vedic tradition. The second, or Buddhist phase, broke sharply with this sacrificial dynamic, but involved an even more radical retreat into acosmic interiority. The third, or Ashokan phase, represents a partial recovery of the notion of a concrete institutional whole, and points beyond itself towards the cosmic asceticism of the Hellenic and Jewish traditions.

In the Upanishads, the mythological discourse which characterized the earlier Vedas gives way to an incipient philosophical theory of the underlying unity of the cosmos, which they term *Brahman*. Some of the texts refer to *Brahman* in personal terms, as a kind of High God.

> Immortal, existing as the Lord,
> Intelligent, omnipresent, the guardian of this world,
> It is He who constantly rules this world (*Upanishads*, Svet. 6.17)

Others speak of Brahman as an impersonal force.

> Verily this whole world is *Brahman*. Tranquil let one worship It as that from which he came forth, as that in which he will be dissolved, as that in which he breaths (Chand. 3.14.1).

The later texts speak of both a formed and formless *Brahman*.

> There are, assuredly, two forms of *Brahman*: the formed and the formless. Now that which is formed is unreal; that which is formless is real (Mait. 6.17, 7).

The Upanishads identify, within each person——indeed within everything existing——what they call *atman*, or soul. This soul is not a separate substance, but rather a participation in the world soul, or *paramatman*, which is itself a manifestation of *Brahman*. By comparison, the manifold world visible to the senses, the world of facts and events, is simply *maya,* or illusion. The goal of humanity, according to the philosophy of the Upanishads, is to attain complete union with *Brahman*.

In this state of *turiya*, world and self are not so much obliterated, as they are united, the self knowing itself only as a part of the world; the world knowing itself only through the self (Chand. 6.8.6).

Thus far the philosophy of the Upanishads represents a considerable advance over the sacrificial ideology of the earlier Vedas. What the Upanishads fail to do, however, is to effect a real break with the system of *varnas* and the sacrificial problematic which governed the Indian tributary social formation. Union with *Brahman* or the universal is possible only through *moksha* (liberation) from involvement in *maya* or the world of appearance or particularity. This is not possible for the members of the lower castes who remain submerged in agrarian production, and thus hopelessly bound up with particularity. For them the only hope is to fulfill the duties associated with their caste, and hope that the process of *samsara* (reincarnation) will elevate them to a higher caste during their next rebirth. In fully developed Hinduism, even the members of the higher castes achieve *moksha* not through productive political or intellectual activity, greater holism and universality of this activity notwithstanding, but rather through ascetic practices. These practices were regarded as a kind of spiritualized sacrifice in which the *tapas* or heat of the self consumed the ascetic, just as the fires of the altar had consumed the old Vedic sacrifices.

The notion of *samsara*, furthermore, served to legitimate the system of *varnas*. A person's station in life was regarded as a result of *karma* accumulated in previous lives.

> Those who are of pleasant conduct here … will enter a pleasant womb, the womb of a Brahmin, or the womb of a Ksatriya, or the womb of a Vaisya. But those who are of stinking conduct here … will enter the womb of a dog, the womb of swine, or the womb of an outcast (Chand 5.10.7).

One care hardly imagine a more effective means of social control.

Buddhism marks a partial advance over the residual irrationalism of the Upanishads. First, it radicalizes the Upanishadic realization of the underlying unity of reality. It is not simply that reality is the manifestation of a unitary underlying principle (*Brahman*). On the contrary, the very existence of *things* is an illusion. The Buddhist principle of "dependent origination" teaches that we exist (or appear to exist) only because we carry with us the false conviction in the reality of the self and permanence of the world. Believing ourselves and the things around us to be real and distinct, we develop cravings for them, and become more and more involved in this world of illusion. Our desires, however, cannot be realized. As a result we suffer.

Now pleasant sensations, unpleasant sensations, indifferent sensations, ... are transitory, are due to causes, originate by dependence, and are subject to decay, disappearance, effacement, and cessation. While this person is experiencing a pleasant sensation, he thinks, "This is my Ego." And after the cessation of this same pleasant sensation, he things "My Ego has passed away (*Digha-Nikaya, Maha Nidana Sutta*)."

According to Buddhist doctrine, reality is a complex of relationships. Individual beings, and their desires, appear real and compelling only because we have failed to grasp the interdependence (dependent origination) of all things. Rejecting the reality of Brahman and *atman*, Buddhists speak not of reincarnation, but rather of rebirth. The appearance of the ego persists so long as we remain ignorant of the principle of dependent origination, and as long as we remain involved with things. What is reborn is not a substantial soul, but rather a particular configuration of relationships to the world (*Milidapanha*).

The Buddha's philosophical innovations had a number of practical implications. First of all, in so far as the goal of humanity was not unity with or participation in the universal substance, but rather recognition of the illusory character of all substance, the attachments of the Brahmin and the Ksatriya, their greater holism notwithstanding, seemed no better than those of the Sudra or untouchable. The Buddha upheld the Law of Karma, but held that birth in a lower caste had nothing to do with past Karma, and that members of any caste could overcome the effects of past karma. Needless to say, this sharply undercut the value of Buddhism as a means of legitimation for the Indian tributary social formation.

Second, the road to *moksha* or liberation was no longer seen as a participation in a vast cosmic sacrifice. On the contrary, the Buddha rejected asceticism as such in favor of the Middle Path of detachment.

There are two extremes, O Almsmen, which he who has given up the world ought to avoid ... a life given to pleasures; this is degrading, sensual, vulgar, ignoble and profitless ... And a life given to mortification; this is painful, ignoble, and profitless (*Vinaya*).

Buddhists seek to avoid attachments, which lead to suffering, but they do not see value in self-sacrifice as such.

While Buddhism was able to overcome those aspects of Indian philosophy which most directly legitimated the tributary social formation, it failed, however, to overcome the acosmism of the Upanishadic tradition. Indian philosophy generally, Buddhist and Upanishadic, represents a *negative* recognition of the "community and interdependence of all things."

It *intuits* the existence of an underlying principle behind the manifold world of facts and events, and, in the case of Buddhism, even grasps intuitively that being itself is not substance but relation,——a network of relations the points of intersection within which appear, to the ignorant, to be individuals. Neither Upanishadic nor Buddhist philosophy, however, was able to grasp the possibility that the underlying unity of the cosmos could be preserved, or even deepened, through the productive activity of human beings, nor were they able to grasp the possibility that human beings might come to an awareness of the "community and interdependence of all things" through labor, through participation in public life, or through scientific research or artistic creation.

Transcendence of this acosmism presupposed the creation of a concrete institutional totality capable of actually promoting the development of human social capacities——and thus the victory of the monarchic principle over Brahminic fragmentation. For the limitations of its warlord form notwithstanding, a unified state of some kind is necessary in order to centralize and allocate resources for development. In the wake of the Macedonian invasion of 325 B.C.E., Chandra-gupta, king of Magadha, had conquered much of the of the Indian subcontinent. It was a particularly brutal conquest, as it involved subduing numerous local dynasties. His grandson Ashoka completed the conquest, adding the Kingdom of Bombay to the already enormous domain. Legend has it that he felt "great remorse" for the suffering which he had caused, and issued a series of edicts intended to rectify the wrong he had done by carrying out a series of social reforms and by offering state sponsorship to the Jain and Buddhist religions.

Behind this public display of remorse, however, was a keen strategic mind which, in the relentless pursuit of power, made important religious innovations. At the center of Ashoka's strategy was centralization of surplus for a large system of public works——including public granaries which provided for the poor in times of famine. The property of the rich (the local dynasties and coteries of nobles he had conquered) was seized, but to the poor Ashoka would lend without interest, and after three years forgive all debts (Sarkisyanz 1965: 54-56). He also provided state funds to Buddhist and Jain monasteries (Sarkisyanz 1965: 28-30).

What did Ashoka gain by this generosity? Buddhism solved one of the principal difficulties facing a prospective emperor in India: the system of varnas which made Brahmin superior to warrior or ruler, and which make all rulers members of a relatively egalitarian caste community in which, up until 500 B.C. a kind of rough internal democracy had prevailed. If all human beings were fundamentally equal, then the claims of the old ruling powers of India were invalid. Second, the Buddhist doctrine of dependent origination taught Ashoka that ultimately his power was rested on the consent and even the support of the people. A king who improved the lot of his people would enjoy their firm support and his kingdom would be secure. In short, Ashoka used a program of public works and public piety to build an alliance with the masses against the local aristocracies who presented the greatest threat to his empire.

Ashoka's insights implicitly pressed beyond the limits of Buddhist doctrine. The doctrine of dependent origination became a guide to political strategy. Active participation in political life became the basis for realizing the principle of dependent origination.

The Mauryan empire eventually fell, but Ashokan state Buddhism became the model for smaller kingdoms in Sri Lanka, Burma, Thailand, and Kampuchea. An entire tradition of lay Buddhism grew up, centered in the *Ashoka-sutras*, which spoke of just kings who had fed the poor, pardoned criminals, and invested the surplus they centralized in public works and the support of the Buddhist *Sangha* (Sarkisyanz 1965: 33). This lay Buddhism, like the monastic tradition of the Buddha himself, pursued liberation from the illusion of selfhood. Its principal means however, was not mediation assisted by monastic withdrawal from worldly attachments, but rather works of *Metta* or charity, through which one gained *karuna*, or a sense of identification with others.   Ashokan Buddhism provided fertile soil not only for the emergence of reforming monarchies, but also, when these monarchies ceased to serve the common good, for the emergence of peasant revolts directed at restoring *dharmaraj* or the rule of cosmic law and setting humanity once again on the path to liberation. And we will see that movements rooted in Ashokan Buddhism, and in structurally similar variants of Indian philosophy, continue to this day to play an important role in humanity's struggle to fulfill its destiny in the cosmos, defining a unique "Indian road" towards synergism.

But if it is possible, on the basis of human reason and experience, to grasp the underlying interdependence of all things, and, on the basis of this knowledge, to order not only one's own life, but whole kingdoms, then we are no longer in the realm of acosmic mysticism, but have crossed over into the realm of cosmic asceticism.

C. Cosmic Asceticism

It would be a mistake to assume with Weber that religiously motivated social historical action is the unique product of the civilizations of Europe and the Mediterranean Basin. We have already seen that mystical movements, in the struggle to bring human society into harmony with the cosmos, necessarily pass over into social-historical action, and, in the process, develop strong rationalistic qualities. At the same time, there remains a very real difference between Taoist, Hopi, *mater/ia*list or Ashokan Buddhist movements which treat cosmic harmony as an end in itself, and the more dynamic outlook characteristic of Hellenic philosophy, Judaism, and their offspring, Christianity and Islam, which, if only implicitly and indirectly at first, regard human civilization as a positive expression of the cosmic order, and a center for the creation of organized complexity. We now need to trace the emergence of this "ascetic" tradition and assess its contribution to the human civilizational project.

1. Hellenic Philosophy

Matriarchal communitarian organization seems to have persisted in Greece longer than in some other regions. By the early iron age (800-500 B.C.E.), however, a network of warlord states had grown up on the Greek peninsula. The economic activity of these *poleis* was centered on the production of wine and oil for export. The *poleis* were dominated politically and economically by a landed aristocracy (*dynastoi*), which gradually reduced a large part of the peasant population to debt peonage. The commercial character of Greek society, however, made it possible for new strata of wealthy landowners and merchants to emerge outside the close hereditary aristocracy. In Athens and certain other cities these *nouveau riches* elements made common cause with the peasantry, and imposed a series of reforms, which opened up political power to all free, landowning citizens, abolished debt peonage, and instituted a system of direct public assistance for the peasantry, thus guaranteeing their survival (Anderson 1974:29-32).

Central to this settlement, however, was the development of chattel slavery, which provided the landed aristocracy with a source of labor other than that of their own peasantry. By the fifth century B.C.E., there were some 80,000-100,000 slaves in Attica, and only 30,0000-40,000 citizens (Anderson 1974: 38). Most of these slaves were owned by a relatively small number of large landowners. This meant that citizens approached the formally democratic public arena with very different economic interests. In order to maintain their sway over the masses, the ruling classes had to develop a whole repertoire of new political techniques. Thus the emergence of schools of rhetoric, where wealthy young men could gain the training they needed in order to maintain their hold on the hearts and minds of the masses, and thus prevail in the law court and the assembly. There had grown up in Athens, in other words, a political-economic system which institutionalized the subordination of the common good——the good of the *polis* as a whole——to the particular interests of the wealthy landowners who manipulated the masses in the assembly. By using the techniques taught them by their sophist mentors, the rich young men of Athens could, quite literally, "make the worse appear the better cause," so that eventually the people began to doubt that there really was any such thing as the Good, the True, or the Beautiful. In this sense, while the Hellenic *polis* opened up qualitatively new possibilities for the development of humans social capacities, by expanding the sphere of political participation and widening the sphere of public discourse, it also made possible a hitherto unthinkable hijacking of the state by private interests, and a hitherto unknown degradation of public morality.

The philosophical tradition was first and foremost a movement of resistance to the ideological degeneration set in motion by slave based democracy.   The   philosophers——faithful   disciples   of   the Goddess——struggled to discover the organizing principle of the universe, so that they could bring human society back into harmony with the cosmic order. Some, such as Democritus, looked for this organizing principle in matter, while others, such as Parmenides, Anaxagoras, and Heraclitus looked in extramaterial forms: in the One, the underlying unity of all things, or in the principle of change which guided the constant flux of matter as it transformed itself from one thing into another.

As important as these early contributions were, however, it was only with Socrates and the Socratic tradition that humanity began to make real progress towards recovering its knowledge of the cosmos as an organized totality. The critical breakthrough in this regard was Socrates' discovery of the dialectic. Unlike the sophists, for whom debate was simply a means by which to sway the opinion of the masses in the service of their clients, Socrates understood that human discourse embodies real, though partial and imperfect truths.   By asking penetrating agitational questions of his interlocutors, Socrates was able to draw out the implications and the contradictions of existing opinion, and struggle towards a higher synthesis.

Both the political context and the concrete political significant of this process are apparent in Plato's *Republic*. The work begins with an account of Socrates, who is returning from a religious festival——a festival of the Goddess——being quite literally way-laid by a group of rich young men——almost an allegory of the captivity of the intelligentsia to the military aristocracy in tributary social formations. He engages them in a debate regarding the nature of the justice and the good. He speaks quite explicitly of the decline of the polis as first military, and later commercial strata gained hegemony (*Republic* 543a ff). The Republic is a polemic against those who teach that "in every city the same thing is just, the advantage of the established ruling body." (Republic 338d), and an attempt to demonstrate that there really is an objective order to the cosmos, and that this order can be discovered by human reason. It is little wonder that Socrates was regarded as a danger to the Athenian state.

Plato built on Socrates' achievement. He identified four distinct levels of knowledge (*Republic* 509c-511c). The first is eikasis or imagination, which grasps only the images of material things. It is not clear whether we are to understand here simply the kind of vague and fading images of things which make up our memory, or if we should think rather of the distorted images of the world embodied in the religious mystifications of the warlord state, which Plato regarded as so destructive. The second form is *pistis*, the knowledge we have of facts and events based on sensory impressions, which constitutes the realm of sound common sense. The third form of knowledge is *dianoia*, rational, demonstrative thought, which allows us to reason on the basis of hypotheses, and thus to grasp the nature of the mathematical objects. Finally, at the highest level, we find *noesis*, or rational intuition, which grasps first principles, and thus confers knowledge of the "community and interrelationship of all things (*Republic* 531d)." Plato, who was more concerned with insight that systematization, spoke of the organizing principle of the universe in various ways, sometimes as the One, sometimes as an integrated, intelligible realm of forms or ideas, and sometimes as the Good, which is beyond being, and which confers organization and thus being on everything in the universe (*Republic* 517c).

Knowledge of the Good, however, presupposes a certain level of human development. The existing structure of the polis is holding back the development of human social capacities and must be transformed. Plato seemed to regard private property, and the private relationships which make up the family, as incompatible with full participation in public life. He thus argued that the leadership (though not the people as a whole) should live in community. The old religion which depicted the gods as dominated by human passions should be abolished in favor of a new myth which spoke of the reward of the just and the punishment of the wicked. Those who were to participate in public life would have to undergo intensive training. Gymnastic would bring the various elements of the body into a proper balance. Through the study of arithmetic, geometry, astronomy, and music, they would develop an understanding of the order and harmony which governs the cosmos. They would gradually penetrate behind the mystifications created by the old religion, and the superficial "knowledge" of facts and events which was the stuff of public opinion, to a rational knowledge of the "forms" or underlying structures which lay behind these facts and events. Eventually, at the end of this process, which Plato called the "journey of the dialectic" they grasp the "community and interrelationship of all things (*Republic* 531d). It was this knowledge which, by focusing human beings on the Good, made them fit to lead. Plato calls on the philosopher, who has grasped the dialectic to return to the dark cave of human society and set his captive fellow citizens free. And Plato himself was actively engaged in a project (however ill-fated) to reorganize the Greek colony of Syracusae.

What is missing in Plato, of course, is any account of development——of how human beings move towards the Good. It was Aristotle who remedied this deficiency. His grasp of the immanent teleology of potency and act, dynamia and *energeia*, product of his work in the new science of biology, represented humanity's first recognition of the lawful and progressive character of change. It also permitted the systematization of Plato's often eclectic and scattered insights, so that humanity was able, for the first time, the comprehend the cosmos as an organized rational totality.

It is not possible to outline this system in detail here. Suffice it to say that Aristotle grasped clearly that matter is first and foremost the potential for organization. This potential is awakened, as it were, by the action of the prime mover, which moves all things while itself remaining for ever motionless and still. The prime mover acts on matter not as an efficient but rather as final cause, the incredible beauty of which draws all things towards itself.

> ... there is something which moves without being moved, being eternal, substance, and actuality. And the object of desire and the object of thought move in this way; they move without being moved ...
>
> On such a principle, then, depend the heavens and the world of nature ... (*Metaphysics* 1072a-b).

It is because of their attraction to the prime mover that the heavenly bodies move around their orbits. It is because of their attraction to the prime mover that seeds grow into plants and embryos develop into complete animals. And it is because of our attraction to the prime mover that we humans develop the latent potential of our intellectual and moral capacities and gradually grow into virtuous human beings. The purpose of the polis is first and foremost the cultivation of virtue, and in the *Politics* Aristotle discusses in considerable detail just what kind of social structure serves best to catalyze the full development of human capacities.

There were, to be sure, limits to the Hellenic philosophical tradition. These limits derive first and foremost from the low level of development of human social capacities. In a predominantly agrarian social order, the vast majority of the population must continue to be absorbed in direct material production——i.e. the cultivation of food crops. The society simply could not support the level of investment in education necessary to make every member of society a philosopher. Plato is thus forced to settle for second best——a society governed by philosopher kings. Aristotle is, if anything, still more pessimistic. Human beings, furthermore, seem to have relatively little role in shaping the order of the cosmos, which appears to have been fixed for all eternity. This dynamic is reinforced by the emergence of patriarchy and the tributary state. Even within the realm of human society, "ordering" and "organizing" appear increasingly as activities wholly distinct from the direct material production (bearing and raising children, the cultivation of food) in which the vast majority of people are engaged. The notion thus develops that order is a pre-existing form imposed on matter from the outside, like the seed sown by the father which gives form to the materia of the woman's menstrual blood, or the laws of a conquering monarch which force a recalcitrant people to submit to forced labor. This is the social basis of idealism.

As a result of these constraints, Hellenic philosophy soon began a process of degeneration. It fell prey to emmanationist doctrines which regarded the material world as at once an expression of, but also a falling away from, the pure Forms which alone were authentically Good, True, Beautiful, etc. Unity with this higher realm was possible only through an intellectual *gnosis*, or through participation in sacred mysteries which joined the believer to the sacrifice of the dying and rising God——a dynamic which Christianity captured and made the religion of all Europe and the Mediterranean basin.

Aristotle's insights regarding the potency and act might have overcome these limitations, had they been developed along the materialistic lines which would later be followed by Avveroes and the Latin Avveroists. But there is a curious ambiguity in Aristotle's system. Immediately after the passage cited above, where Aristotle argues that the prime mover acts as a final rather than as an efficient cause, he adds the following:

> We must consider also in which of two ways the universe contains the good and the highest good, whether as something separate and by itself, or as the order of the parts. Probably in both ways, as an army does; for its good is found both in its order and in its leader, and more in the

latter, for he does not depend on the order but it depends on him (*Metaphysics* 1075a).

Gone is the almost mater/ialist vision of the universe as a cosmic totality moved by love for the goodness of the final cause. Here the universe appears as a kind of army, ordered by a celestial general, who, we must imagine, is simply the philosophical reflex of the great Macedonian generals who paid Aristotle's bills.

It was only under the influence of Judaism, which stressed the value of the material world, and which developed a powerful critique of sacral monarchy, that the philosophical tradition was able to transcend the limits of Platonism and Aristotelianism, and develop for the first time an ontology which regarded human beings as real participants in the self-organizing activity of the cosmos, and a theory of value centered on the development of human creative capacities.

2. Judaism

Where Hellenic philosophy emerged in an urban context, and found its constituency in the great metropoles of the Mediterranean Basin, Judaism emerged directly from the struggles of marginalized peasant communities, and, as it became an increasingly urban religion, carried with it an ethos derived from centuries of peasant resistance. During the Late Bronze Age, the Syro-Palestinian Corridor, where Israel would ultimately emerge, was dominated by a number of city states (really little more than desert fortifications), each with its own warlord, who held sway over the surrounding villages with the assistance of a warrior aristocracy. Land was held in common by the village communities, which redistributed it periodically to individual families for cultivation, according to their needs, but which had to pay tribute to the warlords in the form of rents, taxes, or forced labor. The region as a whole was, in turn, subject to the more or less effective hegemony of the great powers, primarily Egypt, which also extracted a share of the surplus, and no doubt imposed forced labor as well (Gottwald 1979). Bronze technology was used primarily for weapons. This severely limited the development of the surrounding hill country, which could not be cleared and terraced without the use of metal tools.

The prevailing religious form of the Canaanite social formation, at least by the time of the late bronze age, was the cult of *ba'al*, which had, by this point, hegemonized the older, gynocentric cult of the *ashtoreth*. This *ba'al* is generally referred to in the textbooks as a "fertility god," as indeed he was. As we noted above the term *ba'al* means lord, master, owner of land, and ... husband. The term was used for the local warlord, as well as for the deity. Identification of agrarian and human fertility with domination and lordship, and of both with the divine, provided the ruling classes with an especially effective system of legitimation. The Canaanite peasantry, to put the matter starkly, worshiped their landlords.

According to Norman Gottwald, the later Bronze Age witnessed a decline in great power hegemony in the Syro-Palestinian corridor, and corresponding internal strife among the warlords who dominated the Canaanite lowlands. The resulting instability in turn led to an increase in rural unrest, which took the form of social banditry (Hobsbawm 1959). These social bandits——referred to as "*'apiru*" in contemporary sources——were essentially marginalized peasants who had been run off their land, or who had gotten into trouble with their lords, and had (quite literally) taken to the hills, from whence they preyed off caravans, or raided the city states, occasionally entering the service of one or another *ba'al.*

At roughly the same time the collapse of the Hittite Empire to the north broke the monopoly on iron technology, allowing the techniques for production of primitive bloomery iron to penetrate Canaan. Up until this time metal tools had been a ruling class monopoly, protected by royal control of the tin trade——tin being an essential component of bronze, the only metal thus far widely used in the area. This ruling class monopoly on metal tools had in turn held back the development of the hill country, which required metal tools for clearing and terracing. Bloomery iron, while inferior to the bronze used by the Canaanite aristocracy, was superior to the stone tools used by the Palestinian peasants, and could be produced with materials available in the region. The collapse of the Hittite iron monopoly thus put metal tools into the hands of the peasants, removing the obstacle to settlement in the hill country.

The hills were out of the reach of the chariots of the Canaanite warlords, and thus beyond the sphere of Canaanite military hegemony. The *'apiru* groups thus began to terrace and cultivate the hill sides, and their banditry gradually transformed itself into a kind of *guerilla*, or prolonged popular war——the record of which is preserved in the Book of Judges. They organized themselves into *mishpahoth*, or protective associations of extended families. These *mishpahoth* practiced a form of communal land tenure, holding land collectively and redistributing it periodically to individual families, according to need, for purposes of cultivation (Lev 25: 8ff), and also constituted a kind of "popular militia" which helped to defend and extend the "liberated territories" without recourse to a standing army. Israel seems to have provided for a tax of roughly 10% of the agricultural produce to support the Levitical priests.

At the same time, Israelite law insured that the priests could own no land of their own and thus could not degenerate into an exploitative landowning class.

Norman Gottwald has suggested, based in part on the frequency of Egyptian names among the Levites, and their subsequent role as a religious elite, that the Exodus story is in fact the story of the Levites, and perhaps other elements serving in Egypt as forced laborers, who became the carriers of the cult of *yhwh*, and whose flight from Egypt and penetration of Palestine played a critical role in catalyzing the formation of Yahwistic Israel out of the numerous *'apiru* bands. The Exodus is, according to this hypothesis, the story of the vanguard of the Yahwistic revolution.

The emergence of the cult of *yhwh* was at once a product of, and a catalyst for this process. There were both negative and positive dimensions to this process. On the one hand, the cult of *yhwh* broke the nexus between agrarian and human fertility, exploitation and domination, and the divine order. This is the significance of the prolonged prophetic critique of the cult of *ba'al*. The prophets sought to remind the people it was *yhwh* and not *ba'al*, the God of the *guerilla* and not the god of the warlords, who gave Israel her grain, her oil, and her new wine

> Call your mother to account ...
> She does not know that it was I who gave her the grain, the new wine and the fresh oil,
> I who lavished on her silver and gold which they used for the *ba'alim* (Hosea 2:8).

Yahwism rejected the collaborationist priesthood which used rituals as a means of extracting surplus for the warlords but failed to build up the community or promote authentic knowledge of God.

> There is no good faith or loyalty, no knowledge of God in the land ...
> The more priests there are, the more they sin against me ...
> As you have rejected knowledge,
> so I will reject you as a priest to me (Hosea 4: 1, 7, 6).

It was the prolonged struggle of a people faithful to its god, and to the legal norms which it had developed into order to sustain the revolutionary social order, and not rents, taxes, or corvees delivered to a petty desert despot, or orgiastic cults, which insured the prosperity and the power of the peasantry.

Unlike the other gods, who ruled through earthly kings, who were often regarded as their offspring, for Israel *yhwh* alone was king (1 Sam 8:7, Ps

97). Israel achieved an extraordinarily sophisticated grasp of the exploitative character of the tributary social relations which characterized the society out of which it emerged. A king will

> take your sons and make them serve in his chariots and with his cavalry and they will run before his chariot. Some he will appoint officers over units of a thousand and units of fifty. Others will plough his fields and reap his harvest; others again will make weapons of war and equipment for the chariots. He will take your daughters for perfumers, cooks and bakers. He will seize the best of your fields, vineyards and olive groves, and give them to his courtiers. He will take a tenth of your grain and your vintage to give to his eunuchs and courtiers ... He will take a tenth of your flocks and you yourselves will become his slaves (1 Samuel 8: 11-17).

At the same time, the emergence of the cult of *yhwh* marked a new development in human understanding of and participation in the self-organizing activity of the cosmos. Through their own revolutionary activity, through the work of reorganizing the underlying structures of their society, the people of Israel discovered that the self-organizing activity of the cosmos was present not only or even most profoundly in the orderly motions of the heavenly bodies or the apparently unchanging progression of the seasons, nor in the growth, development, and decay of plant and animal life, but also and most especially in the process of human history.

This is apparent from the name *yhwh* itself. The word is apparently the causative form of the verb "to be." Indeed, when Moses asks God who he should say has sent him, the deity answers *"eyeh asher eyeh"*——I am that I am (Exodus 2:14). The emergence of the cult of *yhwh* thus marks, albeit in philosophically undeveloped form, a profound insight into the nature of the divine. God has no particular name, but is rather being itself, the underlying, unifying substratum from which all things derive.

But there is more. There is reason to believe that *yhwh* was originally a name of the Canaanite high god El——specifically *'el yahwi sabaoth yisrael*——"El who brings into being the armies of Israel." It is not at all uncommon for oppressed groups, whose cult previously centered on clan or village deities, to unite around a previously remote High God as the form a confederation to resist oppressors. A similar phenomenon characterized the horticultural societies of North America, which gave birth to a series of cults of the Great Spirit during the period of resistance to European conquest and settlement (Lantenari 1958). This turn to the high god marks the formation of an intervillage, or intertribal unity (Durkheim 1911/1965), but it also reflects an appeal to the underlying unity of the cosmos and of history, against the deformed social relations of the tributary state, represented in the cult of *ba'al*.

yhwh was a god from the mountains and the hills: like the apiru bands he "set forth from Seir" (Judges 5:4). His right hand and strong arm crushed the oppressor, and made the mighty flee in sudden confusion. Israel met its God not primarily in contemplation of the natural order, though it certainly found him there as well (Ps 104), but rather on the battlefields of the revolution.

> *yhwh* is a warrior; *yhwh* is his name.
> Pharaoh's chariots and his army he has cast into the sea;
> the flower of his officers are engulfed in the Red Sea.
> The watery abyss has covered them;
> They sank into the depths like a stone.
> Your right hand, *yhwh*, scattered the enemy.
> In the fullness of your triumph you overthrew those who opposed you:
> You let loose your fury;
> it consumed them like stubble ...
>
> *yhwh*, who is like you among the gods?
> Who is like you majestic in holiness,
> worthy of awe and praise, worker of wonders?
> ...
> In your constant love you led the people
> whom you had redeemed:
> you guided them by your strength to your holy dwelling place (Exodus 15: 3-13).

The liberated zones of which Israel established in the hill country of Judah, Samaria, and the Galil were under constant pressure from both the Canaanite cities of the lowlands, and the surrounding Kingdoms of Edom, Moab, and Ammon. The most dangerous of the enemies of Israel were, however, the Philistines. The Philistines, who appear to have entered the region from the Aegean, possessed iron weapons, and used foot soldiers, and could thus penetrate Israel's mountain strongholds. Interestingly enough, one of their first moves on gaining hegemony over Israel was to prohibit the practice of iron working among the Israelites, for fear that they might beat their plowshares into swords (1 Samuel 13:19-22). In order to mount an effective resistance to the Philistines, Israel was forced to establish first a centralized military command (under Saul) and ultimately a monarchy (under David and Solomon). And the imperatives of centralizing tribute to support the monarchy requires the establishment of a centralized temple cult.

Gottwald regards the establishment of the monarchy as a kind of counter-revolution which restored tributary social relations and made Israel little better than the warlord state which it had supplanted. This seems excessive. Certainly the monarchy was characterized by serious injustices, and the restoration of the temple cult opened the door to a kind of creeping baalism. At the same time the establishment of a monarchy, together with a standing army, permitted the liberation of the entire land of Canaan, and thus a expansion of Israel's rationalizing social project. The Yahwistic revolution permitted the application of iron technology to agriculture, perhaps for the first time, and thus made possible the settlement of the hill country. It also permitted the centralization of the resources necessary for investment in art, science and technology, and philosophical speculation. That Israel's contributions in these fields were limited is due largely to the need to invest most of the surplus extracted in defense against resurgent imperial powers, and to the fact that the Jewish state was destroyed just as authentic science and philosophy were making their appearance throughout the Mediterranean Basin.

Furthermore, even after the establishment of a monarchic state in Israel the Yahwistic tradition made it possible to hold the monarchy accountable for its stewardship of the people's resources. This is the significance of the prophetic movement, which continued the rationalizing activity of the early Yahwists. The prophets denounced the oppression of the poor.

yhwh opens the indictment against the elders and officers of his people:
It is you that have ravaged the vineyard;
In your houses are the spoils taken from the poor.
is it nothing to you that you crush my people and grind the faces of the
poor? (Isaiah 3: 14-15)

Woe to you who add house to house
and join field to field,
until everyone else is displaced,
and you are left as the sole inhabitants of the land. (Isaiah 5:8)

Woe to you who make unjust laws
and draft burdensome decrees
depriving the poor of justice,
robbing the weakest of my people of their rights,
plundering the widow and despoiling the orphan (Isaiah 10: 1-2).

Some of the prophets, particularly in the Southern kingdom of Judah, looked to a reformed monarchy to deliver Israel from both foreign oppressors and from its own internal disintegration. Inevitably they drew their hopes from the memory of David's victory against the Philistines.

Then a branch will grow from the stock of Jesse and a shoot will spring from his roots.
On him the spirit of *yhwh* will rest
a spirit of wisdom and understanding,
a spirit of counsel and power,
a spirit of knowledge and fear of *yhwh*.
He will not judge by outward appearances or decide a case on hearsay
but with justice he will judge the poor
and defend the humble in the land with equity;
like a rod his verdict will strike the ruthless
and with his word he will slay the wicked
He will wear the belt of justice,
and truth will be his girdle (Isaiah 11:1).

Others attacked the institution of the monarchy——and the associated state cult——itself.

... the Israelites will live for a long time without king or leader, without sacrifice or sacred pillar,
without ephod or teraphim ...
After that they will seek *yhwh* their God ... (Hosea 3:4)

In this sense, it is a mistake to regard monarchic Israel as a restored warlord state. Rather, it was a complex and progressive social formation, in which priest and king shared power with the prophetic movement. If the prophets were perpetually unable to realize their vision for the people, for which the necessary social conditions did not yet exist, then the priests and the kings were also unable to rule without making significant concessions to the prophetic movement. The result was a creative tension which guaranteed centralization of sufficient surplus to defend Israel against its enemies, while at the same time subjecting both the process of surplus extraction and the way in which surplus was invested to careful scrutiny to determine its impact on the integrity of the ecosystem and the social fabric (particularly the village communities) and the development of human social capacities.

Gradually the iron weapons technology which had made the Philistines such a potent threat spread to the resurgent great powers——Egypt,

Assyria, Babylon, Persia, and Macedonia. The great size of these empires made it possible for them to field enormous armies and equip them with devastating iron weapons. Israel and Judah resisted, and there is considerable evidence that the tenacity of these small states stunned the advancing armies of the new empires. Thus the Assyrian general Holophernes on the Israelite resistance:

> Tell me you Canaanites, what people is this that lives in the hill-country? What are their cities? How large is their army, and wherein lies their power and strength? Who has set up as king at the head of their forces? Of all the people of the west, why do they alone disdain to come to meet me? (Judith 5: 3-4)

Eventually, however, even Israel was overcome. First the Assyrians, then the Babylonians, Persians and Greeks, and finally the Romans brought Israel under their hegemony. Exploitation was intense, and very little of the surplus extracted was invested in a way which benefited the Jewish masses.

Each of these empires dealt with the Jewish people in a slightly different manner. The Assyrians slaughtered many, and resettled others, replacing them with peasants from other parts of their empire, in an effort to break down Jewish national identity. The Babylonians deported much of the ruling classes, but permitted most of the Judea peasantry to remain intact. The Persians, and later the Romans, on the other hand, permitted significant local autonomy, while insisting that local Jewish authorities keep prophetic elements in check. The Ptolemies followed a policy of gradual and noncoercive Hellenization, while the Seleucids were far more aggressive, and mounted a direct assault on Yahwism.

The Jewish tradition continued, under these circumstances to be a constant source of ferment——and of progressive religious rationalization. Only a few within the Jewish community were actively supportive of the imperial powers. But significant differences emerged not only over the most adequate strategy for resistance, but ultimately over just what it meant to be Jewish, and regarding the role of the Jewish people in what they now realized was a much larger world historical process.

Armed struggle directed at national liberation was, of course one option. There were literally hundreds of armed revolts against nearly all of Israel's imperial overlords throughout the whole period following the collapse of the monarchy and the destruction of the temple by the Babylonians in 587 B.C.E. Another option was simply to collaborate with one or another imperial authority (Persia, Rome) in order to buy space for the re-establishment and conservation of the temple cult, which some elements in

Jewish society understood as the very center of Jewish life.　Eventually these two strategies flowed into one another in the Maccabean revolt of the second century B.C.E., which achieved for Israel a measure of independence, especially in the cultic arena, albeit at the price of an increasingly costly alliance with Rome.

Elements of Jewish society dissatisfied with this settlement continued the struggle for independence, or else withdrew into the wilderness to await some royal and or priestly messiah.　Richard Horsely (1985) has documented the existence of both social banditry (48f) and more eschatologically charged messianic movements.　Royal pretenders and would-be messiahs, including Judas, Simon and Athronges, active after the death of Herod the Great (112-144), Menachem and Simon bar Giora, active during the Jewish War (118-119), and Simon Bar Kochba, active around 132 C.E.　(127ff) all laid claim to the throne of Israel.　The Essenes were a priestly group, and seem to derive from a group of Hasidim who supported the Maccabean revolt, but who could not reconcile themselves to the appointment of a non-Zadokite --Jonathan-- as high priest.　Eventually they Essenes withdrew from Jewish society to form monastic communities, and developed a complex apocalyptic doctrine, which looked forward to not one but two messiahs, one military and the other priestly (Zeitlin 1988:　28).　This latter figure was of paramount importance, and in general the Essenes can be dismissed as a relatively minor group of displaced priestly elements who looked not to the revolutionary transformation of Jewish society, but to the restoration of an independent tributary social formation.　Prophetic figures such as John the Baptist, Theudas and the Egyptian (161ff) led mass movements which attempted to reenact events from the establishment of Israel, such as the giving of the Law, the crossing of the Jordan or the attack on Jericho.

All of these strategies, however, presupposed that Israel could survive——and realize its historic vocation——as an isolated enclave within a larger imperial world order, either constantly struggling for independence, or else trading economic and political dependence for cultic autonomy.　During the long years of the exile and imperial domination the most advanced elements in Jewish society had come increasingly to realize that this would not work.　Israel could survive only in a world which feared its God and loved His law.　This kind of sentiment is already apparent in prophetic writings from the exilic and postexilic periods.

> Listen to me, you coast and islands
> pay heed you peoples far distant ...
> ... *yhwh* has said to me:
> "It is too slight a task for you, as my servant,

to restore the tribes of Jacob,
to bring back the survivors of Israel:
I shall appoint you a light to the nations
so that my salvation may reach the earth's farthest bounds (Isaiah 49: 1a, 6)."

For Israel, however, this hope could never be realized through wars of conquest nor through aggressive proselytization, but rather only through an effort to set a example so beautiful, so powerful, and so compelling, that the world could not help but follow.

This drive to actually fulfill the law, so that the nations might realize its power, and be drawn to worship Israel's God, eventually gave birth to the Pharisaic movement. The Pharisees were an unambiguously patriotic party. They looked forward to the day when the Roman yoke would be lifted and Israel restored. The restoration of Israel did not mean for them simply independence from foreign rule, but the comprehensive transformation of Jewish society in accord with the ethical traditions embedded in the Torah. Neither prophetic denunciation nor armed revolt seemed adequate to this task. The prophets, after all, had developed no strategy to actually transform Jewish society. The Maccabean revolt on the other hand, while winning for Israel a short period of limited independence, left in power a corrupt priestly aristocracy which tended to reduce Judaism to a matter of ritual observance.

The Pharisaic strategy focused instead on the reorganization of Jewish life in such a way as to shift the religious center of gravity from the temple to everyday life, and from ritual observance to ethical conduct, while preserving the ritual law as a mark of Jewish national identity. While the Pharisees did not reject the temple cult or the purity code as such, they did reject the idea that holiness was purely or primarily result of cultic observance, and that it thus belonged primarily or exclusively to the priestly caste which controlled the temple. This involved two critical innovations. First of all, the Pharisees developed the notion of an "oral Torah." As John Pawlikowski has pointed out (1982: 82) the development of the oral tradition was an attempt to give as much specificity to the rather general ethical imperatives of the written law, as the priestly caste had given to the ritual codes.

As the priests had been concerned with the codification of the cultic legislation, so the Pharisees concentrated their efforts on the "codification" of love, loyalty, and human compassion, with the expressed goal of making them inescapable religious duties incumbent upon each and every member of the people Israel .. thus making communion with God possible at any given time or place, with or without the Temple, the priesthood and the sacrificial altar ...

What is clear from this Pharisaic concern with the oral Torah is the necessity for religious persons committed to the ideals of the Exodus covenant to move beyond mere nominal generalization to concrete programs that would correct the injustices being perpetrated by existing social structures (1982: 82-3).

Second, the Pharisees began to create a new institutional structure, centered on the synagogue and the rabbinate, alongside the temple and the priesthood, which served to reinforce the new emphasis on ethical conduct as the center of religious life. The task of the rabbi

resided in the determination of specific solutions to the social problems facing the Jewish society of the day ... In the figure of the rabbi the Exodus understanding of social responsibility as an integral part of genuine religious life received concrete, symbolic expression for the first time (84).

In a very real sense, the rabbinate institutionalized the prophetic function, albeit in the "ordinary" form of a teaching office, making prophetic insight an integral dimension of the day to day life of the Jewish people. The synagogue, by the same token "focused around the congregation as community." Services in the synagogue centered not around prayer or worship, but around "homilies on contemporary demands of the Torah."

For the first time the chief religious institution of any major world religion .. placed concern for social responsibility on a par with ritual worship ... In keeping with their awareness that mere prophetic appeals to conscience were relatively ineffective by themselves in producing the transformation of Jewish society ... the Pharisees enacted a profound structural change in Judaism by gradually replacing the Temple with the synagogue (85).

The Pharisees, furthermore, shifted the center of ritual observance properly speaking from the Temple to the home, and in particular to the arena of table fellowship. This had the effect of displacing ritual power from the priestly aristocracy to the people, while preserving the purity codes which served as a mark of Jewish national identity. In the hands of the Pharisees, the purity codes ceased to be an instrument of priestly hegemony, and became instead a popular institution, through which the Jewish people as a whole preserved their identity (Pawlikowski 1982: 87). The drive to extend the purity codes to the whole people, furthermore, reflected a conviction, radically new in the history of religion, that not only the priest, but also the peasant, merchant, etc. was in some sense "holy." In this sense, the Pharisees took an important step towards breaking down the dualism characteristic of religion in all agrarian societies, which makes a sharp distinction between the priest and the king, who participate directly in the life of God, and the people, who participate only indirectly, by producing food to support the political and cultic leadership.

Pharisaism, in other words, represented an extraordinarily high level of religious rationalization, and a major step forward for humanity in the development of a rational strategy for the reorganization of human society. Where the priestly movements focused on conserving the temple cult and the messianic movements focused on the armed struggle for national liberation, risking the danger of transforming Israel into a nation which was free, but otherwise little different from the surrounding peoples, the Pharisees understood that the original insight of Yahwism——that human beings participate most fully in the life of God through ethical conduct——could be realized only in the context of an ethical civilization. They thus sought to provide ethical guidance to Jews who carried out the will of God by exercising their creativity and initiative in every sphere of human social life. In large part due to their efforts the Jews have served ever since as a kind of leaven for every civilization with which they have come into contact.

Jesus himself must be understood as first and foremost a figure within, or at least at the margins of, the Pharisaic movement. John Pawlikowski and others have documented in considerable detail the clear roots of much of what Jesus taught and did in the Pharisaic tradition. First of all Jesus was clearly a teacher of the oral Torah "reinterpreting the Hebrew Scriptures in ... line with the social setting in which he found himself." Furthermore, the general pattern of his ministry "with its emphasis on teaching and healing" is characteristic of the rabbinic pattern for the period. "He likewise seems to have participated in Pharisaic-type fellowship meals, instituting the Christian Eucharist at the final one he attended (Pawlikowski 1982: 92-93)."[2]

---

[2]. From this point of view any interpretation of the Christian eucharist as a sacrificial rather than a community building act represents a fundamental departure from the actual teachings and practice of Jesus.

The content of Jesus' teaching also shows significant Pharisaic influence. Jesus' answers to the questions posed to him while teaching in the temple (Mark 12:13-34), and particularly the question concerning the great commandment put him more or less squarely in the Pharisaic camp. Like the Pharisees Jesus regarded the imperative to love of God and neighbor (12:28-34) as the greatest of the commandments, he upheld a belief in the Resurrection (12:18-27), and (like many Pharisees) he equivocated when asked whether or not Jews ought to pay tribute to Caesar (12:133-18). The prayer which Jesus taught to his disciples——elements of which are preserved in the so-called "Lord's Prayer" (Matt 6:9-13, Luke 11:1-4), also contains several characteristic Pharisaic elements——a sense of intimacy with God, who is addressed as "Father," a desire for the coming of the Kingdom, for the accomplishment of God's will on earth, a sense of the importance of forgiveness, etc.

From its very beginnings, therefore, the Jewish tradition was characterized by a clearly defined trajectory of religious rationalization. Judaism broke the hegemony of the sacrificial cults of the warlord state, and created among the people of Israel a realization that they participated most fully in the life of God not through military conquest or self-sacrifice, but through concrete ethical action. Fidelity to the covenant might mean conserving the integrity of the ecosystem by respect the law of jubilee, which required, among other things, leaving the land fallow from time to time, or conserving the complex networks of relationships and traditions which constituted the people of Israel. It might mean engaging in productive labor, building and exercising power, composing psalms or works of wisdom——or it might mean the revolutionary transformation of the social order. But the focus remained on concrete social activity which conserved the integrity of the ecosystem and the social fabric and which promoted the development of human social capacities.

Still, there were limitations. Judaism provided a credible means by which human beings could realize the potential latent in their humanity——by which they could lead productive and powerful lives, and gradually grow in knowledge and love of God. It provided a guide towards building an extraordinarily creative and powerful civilization. And it still provides this, as the historic contributions of Jews to the development of our own civilization indicate. At the same time, in return for its early political-military dynamism Judaism made significant concessions to patriarchal social organization and its ideological reflex, idealist metaphysics——the conviction that order is imposed on matter from the outside. The radical transcendence of the Jewish God made human participation in the inner life of the divine all but impossible. And yet we have already seen that every form of matter, because it is relationship, holism, and self-organization, contains within itself the drive to become something more than it already is. For human beings this means the drive to break the bonds of the social form of matter——to become more than human, even to become divine. And this was as impossible from the standpoint of Judaism as it was from the standpoint of the Hellenic philosophical tradition. But as the two traditions began to interact, each tapped into the dynamism of the other, giving birth to extraordinary new possibilities.

3. Towards Synthesis

a) The Hellenistic Jewish Synthesis

One can imagine a trajectory of social development defined by the two rationalizing social projects identified thus far. On the one hand peasant resistance to the degeneration of the tributary state maintains constant pressure on the military and priestly aristocracy, making brutal, exploitative, and wasteful dynasties short, and those which defend the interests of the peasantry, and which invest in the development of human social capacities relatively long lived. At the same time, the rationalizing activity of the prophetic-philosophical intelligentsia undermines the hegemonic sacral monarchic ideology while creating the priests and warriors a relational understanding of power, and a gradually expanding understanding of its own self interest. The tributary state invests less and less in warfare, and more and more in scientific and technological development. Low taxes give the peasantry an incentive to innovate, gradually increasing agrarian surpluses. Over a period of many centuries the kings really do become philosophers. At a certain point what we understand as industrial technology develops, and is implemented within the existing tributary economic structure. The philosopher kings transform themselves into a bureaucratic intelligentsia. A rational, scientific understanding of the underlying unity of the cosmos and its development towards increasingly more complex levels of organization gradually emerges out of the older religious and philosophical conceptions. New means of social communication and transportation make it possible to administer increasingly larger territories, and the new bureaucratic state gradually incorporates its less developed neighbors to heel, leading to the emergence of a single planetary state. A rational, scientific, understanding of the conditions for continued development leads the bureaucracy to keep the rate of exploitation low, and to invest heavily in the development of human as well as machine capacities. This is a society which is on the road to synergism.

It may well be that this is the actual course of social development on many planets. And there is significant evidence that this was precisely the trajectory of development which had been set in motion by the twin rationalizing projects of Judaism and Hellenic philosophy. Of critical importance here was the emerging Hellenistic Jewish synthesis. On the one hand, the Jewish people, now dispersed throughout the great Hellenistic empires which dominated the Mediterranean basin, found in the Greek philosophy a useful tool for drawing out the implications of their own religious insights, and ultimately for systematizing the results. More specifically, philosophy permitted them to demonstrate, in terms which were accessible, and which proved highly credible, to the other peoples of the Mediterranean basin, that the God of Abraham, Isaac and Jacob was in fact the one God of the universe, and that the Law of Moses was nothing other than the law of nature itself. Contact with the philosophical tradition also set in motion a critical dynamic within the Jewish tradition which tended, over a period of time, to restrict the influence of the more backward forms of monotheism (God as heavenly king) which derived more from the period of the sacral monarchy than from Israel's own, unique, historical experience, and to encourage the further elaboration of the concept of God as Being itself. At the same time, the broad masses of the Hellenistic empires found in Judaism an attractive alternative to the otherworldly mystery cults which were proliferating throughout the region. Jewish doctrine was more nearly compatible with the results of the natural sciences, and it offered a clear guide to ethical conduct and active participation in the work of building human civilization. Most important, however, Judaism, unlike the mystery cults, offered real hope that the divine law could in fact be realized in human society. Many actually became proselytes; many more became "god-fearers" who worshiped *yhwh* without seeking circumcision. If the results may have appeared lax from the standpoint of some of the more rigorous interpretations of Jewish practice, they nonetheless represented an enormous extension of Jewish influence, and undoubtedly began to generate some sympathy for the cause of liberating the Jewish homeland from Roman rule.

The potential latent in the Hellenistic Jewish synthesis is already apparent in the Wisdom literature. As Israel reflected on its religious traditions in the light of Hellenic philosophy, it gradually discovered in its God an intelligible principle which orders all things. At the same time, it began to transcend the patriarchal principle which historic Judaism had left largely unchallenged.

In wisdom there is a spirit intelligent and holy, unique in its kind yet made of many parts, subtle, free-moving, lucid, spotless, clear, neither harmed nor harming, loving what is good, eager, unhampered, beneficent, kindly towards mortals, steadfast, unerring, untouched by care, all-powerful, all-surveying, and permeating every intelligent, pure and most subtle spirit. For wisdom moves more easily than motion itself; she is so pure she pervades and permeates all things. Like a find mist she rises from the power of God, a clear effluence from the glory of the Almighty; so nothing defined can enter inter her by stealth. She is the radiance that streams from everlasting light, the flawless mirror of the active power of God, and the image of his goodness. She is but one, yet can do all things; herself unchanging, she makes all things new; age after age she enters into holy souls an makes them friends of God and prophets, for nothing is acceptable to God but the person who makes his home with wisdom. She is more beautiful than the sun, and surpasses every constellation. Compared with the light of day she is found to excel, for day gives place to night, but against wisdom no evil can prevail. She spans the world in power from end to end, and gently orders all things (Wisdom 7: 22 - 8: 1).

At the same time, unlike those strains of Hellenic philosophy which developed in isolation from Judaism, and retreated further and further into otherworldly, gnostic withdrawal, the wisdom tradition regarded philosophy as a means for arriving not at knowledge not only of God, but also of nature——and history.

God .. gave me true understanding of things as they are: a knowledge of the structure of the world and the operation of the elements; the beginning and end of epochs and their middle course; the altering solstices and changing seasons; the cycles of the years and the constellations; the nature of living creatures and behavior of wild beasts; the violent force of winds and human thought; the varieties of plants and the virtues of roots. I learnt it all, hidden or manifest, for I was taught by wisdom, by her whose skill made all things (Wisdom 7: 17-22).

It was this wisdom which made possible Israel's historic victories over her oppressors.

It was wisdom who rescued a godfearing people, a blameless race, from a nation of oppressors (10: 15).

And it was wisdom which would lead to their ultimate vindication.

The high point in the development of the Hellenistic Jewish synthesis, however, is almost certainly the work of Philo of Alexandria. Writing during roughly the period of Jesus' life, Philo explicitly identified the Jewish and the Platonic concepts of God. For Philo, God exists eternally, apart from the material world. We can know God either indirectly, through reason, or directly, on the basis of what amounts to a kind of intellectual intuition. This knowledge of God, however, is limited to his existence. The essence of God, according to Philo, remains ineffable.

The ideas, or the forms of things, exist in the mind of God as a kind of intelligible world. In their latent form Philo called these ideas *sophia* or the divine wisdom. In their active, creative form, they became the *logos*, the word through which all things came into being. Unlike the Hellenic tradition, however, Philo rejects the eternity of matter, and teaches that matter itself was the product of divine creative activity. Thus, while he retains the form/matter dualism which is characteristic of philosophy in all pre-industrial societies (mark of the low level of development of human organizing capacities), he takes an important step towards the recognition of the material world as a realization of, rather than a falling away from, the divine will.

According to Philo, the Law of Moses is nothing other than the law of the cosmos itself, fully accessible to reason and binding on all humanity. On this basis, Philo develops a harmonizing ethics which integrates Jewish and Greek elements, arguing that authentic freedom consists not in citizenship in Rome or some reconstituted Greek city state, but rather in service to the one true God who alone is authentically self-existent. It is the Jews, who know and follow this law, and not the Greeks and Romans, with their devotion to wealth and earthly political power who are the true cosmopolitans. Knowledge of the law flows out from the Jews to the other peoples of the earth who will eventually be united as in a single city, under the one law of the living God (Borgen 1987).

The progressive potential of this vision should be apparent. It may well represent the first claim that it is possible for human beings to fully realize the will of God, and the first call for a unitary world state, which was argued in rational terms accessible to human beings of any nation or religion. It certainly offered humanity a theoretically powerful account of its place in the universe, and a compelling vision of its potential for future growth and development.

The unfolding of this synthesis was soon cut short, however, by the emergence of two new religions, both of which claimed for themselves the heritage of Judaism and of the philosophical tradition, but which interpreted these traditions in profoundly reactionary ways.

b) Socioreligious Deformations

i) Christianity

The Christianity which eventually became the official religion of the Roman empire must be sharply distinguished from the essentially Jewish teachings of Jesus of Nazareth. Jesus found his constituency among the more marginal elements of the Jewish peasantry of Galilee, and his teachings were located clearly within the Pharisaic tradition. Christianity, on the other hand, was a religion of the great urban centers of the empire. It found little support among the principal exploited classes——i.e. the agricultural slaves of the Western part of the empire, or the rural wage laborers or dependent peasants of the East. Nor, for that matter, was it popular with the Senatorial elite of the West, which took great pride in Rome's achievements in the engineering and administrative areas. Rather, it took root among people with an intermediate and even somewhat ambiguous class position: the local representatives of Roman authority, drawn mostly from the equestrian order, the local aristocracy, merchants and artisans, soldiers and veterans, laborers, and urban slaves. Theissen (1982: 69-119) has demonstrated the presence of most of these elements in the Pauline community in Corinth, and we have no reason to believe that the situation would have been too different elsewhere. Most of these cities were also ethnically and religiously heterogenous.

Christianity built on the foundation of the Hellenistic Jewish synthesis, but distorted that synthesis in a way which radically altered its political-theological valence. Christianity took from the Hellenistic Jewish synthesis faith in a single divine creator, identified more or less specifically with the prime mover of the philosophical tradition. It also took the doctrine of the *logos* or *sophia*——God's word or wisdom——again identified more or less explicitly with the Platonic "realm of ideas" and at the same time with the divine law revealed to Moses but also accessible, at least in part, to the gentiles through the exercise of their natural human reason. Finally, Christianity took from the Hellenistic Jewish synthesis its underlying optimism regarding the possibility of salvation——the hope that humanity really could become whole, a real participant in God's plan for the cosmos.

This sane core was, however, subjected to two distinct but related distorting dynamics, both related to the way in which the emerging religion understood its founder and his work. The first of these dynamics is centered on the Christian claim that Jesus of Nazareth was the promised messiah and the incarnate son of God, who brought salvation to humanity "apart from the law," through this crucifixion and resurrection. The term *messiah*, or the Greek equivalent *christos*, means simply the anointed, and originally referred to the King of Israel, who was quite literally anointed during the coronation liturgy. Gradually the term became increasingly bound up with Israel's hope for a liberator, a redeeming king, who would break the yoke of Roman rule. Priestly circles developed their own "messianic" doctrines which flattered the role of the temple priesthood in Jewish life. Messianic doctrines were not, however, looked upon with uniform favor within the Jewish community. We have already seem that some strains in the prophetic tradition rejected the monarchy, and thus messianic hopes outright, and even those currents which looked forward to a redeeming or reforming king stressed the subordination of the king to divine law, and to the prophets who were the ultimate interpreters of that law.

Pharisaic Judaism continued the prophetic critique and refinement of the messianic doctrine. On the one hand, as Jewish patriots, the Pharisees were reluctant to rule out completely the option of armed struggle for liberation. Armed struggle implied military leadership, and successful military leadership was the very core of kingship in agrarian societies. Certainly the leader of a successful liberation struggle, who redeemed Israel from Roman rule, would have some claim to have been "anointed" by *yhwh*. But we have also seen that the Pharisees were skeptical about the transformative potential of armed struggle. The Maccabean revolt, while certainly legitimate and clearly a step forward, had hardly created a society which fully realized the will of God. They stressed that only actual fulfillment of the law could bring true liberation and the authentic reign of God on earth. Thus the notion, popular in Pharisaic circles, that the messiah would come only when the law had been perfectly fulfilled for a day. Jesus' own teaching that the kingdom is at hand, but will come only under the condition of radical *metanoia* is simply an eschatologically charged version of this Pharisaic teaching.

Jewish messianism, therefore, while not the most advanced element in the Jewish tradition, remained resolutely inner-worldly. While messianic pretenders were a dime a dozen in Roman dominated Israel, the claims of these pretenders were always tested against reality. Partisans of the more advanced currents, such as Pharisaism, demanded that they create a society in which the law was actually fulfilled. At the very least a would-be messiah faced the daunting task of liberating Israel from Roman rule. To satisfy many of the Pharisees he would, further, have to lead the people in the whole fulfillment of the law. Not surprisingly, this meant that messianic claims generally failed to stick.

Christian messianism, however, was quite another matter. That messianic hopes should have attached themselves to the figure of Jesus of Nazareth is not surprising. By all accounts he was a powerful teacher who had a profound impact on the lives of those who heard him. Whether or not he himself willingly played the part of royal pretender——a possibility suggested by the story of his triumphal entry into Jerusalem——remains open to debate. But his capture and execution, while in no sense calling into question the value of his teachings, should have put any messianic claims to rest. The early Christians could keep these claims alive only by radically reinterpreting Jewish messianic doctrine. If Jesus is messiah, but Israel remains under the Roman yoke, and the law unfulfilled, then the messianic task must be something other than national liberation and fulfillment of the law.

A similar argument regarding the development of Christianity has been made by Erich Fromm (Fromm 1963). Fromm's arguments focus on Christianity's claims regarding the divinity of Jesus. Early in its history Christianity advanced the claim that Jesus was the adopted son of God. Later, this adoptionist Christology was abandoned in favor of the claim that Jesus was the eternally pre-existing divine *logos*, one in being with the Father, who became flesh in order to redeem humanity. Fromm argues that these two doctrines have a very different political and social-psychological valence.

Adoptionist Christology makes the audacious claim that humanity can become divine, and articulated the rebellion of the Jewish and Hellenistic masses against the oppressive structures of the empire. In the context of this Christology, Jesus appears as the prophet and teacher who, crucified for sedition, nonetheless emerges victorious over not only the empire, but ultimately death itself. Here the resurrection represents a victory for the self-organizing power of matter over the forces of chaos and disintegration. In this sense, claims regarding the divinity of Jesus represent simply a radicalization of the historic Jewish conviction that God is really present wherever the law was being fulfilled, or of the philosophical doctrine that humanity, endowed as it is with the capacity for the intellectual and moral virtues, is a real participant in the life of God. While the scope and depth of Jesus' achievements are hardly sufficient to justify an claim to godhead, it is also clear that the underlying impulse here is progressive. It represents humanity's audacious aspiration to participate fully in the self-organizing activity of the cosmos.

Incarnational Christology, on the other hand is centered on God's paternalistic condescension towards humanity——and on the Son's willing submission to the cross as the just penalty for humanity's sins. God becomes human, representational form of the paternalistic feudal lord. Humanity, in the form of Jesus, obediently does the will of God, even where this means suffering death on the cross. Fromm notes the emergence, in close relationship to incarnational Christology, of the cult of the *theotokos* or mother of God. It is only in the context of the womb, Fromm notes, that the Nicene formulae (two persons, one nature) really make sense. Complete submission to the father leads ultimately to reunification with the mother, a union which, however, far from being fertile, in fact reflects the assumption of an infantile attitude. Here the anger of the masses is turned inward. Humanity's aspiration to become God gives way to radical submission to the will of the divine sovereign.

The gradual emergence of a Christology centered on the doctrine of the crucified messiah, incarnate Son of God, necessarily involved a transformation of the Hellenistic Jewish understanding of the process of salvation. Nowhere is this more apparent than in the classical Pauline formulae:

1) The will of God is manifest in both the natural world (Rom 1:19) and in the Law revealed to the Jewish people (3:2).

2) All have sinned and thus fallen short of the glory of God, meriting eternal death and damnation (Rom 3: 9-20).

3) God's will has, however, through the expiatory sacrifice of Jesus on the cross (Rom 3:25).

4) Justification is available apart from the law through faith in Jesus for all who believe——and only for them (Rom 3:21-22).

5) While those who are justified are also transformed inwardly by the work of the spirit (Rom 6:12, 8:29), Christians are not called to reorganize social structures, which were set in place by God (Rom 13:1). Slaves need not be set free (Philemon), and wealth need not be redistributed (1 Cor 11) so long as Christians treat one another as brothers in Christ, waiting in joyful expectation for the day when they will go to meet their "Lord" in the "air" (1 Thess 4:13).

On the one hand, these formulae preserve enough of the ethical rationalism of the Hellenistic Jewish synthesis to attract a large population of gentile god-fearers. Paul acknowledges that the divine law is accessible through both reason and revelation, and that it is the failure of human beings to fulfill the law——and not their sheer materiality——which is the underlying source of the disorder in human society. At the same time, Paul rejects the Jewish strategy of a protracted struggle to realize the law in favor of what amounts to a spiritualized version of the old cult of divine self-sacrifice. Salvation comes not through the concrete, social historical struggle of human beings to fulfill the law, and to build a civilization which fulfills the law, but through the expiatory sacrifice of the incarnate son of God.

This way of understanding salvation has serious political-theological consequences. First of all, it effectively deprives human beings of all responsibility for actually realizing the will of God, and renders the whole of human civilization theologically irrelevant. The drive to actually fulfill the law gives way to a mere deference to its normative claims, and to an acknowledgement of universal human guilt before the just but ever merciful divine monarch, who seems ultimately uninterested in seeing his decrees actually fulfilled. Indeed, it would not be too much to say that Paul effectively reinstates, albeit at the level of soteriology rather than cosmology, the old doctrine of creation through self-sacrifice which characterized the cults of the warlord states, overturning nearly a millennium of Hellenic and Jewish philosophical-prophetic struggle.

Gerd Theissen has characterized Pauline social ethics quite appropriately as a kind of "love patriarchalism."

> In these congregations there developed an ethos obviously different from that of the synoptic tradition, the ethos of primitive Christian love-patriarchalism ... This love-patriarchalism takes social differences for granted, but ameliorates them through an obligation of respect and love, an obligation imposed upon those who are socially stronger. From the weaker are required subordination, fidelity and esteem (Theissen 1982: 107).

At the same time, the Pauline formulae lay the groundwork for Christian antisemitism. Paul does not simply open up a new road to salvation "outside the law," but actively ridicules the Jewish road to salvation through fulfillment of the law, arguing that it is, in fact, impossible and a dead end (Rom 7). More is at issue here than a bit of sectarian bigotry. Indeed, it would not be too much to say that the rejection of the Jew is a symbolic expression of the larger Christian rejection of participation in the life of God through ethical conduct, and through labor, the exercise of power, and the production of knowledge generally. It should thus come as no surprise that Paul's antisemitism is complemented by an anti-Hellenism. Paul ridicules the Greek struggle for wisdom on essentially the same grounds as he ridicules the Jewish struggle for righteousness (1 Cor 1). In later times this anti-Hellenism would reproduce itself in Protestant anti-intellectualism, and merge with anti-semitism into a virulent hatred of the Jewish intellectual, precisely because of his/her contribution to the development of human civilization.

The resulting ideology played an important role in shoring up the declining Roman Empire, and in facilitating the transition from slave-based production to feudalism. By the middle of the third century the empire was in the midst of a crisis. The closing of the *limes* meant a shortage of slave labor, which had been drawn primarily from among prisoners of war (Anderson 1974: 76, 80). In order to remedy this situation, *servi* were settled on the land, with families (Anderson 1974: 94). This naturally began breaking down the distinction between free and servile labor. At the same time, beginning with Diocletian increasing restrictions were placed on the freedom of movement of the *coloni* (Jones 1974: 302). These economic changes were accompanied by changes in the political and ideological superstructure. The power of the old slave owning municipal oligarchies declined as new elites based in the imperial administration emerged. These elites lived off the revenues generated by the tribute, rather than off the labor of slaves. In effect the ruling class of the empire became less and less an independent slave owning class and more and more a service nobility like those of the empires of the ancient orient. These changes were reflected in Diocletian's effort to exclude the old senatorial order from the most important posts in the state administration, and in the eventual fusion of the senatorial and equestrian order in to a single service nobility, the clarissimate, under Constantine (Anderson 1974: 100).

Meanwhile the emperor Aurelian had proclaimed himself not merely *princeps* but *deus et dominus*, (Anderson 1974: 86). At first this new sacral monarchy took the form of a cult of the imperial genius, which resulted in increasing persecution of the empire's Christian minority. But growing numbers of the clarissimate, and eventually the imperial family itself began to see in Christianity, with its conservative but integrating social ethic, a solution to the cultural crisis of the empire. Christian messianism paved the way for a new kind of sacral monarchy, as the world-rule of Constantine was represented as the final triumph of the risen Christ.

As the knowledge of one God and the one way of religion and salvation, even the Doctrine of Christianity, was made known to all mankind, so at the self-same period, the entire dominion of the Roman empire was being vested in a single sovereign. Profound peace reigned throughout the world. And thus by the express appointment of the same God, two roots of blessing, the Roman Empire and the doctrine of Christian piety, sprang up together for the benefit of men ... (Eusebius Or. Con 16: 4-7 in Ruether 1974: 142).

The emperor himself comes to be regarded in much the manner of a divine king.

The emperor, like the rays of the sun whose light illuminates those who live in the most distant regions, enlightens the entire empire with the radiance of the Caesars as with the far reaching beams reflected from his own brilliance ... Thus he is present everywhere, traversing the entire earth, being in all places and surveying all events.

Having been entrusted with an empire, the image of the heavenly kingdom, he looks to the ideal form, and directs his earthly rule to the divine model and thus provides an example of divine monarchy sovereignty (Eusebius in Cunningham 1982: 51).

Christian love-patriarchalism, meanwhile, offered a way to legitimate and also ameliorate social inequalities——a way which corresponded rather well to the emerging tributary or "feudal" relations of production. As Theissen points out, ancient Greek and Roman society sought to solve the problems of social integration by extending citizenship to an ever greater number of people, "while at the same time providing for the fantastic desires of its ruling classes with an ever larger number of slaves, who were denied even the most basic forms of human dignity (108)." The constantly increasing cosmopolitanism of the empire, and the crisis of the slave mode of production made this solution increasingly untenable. Christianity offered a social ethic which instead stressed mutual obligation between fundamentally unequal parties (109). This ethic corresponded well to the imperatives of new "patriarchal" feudal relations of production, which abolished both the large class of free citizens and the even larger class of subhuman slaves, in favor of a graded series of paternalistic relationships between lord, vassal, and serf.

At least a brief note is in order regarding the basis on which Christianity, which was first and foremost a religion of the service nobility, was sold to the peasant masses who constituted the vast majority of the population of the empire. The key to answering this question lies in the ancient cult of the *Magna Mater* which we have had occasion to discuss earlier in this chapter. The real religion of the Mediterranean peasant masses, this cult too, we will remember, contained a story of a dying and rising son. As missionaries attempted to draw the peasantry into the

Christian fold, Isis, Ishtar, and all of the other names of the goddess were simply assimilated to the figure of Mary, the *theotokos*, the mother of God. Osiris, Tammuz, and their cognates——and even the female Persephone——were assimilated to the figure of Jesus. The church, of course, proclaimed and still proclaims Mary as a "way" to her son, universal mediatrix, but never coredemptrix. For the peasants, however, one gets the impression that it was first and foremost Mary who saved and healed her dead and broken son, and who will save and heal a broken humanity. As one might expect with a religion more interested in gaining the deference and outward submission of the masses than their active collaboration, it mattered little to the clergy what the peasants actually believed, so long as they remained quiescent, and outwardly deferred the authority of the church fathers.

Much the same might be said of the way in which Christianity gained hegemony over the vast intelligentsia of the Mediterranean basin. Especially in Alexandria and the East, but to some extent throughout the region, philosophers were won over the Christianity largely because they had the freedom to reinterpret it in a way which minimized its differences with the Neoplatonism which most of them had previously espoused.

The Christian solution to the crisis of the empire took two distinct forms. In the East, Christianity was actually able to stabilize the empire. The Byzantium preserved a Christian sacral monarchy in the Eastern Mediterranean until the Ottomans completed the Islamic conquest of the region in 1453. And by this point the real center of Orthodox Christianity had already been transferred to Moscow, the "Third Rome," and capital of the eastern Slavs.

At the very basis of the Orthodox settlement were the twin institutions of the village community and the bureaucratic state. Throughout the Orthodox domain, but especially in the Slavic north and east, the village community remained strong. Land was owned in common by the village and redistributed periodically among its member families. As we have noted, the Hellenic and Slavic words for village, *kosmos* and *mir* also mean universe, and order or peace. The villages paid a portion of their produce in the form tribute to the military and priestly aristocracies. Especially in the north-eastern hinterlands of Eurasia, serfdom was imposed in order to prevent peasants from escaping into the vast eastern frontier and depriving the cities of their labor.

But it was not only the peasants who paid tribute. The emerging states of the Slavic north modeled themselves after Byzantium. They were able fairly early on to reduce the smaller warlords to a state of dependency, a process which gave rise to the Grand Duchy of Moscow, and eventually to the Great Russian Empire. Every member of this noble class (like the *decuriones* of old Rome) were required to perform service, initially military, and later bureaucratic, at the course of the *Tsar*. Later on, in the seventeenth and eighteenth century, provision was made for non-nobles who served dutifully in the lower ranks of the military and bureaucracy to themselves obtain noble status. In short, first Byzantium, and later Muscovy, were able to build highly integrated sacral monarchic imperial states.

The ability of the Orthodox East to conserve and construct states of this sort leant credibility to the political theology developed by Eusebius. To use the Neo-Platonic language favored by Orthodox theologians, the state was regarded as an emanation from the divine. As tribute flowed upward from peasant to *boyar* to *Tsar*, divine love flowed downward from *Tsar* to *boyar* to peasant.

The success of the Slavic states, however, came at the expense of the Orthodox church. If the state is an emanation from the divine, then there is but little room the church to carry out its prophetic functions. The Orthodox Church, at least by comparison with its Catholic counterpart, tended to remain subservient to the state, and to confine itself to priestly functions. It was content, in the context of the sacred liturgy, it claimed to create a "heaven on earth," and did not trouble itself too much about the reorganization of the social order in accord with divine law. The result was a social formation characterized by greater unity, but also by less dynamism———at least until the penetration of market relations unleashed the great upheavals of the late nineteenth and early twentieth centuries.

In the Western Mediterranean and in the hinterlands of Northwestern Europe, on the other hand, the empire rapidly succumbed to internal disintegration and to the pressure of invading Germanic tribes. This undermined attempts to claim for the Western emperor the quasidivine status enjoyed by his Byzantine counterpart. Indeed, by the end of the fifth century of the common era, the Western empire had collapsed, and the Roman Catholic church was the only remaining supranational institution in Western Europe. A long period of struggle ensued between Germanic warlords who attempted to rebuild the empire in the west, and the Bishops of Rome, who also claimed for themselves the heritage of the Caesars.

The resulting stalemate at once created the social basis for, and was reinforced by the Augustinian theology which became dominant in Western Europe. Augustine wrote from a unique perspective. As Catholic Bishop of Hippo in North Africa during the late fourth and early fifth centuries of the common era, and a son of the provincial aristocracy, he was a friend of neither the empire, which had ruthlessly exploited his homeland for centuries, nor of the Donatist peasants who raided the estates of the wealthy, wielding large clubs and crushing the skulls of the landowners while shouting *Deo laudes!* (Praise God!). As a member of the exploiting classes he looked to the public authorities to maintain order. As a member of an oppressed people he rejected the empire's claim to sanctity.

Drawing on the Neoplatonic philosophy of the late empire, Augustine argued that the highest good is order, which exists perfectly only in God himself. Human beings participate in this order to the extent that their love was directed towards God. Humanity is divided into two cities, the City of God, composed of those who loved God, and the City of Man, composed of those whose love was disordered, and directed towards the creature rather than the creator. Augustine was, however, able to identify a certain limited order within the city of Man. Those who love honor, he argued, are generally more disciplined, and are able to prevail over, those who love pleasure. This, he argued is the foundation both of slavery and of the state, which derive their legitimacy from the capacity of lovers of honor to raise lovers of pleasure to a higher degree of order. Neither institution, however, really participates fully in the life of God. Full participation in the life of God is possible only by divine grace, which orders the love of the believer and incorporates him into the City of God.

Augustinian theology provided sufficient legitimation for the warlord states to permit them to gradually establish a semblance of public order throughout Europe, but it inhibited the formation of a unified European tributary state, which would have required a fully developed sacral monarchic ideology. It guaranteed the superiority of the church, while imposing radical limits on the church's claim to secular authority and thus its own claims to empire.

The resulting complex of conditions——a religion which affirmed, if only a limited and partial way, the meaningfulness of the material world and humanity's place therein, and a weakened and internally divided warlord class unable to horde the whole social surplus product for itself——permitted the development of an unusually productive and progressive variant of the tributary social order in Europe as the Hellenistic Roman civilization of the Mediterranean basin declined (Amin 1980, 1988). Feudal land tenure patterns, the division of land into demense, virgate, and commons, gave the peasantry a stake in increasing productivity. Feudal Europe made great strides in the transition from human to animal power, invented the three field system, the alpine plough, etc. The period between collapse of the Roman Empire and the middle of the 12th century saw agricultural yields increase from 4:1 to 9:1, the first real increase since the agricultural revolution (Anderson 1974). Higher agricultural yields made possible the development of a rich culture centered at first around monasteries and cathedrals, but eventually around universities which absorbed the intellectual wealth of the more advanced Islamic civilization to the south and east.

Augustinian theology, however, placed real limits on the development of human social capacities. The sense which pervades Augustinian theology, that a real interest in any particular system represents a kind of apostasy, and threatens the order and stability of the whole, hardly favored the development of a rich and complex social fabric, the investment of energy and resources in technological innovation, economic development, statecraft, the arts, or the special sciences. We will see that as European civilization developed, it gradually restricted the role of the Augustinian element in the Catholic tradition in favor of elements drawn from the philosophical tradition. But first we need to analyze the emergence of the third great monotheistic religion, Islam.

ii) Islam

Islam represents a second type of deformed monotheism. Where Christianity served to shore up the fortunes of a decaying empire, and symbolically, at least, exalted the priestly over the warrior aristocracy, Islam was the ideology of an emerging empire, which exalted the calling of the warrior above all others, and effectively eliminated the priestly aristocracy and its claims on surplus altogether.

In order to understand Islam, we need to examine the social matrix out of which it emerged. The Arabian peninsula was located on the periphery of the great civilizational complexes of the Mediterranean basin. Largely desert, the greater part of the peninsula was given over to nomadic Bedouin tribes. These tribes had perfected the technology of camel nomadism, which enabled them to travel long distances across the barren desert, driving their flocks from one oasis to another, or else organizing merchant caravans or raiding parties which permitted them to garner for themselves a share of the surplus produced by the more developed centers of the Eastern Mediterranean. Along the southwestern coast of the peninsula was a string of small urban centers——Yathrib, Mecca, Aden——referred to collectively as the Hijaz. These urban centers were themselves still organized on a quasi-tribal basis, never having developed a stable state apparatus, even at the city state level. Wealthy merchant clans dominated the cities, constituting a kind of urban, mercantile patriciate. Like most societies which have yet to develop a strong centralized state apparatus, Arab society was polytheistic, though Judaism and Christianity both exercised influence, and there is some evidence for the existence of indigenous movements towards monotheism (Sourdel 1983).

Mohammed himself was a member of the relatively poor Hashemite clan, of the Quraysh tribe——the ruling Meccan tribe in this period. He married a wealthy widow and became a successful merchant. At age 40 he underwent a religious crisis, and began preaching a radical monotheism and a program of social transformation which looked to replace the complex tribal structure of Arab society with a single community of believers——the *'ummah*——united by a common faith and governed by a single divine law——with him as its leader. Not surprisingly, Meccan elites rejected his program, but in 622 C.E. a delegation from Yathrib, the future Medina, which was experiencing serious intertribal feuds, invited Mohammed to come to the city as an arbiter. There he established a state based on his political-religious program. Over the course of the next ten years, he conquered most of central Arabia, and transformed Arab society into a dynamic political-military-ideological force, capable of mounting a real contest for power with the surrounding empires (Esposito 1984: 3-4).

After Mohammed's death in 632, the Islamic community or " *'ummah*" was led by a series of *caliphs*, or "successors" of the prophet. The caliphs were able to consolidate Islamic control of the Arabian peninsula and extend it to Byzantine Syria and Sassinid Mesopotamia (636-641), Sassinid Iran (642-650), Egypt (674), the Maghreb (670-707), Spain (712-714), Southern Gaul (717-732), and Transoxiana and Sind (711) (Sourdel 1983: 16).

It was during this period of rapid state building and imperial expansion that the basic character of the Islamic religion was forged. Jewish monotheism emerged out of a radical rejection of the claims to divinity advanced by the warlord state, and developed in the context of a protracted struggle to build a civilization which authentically realized the law of God. It was on the basis of this experience that the Jewish people came to recognize in *yhwh* not only "he who brings into being the armies of Israel," but the one true God, creator of heaven and earth. Christianity reverted to the more common pattern of advanced tributary states, in which a "high god" presides over a pantheon of lesser deities, with the one difference that in the Christian religion, only the high god is accorded fully divine status (just as in the Roman Empire, only the Emperor really organized). As the empire broke up, and a feudal structure emerged in its place, the saints not surprisingly recovered something of their older divine status. Islamic monotheism, on the other hand, was at once the condition for and the religious reflex of a process of rapid imperial expansion. Allah is the only god, because Mohammed, or his successors the caliphs, are the sole organizers and directors of the Arab empire.

Where Judaism demands actual ethical conduct, and Christianity a kind of deference to divine authority (religious reflex of a deference to a decaying empire), Islam demands radical submission. Indeed, the word Islam itself means the "peace which comes from submission to God," and a Muslim is one who has so submitted. Concretely this submission involved a number of acts denoting religious submission: reciting "There is no god but Allah and Mohammed is his prophet," praying, fasting during Ramadan, and making a pilgrimage to Mecca. But it also involved participation in the *jihad*, and payment of the *zakat*, often translated as almsgiving, but in fact the foundation of the Islamic taxing system.

The term *jihad* refers, in its most general sense, to the struggle to realize the will of God. If Christian observers have sometimes been too quick to construe *jihad* as a *carte blanche* for religious violence, Islamic reformists have often tended to deny its military dimension entirely, or to restrict it to defensive struggles. Neither approach is really accurate. The concept of *jihad* probably arose out of an attempt to control the divisive intertribal warfare which was preventing the formation of a unified Arab state out of the various tribes of the Arabic peninsula, without, however, opting for a pacifism which would blunt the dynamic expansionism of a potential Arab empire. Thus the distinction between just (*jihad*) and unjust (*fitna*) war (Pipes 1983: 39). Theologically, just war was understood as war which helped to realize the will of God. In real terms, this meant war which did not disrupt the unity of the emerging Arab nation, but rather contributed to the process of Arab state formation and empire building.

The Islamic understanding of what it meant to realize the will of God required the establishment of an Islamic state, for only an Islamic state could enforce Islamic law (*shariah*) and ensure that human beings submitted to the will of Allah. The world was divided between *Dar al-Islam* (the abode of peace) and *Dar Al-Harb* (the abode of war). Islam necessarily involved armed struggle directed at the spread of Islamic rule. Those who were conquered were offered three alternatives. They could convert to Islam. If they were members of other monotheistic religions——Judaism, Christianity, Zoroastrianism, Sabaenism——and were willing to accept "protected" or "*dhimmi*" status, they could continue to practice their own religions, and live under their own law in return for paying additional taxes, suffering certain civic disabilities, etc. Those who resisted, or who clung stubbornly to polytheistic religions, were either enslaved or executed.

Closely related to the extension of Islamic rule was the extension of the Islamic taxing system. Muslims paid the *zakat*, a tax on wealth as well as revenue. *Dhimmis* paid additional taxes as well. In practice, non-Arab Moslems were taxed more heavily than Arabs (Esposito 1984: 8). All revenues were collected and administered by the centralized state apparatus. This was in sharp contrast to the complex network of taxing authorities which characterized feudal Europe. A significant portion of the surplus produced by the population became available to the state to support further conquests or to support the development of an Islamic cultural apparatus. The political-military authorities entirely displaced the priestly aristocracy. The Islamic clergy, or *'ulamma,* was reduced to a corps of legal interpreters and advisors with little or no political or cultic responsibilities. In this sense Islam represents perhaps the most extreme manifestation of the warrior cult which characterized the religious structure of tributary social formations.

The result was a highly unified and rationalized tributary state. The great empires which emerged from the Islamic conquests represented, from a global perspective, one of the high points of human civilization. This is true on several fronts. At the economic level, the conquests created a unified civilizational sphere throughout which people and products could spread with unprecedented rapidity, so that new technologies developed in China, or in the Islamic world itself, soon found their way even to the most remote backwaters of feudal Europe. The Islamic state provided an unusually efficient taxing mechanism which permitted the centralization of a large surplus without imposing oppressive burdens on the peasantry of the conquered regions, who often greeted the Arabs as liberators, and willingly adopted the Islamic religion.

Islam took up the artistic, scientific, and philosophical work of the Hellenistic and Roman social formations, and extended it in important ways. Cordoba, Baghdad, Cairo, Nishapur and Palermo all outranked the emerging cities of Europe as intellectual and cultural centers. Arab contributions in the fields of mathematics, astronomy, and philosophy are especially noteworthy. Arabic philosophy is the mother of European philosophy and thus forms a key link in the whole journey of the dialectic. As Islamic philosophers began to translate, read, and comment on the philosophical texts of their Hellenistic predecessors, and to produce philosophical works of their own, they began to interpret Islam in a way which gave priority to its rationalizing tendency, and which down played the demand for radical submission to the will of a single divine sovereign (Sourdel 1983: 72-79, 91-96). Mystical currents emerged which sought a kind of union with God which Islam, with its emphasis on divine transcendence and sovereignty, had hitherto discouraged, in some cases even speaking of the inherence of God in the human soul (Sourdel 1983: 87-91, esp. 88).

Islam, furthermore, permitted a relatively high degree of religious and philosophical pluralism. Indeed, in so far as *dhimmis* paid higher taxes than Muslims, Islamic governments actually had an incentive not to pressure their subjects to convert to Islam (Pipes 1983: 53). Islamic spirituality, whatever its limitations, *did* help to promote, even more clearly than Hellenistic cosmopolitanism or Christianity, the realization that the planet was inhabited by a single humanity.

At the same time, Islam set definite limits to the development of human social capacities. Radical monotheism blocked the formation of the multiplicity of interests which characterizes complex societies. While Islam permitted other monotheistic religions, the whole process which gave birth to Islam effectively annihilated the tremendous diversity of pre-Islamic cultures which had developed from the Maghreb in the West to Persia in the East, destroying Berber, Egyptian/Coptic, Syriac, and Mesopotamian civilization and partially Arabizing various African, Turkish, Persian, Indian, and Mongolian cultures as well.

A spirituality centered on conquest and submission, furthermore, leaves human beings little scope to actually begin to understand themselves as active participants in the self-organizing activity of the cosmos. And the self-organizing activity of the cosmos itself is apt to be (mis)understood as the activity of a single divine sovereign who acts on inert matter in much the same way the *caliph* acts on a subject population. Human creative activity, when it understands organization in this way, cannot help but taken on authoritarian characteristics.

Closely related to this problem is the uniquely Islamic form of women's oppression. Repression of sexuality, and of the bond with the mother, is the cornerstone of the socialization process in tributary social formations. This repression undermines the spontaneous sociality of human beings in favor of submission to paternal, or by extension royal, and ultimately divine, authority. This dynamic works in different ways, however, in different societies. Christianity denied female sexuality and thus female creativity, making the creative act something performed by the male God-head alone. This is the significance of the doctrines of the Virgin Birth and of creation *ex nihilo*.

Islam, on the other hand, perhaps because the Arab tribes were still closer to a communitarian mode of social organization, or perhaps because the Arab domain included so much of the ancient homeland of the Mediterranean cult of the *Magna Mater*, regarded female sexuality as something very real and very threatening to the patriarchal authorities led by the caliph, of whom Allah was simply the representational form. Daniel Pipes notes that Islam regarded female beauty as something so immensely attractive that it could compete with God himself (Pipes 1983: 176). He might have written that Islam feared the radically different understanding of the divine nature, and the very different path to God, which would be opened up by the liberation of women and of female sexuality——an understanding of, and path to, God centered on tenderness and nurturing, rather than on conquest and submission.

c) A Second Attempt at Synthesis

Their very real limitations notwithstanding, both Catholic and Islamic theology unleashed significant new progress in the development of human social capacities. Rapid increases in agrarian productivity, the cultivation of previously vacant land, the rapid proliferation and expansion of towns and cities, the development of new methods of extracting and centralizing surplus, and the establishment of the planet's first independent scholarly and scientific centers, the universities, gradually created within humanity a sense that it was, in fact, an authentic and even an important participant in God's plan for the universe.

The Catholic Church played a critical role in these developments. Not only did the church centralize and allocate resources for human development, while the secular warlords were busy fighting over which villages they could tax. The church also endowed the human civilizational project with theological significance, proclaiming it a real, if only partial and finite, participation in the life of God. When Pope Innocent III claimed for himself "authority over nations and kingdoms to uproot and pull down, to destroy and to demolish, to build and to plant (Jeremiah 1:10) he effectively institutionalized the ancient prophetic function, the work of critically assessing the development of human society from the standpoint of its impact on humanity's larger spiritual vocation, and where necessary, intervening to reorganize the social order. The implication was that the organization of human society affects the realization of God's plan for the universe, that the human civilizational project has real cosmic significance. The Catholic Church was the first human institution to set itself above and outside the organization of any particular society, and to claim the authority to judge and reorganize the social order on the basis of transhistorical criteria. In this sense, the Catholic Church is the predecessor of the communist movement.

The work of the great Islamic empires in many ways overlapped with that of the Catholic Church——rationalizing the centralization and allocation of resources, building great urban centers, establishing universities etc. Islam did not, to be sure, support the existence of an autonomous prophetic institution such as the church. But what Islam lost in the development of the prophetic function it gain in its exercise of royal authority. Islam did not invent the idea that raw political power could be used as a force for doing God's will on earth, any more than the Catholic Church invented the prophetic office. Both are part of the legacy of Judaism. But it was Islam which first institutionalized this idea, and applied it systematically on a world historical scale. In this sense Islam as well was a predecessor of the communist organization.

In the long run, these developments could not help but have an impact on theology itself, as the most advanced elements in Europe and the Mediterranean basin searched for a theory adequate to humanity's growing sense of its own power and significance. At the center of this effort was the struggle to resolve what we have identified as the principal contradiction of the Hellenic philosophical tradition: the notion that matter is, on the one hand, the potential for organization and, on the other hand, simply the passive capacity to receive form, which comes to matter only and always from the outside. Given the inherent instability of this position, it should come as no surprise that the "hylomorphic" center tended to continuously give rise to radical idealist and materialist extremes. Avveroes, the great Arab interpreter of Aristotle, for example, rejected the idea that form came to matter from the outside, and argued instead that

> the primary, unformed matter contains potential forms as 'seeds'. Forms are therefore not extrinsic to but immanent in matter. If the forms were to come to matter from the outside, this would be a sort of *creatio ex nihilo*. The forms are as eternal and uncrated as is matter. God creates neither matter nor form. The task of the "prime mover" is to convert possible forms into actual forms, i.e. to develop the seeds contained in matter … Prime matter is universal potency that hides the seeds of the forms; the prime mover does nothing but turn potency into act (Trachtenberg 1957: 66).

Similar positions were adopted in Europe by Siger of Brabant, and later by Giordano Bruno.

Other philosophers, such as Amalrich of Bena, took precisely the opposite strategy. Reasoning that it was form, not matter, which characterized actual being, Amalrich argued that in a very real sense everything is form, and thus implicitly, at least, everything is God (Dahm

1988: 94). This is the line of reasoning which ultimately gave birth to the panentheism of Meister Eckhart and the mystical tradition associated with him.

What should be apparent, of course, is that both of these lines of reasoning lead to the same, or at least to a very similar conclusions, conclusions which were ultimately reached by David of Dinant.

> The philosophy of David of Dinant is ... a materialist pantheism ... The basic ground of this philosophy is the pantheistic unity of the material, spiritual, and divine principles ... this unity lies not in the empirical world, and not in the reason of the individual, and not in the matter of single things, but in a higher realm, where reason *as such* melds into God and "prime matter (Trachtenberg 1957: 96-97)."

This is an extraordinary doctrine which indicates at least a preliminary grasp of the self-organizing character of matter, and prefigures many of the later claims of the dialectical tradition. And its political-theological implications are revolutionary indeed. The entire material world, and humanity with it, represents a real participation in the life of God, a gradual working out of the divine form latent in matter.

The revolutionary implications of this new orientation are perhaps most important apparent in the work of the Calabrese Abbot, Joachim of Fiore. Joachim was born around 1135 in Celico, in the region of Calabria. Some have said he was of Jewish descent. After many years as a Benedictine, rising to become abbot of Corazzo, he received permission to found an order of hermits dedicated to ascending the mountain of contemplation in remoteness and austerity, and retired to the community of San Giovanni in Fiore, where he began to produce a series of commentaries on the scriptures. That he wrote earlier than the Latin Avveroists, and almost certainly worked without benefit of direct access to philosophical texts only confirms that his vision, like theirs, reflected a deeply rooted revolutionary dynamic at work in medieval society.

Through his long years of prayer and study, Joachim arrived at a new understanding of the doctrine of the Trinity, and of its relationship with salvation history. While the scholastics, following Peter Lombard, held that the essence of God existed separately from the persons, Joachim argued that the unity of God could be understood only as an interpenetrating community of persons, which was the perfect realization of divine charity--a kind of prefiguration of the classless and communal society of the Kingdom of God. His doctrine is, in this regard, more in continuity with certain strains of the Orthodox tradition (cf Osthathios 1979:59-79) than with Roman Catholic doctrine, and some have argued that this reflects the persistence of Greek and Albanian influence in the *Mezzogiorno* (Reeves 1976:3). At the same time, unlike the Orthodox, Joachim regarded the historical process as the progressive and gradual realization of the communal reality of the Trinity. Each person was, he argued, especially expressed in a particular "*status*," or state of human history. These states are not precisely ages, but rather successive but overlapping stages in the development of human social existence. Thus the first *status*, which corresponds to the activity of the Father, is governed by the order of married men, and is thus characterized by patriarchal rule. The second *status*, that of the Son, is associated with the clerical order and with rule by priests and kings, of whom Christ is the perfect exemplar. The third *status*, that of the Holy Spirit, is associated with the monastic order.

During the third status, all will live in community, under the direction of a Spiritual Father, who resembles an abbot more than an Emperor or a Pope.

> They will have their own houses, but food and clothing will be in common. . . the surplus will be taken from those that have more and given to those who have less so that there may be no need among them, but all things held in common (in McGinn 1979:114-48).

The transition to this new *status* was to be catalyzed by two new religious orders: the first an order of contemplatives, the second an order of preachers.

Joachim resolved, or at least mitigated, the political-theological contradiction implicit in the claim that Jesus was the messiah, and laid the groundwork for a movement to reorganize the church into an effective force for promoting social transformation. First of all, he effectively displaced the drama of the cross and resurrection from the center of Christian doctrine in favor of a theory of redemption through concrete social-historical transformation. Indeed, Joachim's effort to penetrate behind the symbolic forms of religious language to the real historical content which these symbols at once disclose and conceal——to see *in plenitudo intellectis,* rather than through the dark eyes of faith——laid the groundwork for a progressive rationalization of Christian doctrine, and ultimately for the creation of a scientific theory of human history. In this sense Joachism must be regarded as a precursor of historical materialism.

Unfortunately, humanity had not yet developed the organizing capacities necessary to carry out the kind of transformation which the Avveroists and the Joachites envisioned. Indeed, there were powerful forces at work which eventually undermined not only these revolutionary trends, but ultimately the larger Catholic synthesis of they were simply radical manifestations. For the Avveroists the most immediate problem was a lack of scientific evidence to support their incipiently dialectical ontology and cosmology.

The difficulty, of course, is that medieval science could not supply the empirical data to support this doctrine. Indeed, the trend in the special sciences was towards atomism and positivism, and it is only in the present century, with the emergence of relativity, quantum theory, complex systems theory, and postdarwinian evolutionary biology that the basis really exists for the sort of claim first advanced by the Avveroists. It is not surprising that they were crushed.

Joachim's ideas, meanwhile, were picked up by the left wing of the Franciscan order——the so called Spirituals——who integrated them with radical interpretation of St. Francis' teachings on poverty. Joachim's vision had been firmly located in the Benedictine tradition, which was centered on communitarian work and prayer——work which helped to build up human civilization, and prayer which situated that civilization in its larger cosmic context. In the Benedictine tradition "poverty" meant first and foremost a rejection of the private accumulation of wealth with its attendant wastefulness and neglect of the needs of the community. There was no glorification of the lack or suffering of the poor. On the contrary, the Benedictines valued beauty, concrete physical beauty, immensely. Joachim's radicalism consisted not in a renunciation of the fruits of human labor, but rather in his desire to "Benedictinize" Europe——to reorganize European society as a network of Benedictine monastery communities.

The Franciscans, on the other hand, had little use for labor. And it was precisely the lack and suffering of the poor which the Franciscans valued in "Lady Poverty," a lack and suffering which they saw as an imitation of Christ's sufferings on the cross, and thus a type of the self-sacrificial love which they believed was the only possible route to wholeness and salvation.

This identification of Joachism with Franciscan poverty effectively drained the doctrine of its progressive content. Where Joachim had radically relativised the role of Jesus in the process of salvation, and largely negated the doctrine of salvation through self-sacrifice, the Franciscan Joachites reinstated this doctrine with a vengeance. The practical results were disastrous. Franciscan Joachites became involved in a long series of flagellant movements and hopeless uprisings. Not infrequently they directed the anger of the peasants and the *popolino* or urban poor against the Jews, who they accused of being usurers and "Christ killers," rather than against the landlords or urban patriciate. And even when they took aim at their real adversaries they lacked a compelling vision of the future.

Both Averroism and Joachism were ultimately the victims of the emerging market system. The marketplace undermined humanity's ability to know and love the cosmos, making the universe seem more and more like a system of only externally related atoms, and human society an arena of unending conflict. And it was the marketplace which would ultimately doom the Catholic synthesis as a whole. But before this happened, Catholic philosophy experienced on last flowering, and it produced a system which, while it did not go so far as the Avveroists and Joachites, nonetheless affirmed in a powerful new way the participation of matter generally, and humanity in particular, in God's plan for the universe. I am speaking of the system of Thomas Aquinas, which not only survived attacks from the Augustinian right, but which eventually became the official doctrine of the Roman Catholic Church. Let us sketch the Thomistic system at least briefly, and show how, its limitations not withstanding, it represented a real step forward in realizing the progressive potential which had always been latent in Christianity.

According to Aquinas there is an eternal law which exists in God as a kind of exemplar for the cosmos as a whole. Human beings, endowed as we are with reason, are able not only to achieve knowledge that this eternal law exists, but also to grasp a certain portion of it——the part Aquinas referred to as the natural law. This natural law provides a guide to virtuous human conduct and serves as the basis for the development of human laws, which are essentially applications of natural law to particular situations, based on a prudential estimate of particular conditions (Aquinas, *Summa Theologiae*, I-II: 90, 91, 93, 94, 95, 99).

Human beings, created in the image of God, are naturally good, and are capable of participating, at least in a limited way in the divine life regulated by the eternal law. If we are incapable of achieving salvation——i.e. full participation in the life of God——by our own powers, this is only because as finite beings we cannot grasp the infinite goodness, truth, and beauty of God in its essence, and thus cannot love God as he really deserves to be loved. (Aquinas, *Summa Contra Gentiles*, Third Book, Chapter II). Thus the necessity of revelation, which makes known the divine law; the incarnation, by which God joins the divine nature to the human and infuses it with supernatural capacities which permit us to participate fully in a life which would otherwise be beyond us; and the church, the body of Christ, which mediates supernatural grace to the individual believer through the sacramental system. Humanity, as essentially social being, is redeemed through participation in community. The Thomistic system goes a long way towards rectifying the distortions introduced into the Hellenistic Jewish synthesis by Christianity. Gone is the focus, characteristic of both Pauline and Augustinian theology, on the impossibility of knowing and fulfilling the law. Aquinas interprets the limitations to human knowledge and goodness in ontological rather than moral terms. And gone is the emphasis on the messianic nature of Jesus and the substitutionary interpretation of his saving activity. The incarnation, for Aquinas, marks the beginning of a new sociohistorical process, centered on the Church, which, through inner-worldly communal activity brings finite human society into communion with the divine. Aquinas thus not only creates significant space for theologically meaningful participation in the human civilizational project. He expands our understanding of that project so that it becomes, fully and explicitly, a real participation in the life of God.[3] The fullness of salvation, involving as it does participation in the eternal life of an infinite God, by its very nature transcends anything which can be accomplished within the confines of a finite social historical project. But all human endeavors, at least in so far as they develop the social capacities of humanity, are at least a partial participation in this same divine life.

There are, to be sure limitations. Thomism clearly reflects the still very low level of development of human organizing capacities. Aquinas' conception of natural law is static, based as it is on the science of his day, which was limited to arithmetic, algebra, geometry, (Ptolemaic) astronomy, (Aristotelian) physics, music, and anatomy. Prior to the democratic revolutions, a science of history, and thus an understanding of the natural law as a self-reorganizing logic of matter, was not yet possible. More broadly, Aquinas fails to resolve in a fully satisfactory way, the contradiction between matter and form which lies at the heart of the Hellenic philosophical tradition. The sharp distinction between matter and form, nature and grace, finite human and infinite divine life is a kind of theological reflex of the enduring contradictions between women and men, and between the peasantry, on the one hand, which participates only indirectly in the human civilizational project (by producing food), and the clergy, which moves in the realm of the universal. And the Christian residues within the Thomistic system meant that Aquinas himself was not immune to antisemitism.

---

[3]. I say "gone is the emphasis" because it is, of course, possible to find places in Aquinas' writings where he explicitly defers to Catholic dogma on these points. This is clearly a weakness. More important than the letter of the text, however, is the logic of the system, which clearly breaks with the Pauline and Augustinian pattern and suggests a very different road to salvation.

But on the whole the progress is enormous. By endowing human labor, human political organization, and human scientific research with real theological significance, Scholastic philosophy helped create the conditions for the industrial, democratic, and scientific revolutions. A Thomistic theology enriched by the experience of these revolutions might have had far more understanding of the self-organizing character of matter itself, something which would have gradually softened the form/matter dualism. This is the road which leads towards the dialectic. It would certainly have had a greater appreciation of human organizing capacities, and thus of the full extent of humanity's natural capacity to participate in the life of God. And indeed, at least at the philosophical and theological level, such a trajectory of development exists. It is the tradition which leads from Aquinas through Eckhart and Cusa, Spinoza and Leibniz, to Hegel, Marx, and Bogdanov——and ultimately to our own emerging synergistic theory.

Once again, however, the process of socioreligious evolution was derailed. Late medieval, Reformation, and Counter-Reformation theology rejected the enormous achievement represented by the Thomistic synthesis, and reinstated in even more radical form the reactionary, patriarchal, antisemitic theology represented by the Pauline and Augustinian tradition. The reasons for this development, of course, lie in the reorganization of European society set in motion by the industrial revolution and the emergence of capitalism——something we will discuss in detail in the next chapter.

CHAPTER FIVE:
The Mandate of the People

Its very real limitations notwithstanding, the socioreligious synthesis effected by the salvation religions opened up an extraordinary new period in the history of humanity and in the cosmohistorical evolutionary process. Scholastic philosophy, and parallel tendencies in the other socioreligious traditions, by grasping humanity as a real participation in the life of God, opened the way for powerful innovations in the technological, economic, political, artistic, scientific, and religious arenas. Three great revolutions swept across the face of the planet, transforming everything in their wake. The first of these was the industrial revolution, which catalyzed a qualitative leap forward in the development of human productive capacities. The second was the democratic revolution, which marked the recognition that human society was a human social product, subject to rational reorganization just like any other form of matter. The third was the scientific revolution which began, for the first time, to unlock the mysteries which governed the complex organization of the cosmos. By vastly increasing human organizing capacity, these revolutions also increased the real level of human participation in the self-organizing activity of the cosmos. Human beings came increasingly to understand themselves as the subject of an organizing activity which, if not at the center of the cosmos, was, at the very least, among its most highly developed products and an important agent of its future development.

These developments had particular significance in the light of the philosophical tradition which Europeans had synthesized out of Hellenic and Jewish elements. The cosmos was not simply ordered——it did not simply follow certain rationally knowable mathematical laws——nor was it simply organized. Its structure actually developed through time, in order to more adequately fulfill the designs of its creator. And human productive, political, and scientific activity was precisely the locus of this progressive activity. Human beings had both the capacity, and the awesome responsibility, to fulfill God's plan for the universe. The ancient teaching that humanity had been created in the image and likeness of God thus took on a radically new significance.

And yet the full potential of this synthesis was never achieved. As human society became increasingly complex, it became ever more difficult to organize, develop, and allocate resources using the bureaucratic methods which had been developed by the tributary state. Bureaucratic state structures, even under the rationalizing impact of the salvation religions, simply couldn't comprehend the tremendous diversity of interests to which the emerging social order gave rise. Indeed, if anything, the state became an obstacle to development, as heavy taxes withdrew capital from progressive initiatives in order to support luxury consumption and military adventures. Many of the most exciting of the new initiatives took place outside the state sector, drawing on resources centralized and allocated by the spontaneous mechanisms of the marketplace.

The marketplace played, from the beginning, a contradictory role in the development of human social capacities. On the one hand, market relations were able to organize far more complex activities, and to mediate a far more complex level of interdependence than the centralized, redistributional states of the tributary social formations. The complex, intersecting and overlapping network of relationships between individuals made possible by the marketplace permitted the development and realization of a far wider range of interests than could be successfully planned by any central authority. It is, of course, for this reason that the liberal tradition argues that the market serves as a catalyst for the development of human social capacities (Hayek 1988).

But even as it was catalyzing development of the highest level of productivity ever known by humanity, the marketplace was rapidly undermining the complex fabric of relations which held society together and which nurtured the development of human social capacities. The marketplace has no access to information regarding the impact of various activities on the qualitative complexity of the ecosystem or the development of human social capacities. Indeed, the marketplace permits development of a rich variety of relationships precisely because it acts as a solvent which breaks down traditional bonds of family and village community——bonds which, we have seen, played a critical role in developing the capacity for work, for the exercise of power, for understanding the unity and interrelatedness of the cosmos, and especially the spiritual capacity to take an interest in the cosmos as a whole. And the marketplace allocates resources not on the basis of their potential for contributing to the development of human social capacities, and thus to the organization of the cosmos in general, but rather on the basis of their relatively profitability. Development thus took on a contradictory character: on the one hand unprecedented progress, and on the other hand unprecedented destruction. It is this contradictory pattern of development, born of the industrial, democratic, and scientific revolutions, but organized and directed by market mechanisms, which we call capitalism. In this chapter we will trace the history of capitalist development. We will begin by analyzing the industrial, democratic, and scientific revolutions, in order to determine the specific contribution made by each to the development of human social capacities. Second, we will analyze the mechanisms by which the marketplace breaks down the complex organization which had been built up during the course of thousands of years of agrarian civilization, and the ways in which it actually holds back the progress of the industrial, democratic, and scientific revolutions. Finally we will examine the development of capitalist civilization as a whole. We will analyze three very different "roads" to capitalist development and demonstrate that, their differences not withstanding, it was in every case the interventionist state, centralizing and allocating resources for investment in infrastructure, education, research, and development, which served as the real agent of development, and that the marketplace was always and everywhere a drag on development, draining off resources for unproductive consumption. We will examine the internal contradictions of the capitalist system, showing that the operation of the marketplace inevitably tends to draw resources away from high technology, high wage activities and redeploy them to low technology, low wage activities on the periphery of the system, Finally, we will demonstrate the inadequacy of reformist approaches to resolving these contradictions and demonstrate the necessity of a wholesale break with market relations.

I. The Industrial, Democratic, and Scientific Revolutions

A. The Industrial Revolution

Industrial production has traditionally been defined as the application of non-human energy sources to the production process, coupled with a rationally developed social and technical division of labor which makes possible the mass production first of consumer goods, and later of the capital goods (the machines) which are used to produce those consumer goods. This definition does provide a more or less adequate *empirical* marker by which we can differentiate industrial from preindustrial production. But in order to grasp the *concept* of industry, we need to analyze the actual process of industrial production in the context of its impact on the qualitative complexity of the ecosystem.

Industry proceeds by breaking down the organization of both the natural and the social factors of production in order to release the energy which they contain, and then harnesses that energy to reorganize matter into more complex forms to serve human purposes. Industrial generation of energy, for example, breaks down the chemical (electromagnetic) bonds of various fossil fuels, or the nuclear bonds of certain heavy metals and uses the resulting heat either directly to drive machinery, or to produce steam, and eventually electricity, which in turn drives machinery. These machines then combine various elements mechanically, chemically, or biologically to produce complex structures which serve human purposes. Similarly, industry breaks down pre-existing social organisms (family, clan, village), and harnesses the social capacities of the individuals through a complex social and technical division of labor, enabling them to create products more complex than any individual worker could alone.

There are several distinct stages in the development of industrial production. These stages are defined by progress in the development of new energy sources and in the organization of labor power. Let us consider each of these dimensions separately. The development of energy sources is fundamentally a matter of harnessing the four fundamental forces of the universe. The first steps in this process were taken prior to the industrial revolution: the discovery of air, water, and animal power, which, in effect, harness thermodynamic gradients, the force of gravity, and the complex organic capacities of other mammals to perform useful labor. With the scientific revolution, however, it became possible for humanity to begin to harness the electromagnetic force, at first indirectly through the oxidation of various compounds to produce heat and drive steam and later internal combustion engines, and later directly with the development of the electric motor. Finally, during the middle of the present century, we learned how to harness (albeit only very imperfectly) the strong and weak nuclear forces. The logical conclusion of the replacement of human by nonhuman energy sources is, of course, the elimination of direct manual labor, or at least routine and uncreative direct manual labor, altogether through automation. Automation may, in fact, not be possible within a still largely industrial mode of production, in that it requires the development of a kind of holistic, systems thinking which at the very least pushes at the limits of industrial technology and the analytic science on which it is based.

The organization of labor power has also proceeded through various stages. The first of these——the social division of labor or the development of specialists in various crafts——was largely completed within most advanced tributary social formations. The second stage consists in dividing the production of a particular object into a constantly increasing number of more or less routine tasks, which can be performed by relatively unskilled (and therefore relatively cheap) labor. This is what we call the technical division of labor. Integral to this stage is division between the mental and manual dimensions of the production process. One worker or team of workers (initially the inventor or entrepreneur, later a team of salaried engineers) designs the object and the production process; another team actually executes the process. Third, the individual tasks involved in production are subjected to detailed analysis (time motion studies, etc.) and the whole complex of bodily movements of the worker are subjected to a scientifically developed discipline designed to increase the efficiency of the production process, transforming the worker, in effect, into simply one part of a larger machine——and not, by any means, its most intelligent part.

Industry involves the vast majority of the population in the civilizational project in a way which was not previously possible. Much, though not all, of the industrial working class is involved in producing machines, building structures, etc., which become a semipermanent part of the human environment, and which contribute not simply to the *reproduction* of humanity, but to its actual *development*. And the high level of productivity characteristic of industry makes it possible to release an increasingly large part of the population from direct production altogether.

At the same time, industrial uses of energy and natural resources run up inevitably against certain definite ecological limits. It is not at all clear that industry really increases the overall complexity of the ecosystem, and if it does, there seems to be a real point of diminishing returns. There are, furthermore, serious limits to the ability of industrial production to tap fully the self-organizing potential inherent in matter itself. Rather than catalyzing the development of complexity which already exists in matter (physical, chemical, biological, social), industry breaks matter down into its component parts and reorganizes it in a way that serves human purposes. For the most part, it uses only the raw energy stored in material structures, not the organizing dynamic which created those structures. This is most apparent in the way industry uses the social form of matter. Industrial strategies for increasing productivity reduce the vast majority of workers (including, increasingly, intellectual workers) to mere cogs in a wheel, simply moving physical, chemical, biological (or, in the case of intellectual workers) social matter from one place to another. Industry has no way to tap into the skill or creativity of workers, and no way to recover the waste that derives from all of the potential Mozarts and Einsteins who have passed their lives painting tail fins for Cadillac. Both of these limits are rooted in the "exploitative" character of industrial production, an exploitation which is prior to the marketplace or any other means of surplus extraction. In this sense, it is necessary to regard industrialization as an important, but also very ambiguous, step forward in the antientropic vocation of the social form of matter.

B. The Democratic Revolution

Throughout most of human history the organization of society has been regarded as an integral part of the natural order: something fixed eternally, as it were, by the mandate of heaven. This did not, to be sure, prevent human beings from regarding their particular societies as somehow disordered, or from struggling, often quite effectively, to rectify that disorder. On the contrary, we have seen that from the very earliest stages of human civilization, both the peasantry and the intelligentsia have struggled, in different ways, to resist or restrict the emergence of the warlord state, and to rationalize the extraction and investment of the social surplus product. In this sense, Laclau and Mouffe (Laclau and Mouffe 1985) are wrong to regard the conscious resistance to oppression as a product of the democratic revolutions. At the same time, we have also seen that the great socioreligious movements of tributary social formations always conceived their struggle as ultimately the work of God, and almost always as a struggle to restore a divinely ordained (or naturally pre-existing) social order which had been decreed before the foundations of the world.

The democratic revolutions of the past four hundred years mark the realization that the organization of human society is, immediately at least, a human product, and that human beings can, through rational, collective action, reorganize human society in much the same manner that they can reorganize any other form of matter. On the theoretical side, this realization developed only very gradually. At first, human society was regarded as an artifact the purpose of which was to defend either consumer interests (Hobbes 1962) or else the "natural rights" (Locke 1967) of individual human beings. Gradually this focus on the defense of individual interests and rights gave way to a generalized utilitarianism (Mill 1965). States which failed in these tasks were illegitimate, and could be overthrown, if necessary, by force of arms. Eventually, the development of human society was recognized as a uniquely social process, which brought into being the whole realm of property, law, morals, and religion, and which transformed human beings from narrow, selfish creatures into citizens capable of devotion to the common good (Rousseau 1962). It was only after a rather extensive revolutionary experience that the reorganization of human society became an object of scientific analysis (Marx 1848), and that human beings began to identify the laws which govern the process of social reorganization, mastery of which was as much a precondition for realization of the democratic revolution, as the mastery of physical, chemical, and biological laws was a condition for the realization of the industrial revolution.

It is possible to identify three stages in the development of the democratic revolution. During the first stage, it was simply the political sphere——the means of exercising power——which was subject to reorganization. The economic structure and the kinship structure, which regulates the socialization process, continued, at this stage, to be regarded as "natural" and not subject to reorganization through rational collective action. There are two dimensions to this process. On the one hand, the residues of old imperial structures (such as the Roman Catholic Church, heir of the Roman empire) were cast off, and nation states were organized on the basis of existing economic networks (development of a unified national market), and political and cultural identities. On the other hand, irrational claims to authority based on hereditary privilege were swept away, and leaders were increasingly forced to legitimate themselves on formally or (in rare cases) substantively democratic grounds. By formally democratic legitimation we mean legitimation on the basis of election or other popular manifestation of support, which confirms the principle that human societies are a social product. By substantively democratic legitimation we mean legitimation on the basis of the leader's ability to actually increase the organization, and thus the power, of the society (cf. Weber 1921/1968).

This stage of the democratic revolution is often referred to as "bourgeois" in character because of the leading role played by the bourgeoisie during this phase of the revolution in European societies, and because the failure to subject economic structures to rational reorganization leaves the power of the bourgeoisie unchecked. The process of rationalizing political structures itself however, neither presupposes bourgeois political power, nor is it particularly conducive to the persistence of that power. On the contrary, we will see that completion of the democratic revolution, even in its political manifestation, calls the authority of the bourgeoisie radically into question, and is, in turn, held back by the mediation of interdependence through market institutions.

During the second phase of the democratic revolution, economic as well as political institutions were subjected to reorganization, and an attempt was made to secure as much as possible of the social surplus product for investment in the all sided development of the species. This stage of the revolution, customarily called "socialist," has generally been led by the intelligentsia with support from the proletariat and the peasantry. Even more so than the first, political phase of the democratic revolution, the second, economic phase, calls market relations radically into question.

Finally, during the present century, cultural institutions have been subjected to the democratic revolutionary dynamic. There are two dimensions to this process. On the one hand the structure of the family and the organization of the socialization process has come under increasing scrutiny (Freud 1930, 1933), as scientists have recognized that realization of the democratic and socialist project presupposes reorganization of the human personality as well as of social institutions (Fromm 1941, 1947, 1955, 1973, Marcuse 1955, Reich 1977). This realization has led to a broad range of movements to transform human personality from mass political campaigns (the Chinese Cultural Revolution) to the emergence of vast networks of quasi-religious psychotherapeutic sects. At the same time, the advent of artificial contraception has liberated women from the burden of pregnancy and childrearing, and permitted them to call into question the structure of the nuclear family, which seems increasingly to be the locus of their oppression (Rubin 1975, Chodorow 1978). The women's movement has become the principal constituency for efforts to reorganize the socialization process.

The democratic revolution thus represented a tremendous leap forward in the development of human social capacities. On the one hand, human beings recognized, for the first time, that they could systematically reorganize the structure of human societies. And they recognized that building the power to carry out this kind or reorganization——or for that matter any other kind of complex social activity——meant comprehending and engaging the interests of thousands or even millions of people. In a very real and very profound sense all power does, ultimately, come from the people.

As in the case of the industrial revolution, however, there were real limitations. Democracy fails to grasp the drive towards organized complexity and holism implicit in the revolutionary movements——a drive the trajectory of which defines a moral imperative which transcends anything the people generally, or even the most advanced elements in the leadership, can know or understand. The democratic revolutions put human societies at the mercy of the empirical interests of population at large, interests which are, for the most part, relatively narrow and underdeveloped. This problem becomes even more serious in market societies where the disintegration of the social fabric undermines the ability of the people to comprehend the relational and interdependent character of the cosmos, and transforms all activity into merely a means for realizing individual consumer interests.

A correct understanding of the democratic revolutions thus distinguishes sharply between the authentic contributions of democracy——a recognition of the collective capacity of humanity to reorganize society, and the development of a relational understanding of power——and fundamentally reactionary ideas which would subject society as a whole to interests of backward elements, expressed through such mechanisms as the plebescite, representative government, etc. Authentic democracy consists in the capacity of the most advanced elements in the society to mobilize the talents and interests of the majority in the work of their own self-development.

C. The Scientific Revolution

The scientific revolution was at once a precondition for and a product of the industrial and democratic revolutions. It is only on the basis of a rational understanding of the organization of matter that we are able to reorganize it in the systematic, rational manner which characterizes industrial production and democratic revolutionary politics. At the same time, it is precisely the sort of engagement in reorganizing the world which provides both the occasion of insights into the organization of matter and the opportunity to test out new theories in practice.

Precisely because it participates in such an advanced way in the self-organizing activity of the cosmos, humanity is able to reflect in itself the complex organizing principle which gives birth to the various forms of matter (physical, chemical, biological, social). Initially this capacity takes the form of an intuitive perception of pattern, order, structure, organization, and system, coupled with a rigorous exploration of the logical implications of these concepts. This intuitive capacity formed the basis of the first true science, mathematics, which for the Greeks included arithmetic, geometry, astronomy, and harmonics, as well as the protoscientific theories embodied in Aristotelian physics, biology, politics, etc. and comparable doctrines.

What these earlier scientific theories lacked was the capacity to break down the complex organized totality which constitutes the cosmos into its component parts, and isolate the specific relations, structures, processes, etc. which characterize the various forms of matter. It was precisely this analytic thrust which came to the fore during the scientific revolution which began in the seventeenth century, and which has continued into the present period, unlocking the secrets of progressively more complex forms of matter: physical, chemical, biological, and social. Closely associated with this analytic thrust is the development of the experimental method which permits us to isolate one specific relationship from the complex manifold of reality, and test hypotheses concerning the nature of this relationship in order to determine whether or not our intuitive perceptions of order are

206 *Towards Synergism*

actually born out.

Beginning with Copernicus, Galileo, Kepler, and Newton, humanity began to unlock the four fundamental forces by which matter is constituted: first gravity, then the electromagnetic force, and in our own time the strong and weak nuclear forces. We have uncovered the process by which these forces interact to produce what previously were regarded as distinct substances (the elements) and the way in which these elements combine to form more complex compounds. We have used this chemical theory to analyze many of the processes which characterize living matter. And we have begun to understand some of the principal processes which define the social form of matter, as well as the structures which regulate those processes. The resulting body of knowledge makes it possible for us to tap into physical, chemical, biological, and social processes in order to tap into the self-organizing dynamic present in matter, and reorganize matter in new and increasingly more complex and powerful ways. In this sense the scientific revolution has become the driving force in humanity's technological, economic, and political development.

It would be a mistake, however, to ignore the limitations of the scientific revolution. Science as it has developed over the past 400 years is largely a compendium of separate theories describing countless relationships and processes. Our ability to grasp whole systems remains largely undeveloped. And since, as we saw in the first chapter, it is the whole alone which is really real, we have not yet advanced to the point of authentic knowledge. The irreducibly analytic character of bourgeois science is most apparent in the dominant role which classical analytic mathematics (the calculus) continues to play in most of the sciences. But it is true even of the biological and social sciences where holistic thinking has, from the beginning, been more common. Most biological theory tends to break the life process down into a series of simpler chemical processes. Biology still cannot tell us what life is, and even has difficulty determining which forms of matter are living and which are not. What about viruses? or Sheldrake's self-reproducing crystals? Even sociology has difficulty conceptualizing rigorously the whole process of human social development. This limitation is exacerbated in market societies, where the social disintegration engendered by the marketplace has undermined humanity's ability to grasp the relational and interdependent character of matter. The result has been the emergence of atomistic and positivistic theories which deny the relational, holistic, self-organizing, and teleological nature of the universe.

The full development of the human capacity for knowledge requires that we transcend this limitation, and that we advance beyond analytic thinking to the synthetic. Partly this change must take place in the specialized sciences themselves——a process which is already underway with the emergence of unified field theories, complex systems theory, postdarwinian evolutionary biology, dialectical sociology and anthropic cosmology. Partly it will involve a rebirth of philosophy.

For now, though, we need to examine in more detail the complex interaction of the industrial, democratic, and scientific revolutions, with the regulating mechanism of the new capitalist society: the marketplace.

II. The Development and Penetration of Market Relations

Trade of various kinds has existed from the very earliest stages in the development of human civilization. Even the communitarian societies of the Pacific and of North America engaged in trade with their most distant neighbors. With the partial exception of advanced tributary social formations, however, this trade was always strictly subordinate to the production of use values. People traded in order to secure goods which they could not produce themselves. They did not produce in order to trade. Even in advanced tributary societies production for trade formed a strictly limited sector within a larger economy still dominated by the production of use values. It is only with the advent of capitalism that the dynamics of the marketplace become the dominant, regulating structure of the social formation, and that production of commodities (production for exchange) becomes the principal form of economic activity.

How does this come about? Partly the emergence of markets is simply a result of the fact that the internal differentiation of human society outstripped the organizing capacity of the bureaucratic state apparatus. So long as only a small ruling class, together with a small cadre of craftsmen, scribes, etc., has been withdrawn from agrarian production, it is possible to provide for their needs through the mechanisms of the centralized redistributional state, assigning the taxes from various lands and villages to high officials, and providing a small salary or benefice for craftsmen and minor functionaries. But when the vast majority of the population is engaged in producing either manufactured goods or non-food crops this is no longer possible. Goods and services begin to be exchanged, and resources for investment centralized and allocated, spontaneously, outside the state sector.

For an authentic market economy, however, something more is required. It is necessary, on the one hand, for the capital necessary for industrial production to be centralized in the hands of a relatively small class of investors, so that entrepreneurs must compete with each other for capital, constantly bidding up the rate of return on investments, which becomes the criterion by which investors evaluate the relative merit of various enterprises. It is also necessary for the direct producers to be alienated from control over the means of production. So long as people can grow their own food and provide their own shelter and their own clothing, even if they are forced to pay relatively high rents and taxes, and to performed forced labor of various kinds, they are unlikely to sell their labor power or to submit to the discipline of factory or office work. So long as a craftsman can make ends meet plying his customary trade, exercising control over the production process (and over the way he spends his time) he is unlikely to submit to the technical division of labor. Neither the peasant nor the craftsmen, furthermore, is likely to relinquish control over his means of production without first mounting some rather sustained resistance. For this reason, development of market relations presupposes a violent struggle between the bourgeoisie, which seeks to create a pool of "free" wage labor, and the peasants and the artisans, who seek to defend their traditional way of life, as well as an parallel struggle, perhaps less violent, within the bourgeoisie between industrialists and *rentier* elements.

In this section we will examine the development of market relations in some detail, focusing in particular on these early, violent stages of the process, and reserving discussion of the contradictions of the developed market economy for the next section.

A. The Destructive Impact of Market Relations on Communitarian and Tributary Civilization.

1. The Penetration of Capitalism into the Countryside

Everywhere, almost without exception, the penetration of capitalism into the countryside has involved three definite stages:

a) conquest by an emerging capitalist power,

b) the enclosure of common lands and the dispossession of the peasantry, together with the conversion of land from subsistence to cash crops, and

c) the internal differentiation and disintegration of the village community.

If the development of industrial forces of production was the basis for the development of capitalism, then clearly the European conquest of Asia, Africa, and the Americas was the leading factor. This conquest laid the foundation for the creation of a unified world market——a process which is only now reaching completion. From the beginning Europeans sought to transform the complex civilizations of these continents into means for their own enrichment. Thus Columbus, on his first meeting with the Arawak Indians.

> They brought us parrots and balls of cotton and spears and many other things, which they exchanged for the glass beads and hawk's bells. They willingly traded everything they owned ... They were well-built, with good bodies and handsome features ... They do not bear arms, and do not know them, for I showed them a sword, they took it by the edge and cut themselves out of ignorance. They have no iron. Their spears are made of cane. They would make fine servants ... With fifty men we could subjugate them all and make them do whatever we want (in Zinn 1980: 1).

When conquered populations could not be enslaved or put to forced labor, they were simply exterminated. There is good evidence that the Spanish slaughtered between 12 million and 15 million Indians. Millions more died from disease, from starvation induced by the disruption of delicate horticultural adaptations to American ecosystems, or from the rigors of forced labor (Zinn 1980).

In areas where Indians successfully resisted subjection to forced labor, African slaves were imported to work the lands which had been seized from them.

> As the slaves come down to Fida from the inland country, they are put into a booth or prison ... near the beach, and when the Europeans are to receive them, they are brought out into a large plain, where the ships surgeons examine every part of everyone of them, to the smallest member, men and women being stark naked ... Such as are allowed good and sound are set on one side ... marked on the breast with a red-hot iron, imprinting the mark of the French, English or Dutch companies ... The branded slaves after this are returned to their former booths where they await shipment, sometimes 10-15 days (Zinn 1980).

On the ships

> the height, sometimes, between decks, was only eighteen inches so that the unfortunate human beings could not turn around, or even on their

sides, the elevation being less than the breadth of their shoulders; and here they are usually chained to the decks by the neck and the legs. In such a place the sense of misery and suffocation is so great that the Negroes ... are driven to frenzy (Zinn 1980).

Zinn summarizes the situation in this manner.

By 1800, 10 to 15 million blacks had been transported as slaves to the Americas, representing perhaps one third of those originally seized in Africa. It is roughly estimated that Africa lost 50 million human beings to death and slavery in those centuries we call the beginnings of modern Western civilization, at the hands of slave traders and plantation owners in Western Europe and the Americas, the countries deemed the most advanced in the world (Zinn 1980).

Much the same process was played out in Asia, limited only by the greater capacity of the more powerful Asian civilizations to resist. Thus Hauser

Vasco de Gama's second voyage (1502-1503) at the head of a veritable war fleet of 21 vessels, resulted in the replacement of the Egyptian-Venetian monopoly of the spice trade by a new monopoly. It was punctuated by horrible atrocities ... Arson and massacre, destruction of rich cities, ships burnt with their crews in them, prisoners slaughtered and their hands, nose and ears sent in mockery to the 'barbarian' kings, these were the exploits of the Knight of Christ; he left alive, after mutilating him in this way only one Brahmin, who was given the task of conveying these horrid trophies to the local rulers (Hauser in Mandel 1968: 108).

Mandel adds that

the new commercial expansion remained based on monopoly ... Dutch merchants, whose profits depended on their monopoly of spices obtained through conquests in the Indonesian archipelago, went over to mass destruction of cinnamon trees in the small Islands of the Moluccas as soon as prices began to fall in Europe. The "Hongi voyages" to destroy these trees and massacre the population which for centuries had drawn their livelihood from growing them, set a sinister mark on the history of Dutch colonialism, Admiral J.P. Coen not shrinking from the extermination of all the male inhabitants of the Banda islands (Mandel 1968: 108).

Thus, if capitalism developed human social capacities (which it did, and in some remarkable ways), it did so only when this served individual consumer interests——specifically the interests of the small minority whose wealth governed the operation of the marketplace. It was also capable of great destruction, and its development actually presupposed the destruction of all of the other great civilizations on the planet.

Lest we be tempted to draw the conclusion that the conquest and plunder which accompanied the development of capitalism was a function less of developing market relations, than of some will to power embedded in European culture, or that the victims of this process were exclusively Asian, African, and American, we should note that the development of capitalism in Europe was itself marked by a series of quasi-imperial conquests. The "unification" of England, Scotland, Wales and Ireland under English rule, the formation of a unified French state, of the Prussian, Russian, and Austro-Hungarian empires all involved the conquest, and subsequent impoverishment of rich agricultural regions by developing commercial and industrial centers. Perhaps the clearest example of this pattern of conquest within Europe, however, is the Italian *Risorgimento*, which transformed the essentially progressive and popular struggle to liberate Italy from feudal domination into a colonization of Sicily and the South by the Piemontese bourgeoisie.

> The policies of the first 40 years of unification effected the industrialization of Lombardy, Liguria and Piemonte, allowed the development of an advanced capitalist agriculture in Romagna, which had been one of the most desolate regions of the Papal States, and was large with benefits for Tuscany. Into these regions the ruling classes of which governed the country as a whole, flowed the better part of the disposable resources of the nation, while the *Mezzogiorno*, at the close of the nineteenth century was reduced to an outlet for the products of Northern industry ... Unity, in assigning to the South a colonial function, had made it an underdeveloped country where conditions are worse than in many parts of the Third World ... a country the hopes of which could only find refuge in the American dream (Zitara 1971:76).

Dispossession of the peasantry was everywhere a precondition both for the development of a pool of "free" labor and for the development of a capitalist agriculture devoted to the production of industrial raw materials (wool, cotton), wage foods (cheap grain) and luxuries (cocoa, coffee, etc.).

In the developed areas of the Americas this process was gradual. Initially the *conquistadores* simply took the place of the old Aztec and Inca ruling classes, imposing various forms of forced labor and tribute on the population, but leaving at least a very significant portion of the landholdings of the Indian communities intact. It was not until the middle of the nineteenth century, with the great Liberal revolutions, that a fully capitalist agriculture emerged, based on the production of coffee, cotton, and other export crops, and that a serious effort was made to expropriate the peasantry. The case of Mexico was typical in this regard. "The Law of Expropriation (*Ley de desamortizacion*) of June 25, 1856 held that (Wolf 1969: 12)"

> no civil or ecclesiastical corporation could acquire or administer any property other than the buildings devoted exclusively to the purpose for which that body existed. It provided that the properties then owned by such corporations must be sold to the tenants or usufructuaries occupying them and that properties not rented or leased would be sold at public auction (Whetten 1948: 85).

Eric Wolf points out the implications of this law, apparently directed at the Church and the large landowners, for the Indian communities.

> Freedom for the landowner would mean added freedom to acquire more land to add to his already engorged holdings; freedom for the Indian——no longer subject to his community and now lord of his own property——would mean the ability to sell his land, and to join the throng of landless in search of employment (Wolf 1969 :13).

The law also effectively prohibited educational, scientific, and religious institutions from establishing an autonomous economic base, rendering them dependent on support from the bourgeoisie or the bourgeois state.

The resulting concentration of land ownership should come as no surprise.

> ... at the end of the Diaz period there were 8,254 haciendas. Three hundred of them contained at least 10,000 hectares; 116 around 250,000, 51 possess approximately 30,000 each, 11 measured no less than 100,000 ... one hacienda owner might own more than one hacienda ... (Wolf 1969: 16).

Similar conditions developed throughout Asia. Barrington Moore reports that at the end of the period of British rule, 20% of Indian households were landless, and another 50% owned only 2% of the total arable land. The richest 10%, on the other hand, controlled fully 48% of the land. Meanwhile, the ranks of the rural proletariat swelled. In 1891 only 13% of the Indian population could be classed as landless agricultural laborers.

By 1931 this figured had risen to 38% (Moore 1966: 368).

Sarkisyanz reports that in Burma, prior to the British conquest, 80% of the people owned their own land. By 1920 only 50% owned land, and only 22% were able to rent it. Fully 27% had been proletarianized. Wages fell by 20%, between 1870 and 1930, and rice consumption declined by 25% between 1921 and 1941 (Sarkisyanz 1966: 137-141).

Penetration of market relations led to the effective dispossession of the Chinese peasantry, even in the absence of foreign conquest.

> Studies of four *hsien* or districts in north China in 1936 showed that landlords who formed 3 to 4 percent of the population possess 20 to 30 percent of the land; poor peasants formed between 60 to 70 percent of the population but controlled less than 20 to 30 percent of the land ... between 40 to 50 percent of the peasant families did not have enough land to provide them with food (Wolf 1969: 134).

French rule in Indochina led to a similar phenomenon.

> By 1938 ... nearly half [of the land] was in the hands of 2.5 percent of all landowners. Seventy percent of all landowners owned only 15 percent of arable land. Still larger was the class of landless tenants in the South, numbering some 250,000 families, and constituting about 57 percent of the rural population (Wolf 1969: 166).

Once again, the European peasantry suffered much the same fate as the rural population of Asia, Africa, and the Americas. In England, this process began around 1580 with "encroachments made by lords of manors or their farmers upon the land over which the manorial population had common rights or which lay in the open arable fields (Tawney 1922: 150)." As the English textile industry expanded the market for wool grew too. And wool production was less labor intensive than grain cultivation, and thus far cheaper. "Progressive" landowners thus found countless was, some legal and some not, to deprive the peasantry of access to the commons or to the open arable fields, so that this land could be given over the pasture for England's rising population of sheep (Moore 1966: 9).

In Italy, the *Risorgimento*, which, we have noted, amounted to nothing less than the Piemontese conquest of the Papal States and the ancient kingdom of the two Sicilies, led to the more or less complete expropriation of the Southern peasantry. Consider the following statistics from the *Agro Romano*, the rural hinterland of Rome. Prior to unification 55% of all arable land belonged to the nobility, 30% was in mortmain (i.e. was church or demense land) and 15% was held *in borghese* (i.e. as private property, mostly by bourgeois landlords, more rarely by rich or middle peasants). After the "land reform" 53% of the land remained in the hands of the nobility, only 7% remained in mortmain, and fully 40% was in the hands of the bourgeoisie (Sereni 1968: 144). Lands held in mortmain were either demense lands administered by the *communi* and used as commons where the poorest peasants could pasture their flocks, gather wood, and forage for the wild asparagus and cactus fruit, mulberries, acorns and chestnuts which formed the greater part of their daily fare, or else belonged to the church and the so called "*enti morali*," or charitable institutions. Revenues from these latter lands supported hospitals, schools, universities, and other activities which contributed to the development of human social capacities. Capitalization of landed property proceeded at the expense of both the subsistence needs of the peasantry and the development of society as a whole (Sereni 1968: 139).

The capitalization of landed property was, furthermore, accompanied by a deterioration of the terms of agrarian contracts, as the ancient *mezzadria* and *metateria* contracts, both of which involved a roughly equal division of the harvest between the landowner and the peasant, gave way to the *parspolo* and *compagno e padrone* contracts under which the landlord received 2/3 and 3/4 of the harvest respectively, the hated *terragio* contract, under which he received a fixed portion of grain per hectare, the quality of the harvest notwithstanding, or *retrometateria*, under which the peasants harvested sown land in return for as little as 10% of the harvest. Many peasants were driven off the land entirely and forced to work as *braccianti* (wage laborers) (Renda 1977: 62-64, Romano 1959: 77-80, Sereni 1968: 150, 257).

In Eastern Europe, which, like the South, had experienced a seigneurial reaction after the defeat of the peasant revolts of the late feudal period, "emancipation" of the still enserfed peasants was carried out in such a manner as to guarantee their effective expropriation. Peasants were forced to buy the land which they had worked for years. Wolf reports that in the fertile black-soil regions of Russia, "the average allotment before the reform was 9.18 acres; after it was 6.75 acres," as the landlords kept as much as possible of the best land for themselves.

In the non-black-soil industrial provinces, on the other hand, where *obrok* still dominated, the reverse was true. The landlords benefited by ridding themselves of unproductive land, transferring this to the peasants on the basis of excessive valuations. In eight such provinces, the average pre-reform allotment per person had been 10 acres; after the reform it was 11.6 (Wolf 1969: 56).

This poor land cost the peasants dearly. They had to come up with a down payment of 20 percent of the land, and repay the remainder to the state at 6% interest over the period of 49 years. Nearly all fell behind in payments or were impoverished as a result of meeting them; many lost their land altogether.

Dispossession of the peasantry was accompanied, in most cases, by conversion of the land to production of cash crops. Often, particularly in tropical and subtropical regions, this meant the abandonment of basic food (grain) crops altogether, as the triumphant agricultural bourgeoisie forced the peasants to plant sugar cane, cotton, coffee, cocoa, or any of a long succession of export crops which served as raw materials for the developing industrial centers. In other cases——as for example in Burma and Indochina——it meant the export of valuable food crops at a time when domestic consumption of wage foods was actually dropping.

Once again, this process was not confined to Asia, Africa, and the Americas. In Italy, after the imposition of a policy of free trade by the Piemontese bourgeoisie, southern wheat production suffered terribly, especially after 1870, when American grain flooded the world market and new technology made it possible to ship, store, and make pasta from cheap soft American wheat, thereby eroding the comparative advantage long enjoyed by Sicilian durum. Free trade forced conversion of arable land to exportable cash crops, such as olives, wine, fruit and especially citrus, for which the coastal lowlands and the eastern part of Sicily generally were well suited. This meant displacement of peasant subsistence agriculture, and a consequent deterioration in the peasants' diet, as vineyards and cattle pastures displaced grain fields, and citrus trees encroached on acorn, chestnut and mulberry groves, traditional sources of food for the poorest *contadini* (Zitara 1971:59).

The result of this process was the physical degradation, and in some cases even the destruction of the peasantry——a sobering thought for those who extol the virtues of the marketplace as a force for social progress and for the "development of the productive forces."

The physical degradation of the peasantry was accompanied by the disintegration of the village community, itself one of the most creative "productive forces" ever developed by humanity.

In Mexico, entire villages were swallowed up by the haciendas.

... in six states (Guanajuato, Michoacan, Zacatecas, Nayarit, Sinaloa, more than 90 per cent of all inhabited places were located on estates; in eight more states (Queretaro, San Luis Potosi, Coahuila, Aguascalientes, Baja California, Tabasco, Nuevo Leon) more than 80 per cent of the rural population lived in estate communities ... (Wolf 1969: 18)

This leads one author to suggests that "the plantation had absorbed not merely the land but the self-directing life of the communities, and had succeeded in destroying their mores (Tannenbaum 1937: 193)."

Eric Wolf reports that in Cochin China (the southern region of Vietnam), where penetration of market relations was most advanced,

> Villages lacked the historical depth of association among fellow villagers characteristic of areas further north. Attachment to patrilineages and lineage ancestors was less functional; the role of the *dinh* (the communal temple) in communal life was less central ... (Wolf 1969: 177)

Barrington Moore argues that "the enclosures were the final blow that destroyed the whole structure of English peasant society embodied in the traditional village (Moore 1966: 21)." In 1906 Russian Prime Minister Stolypin enacted legislation giving peasants the right to claim as private property the lands to which they were entitled in periodic redistributions, and converting to private property the lands of those communes which had abandoned regular redistributions. As a result some 3 million peasants abandoned the village community; of these at least 900,000 sold out entirely and went to the cities to look for work.

In short, the penetration of capitalist relations of production into the countryside was accomplished, for the most part through force——in many cases through a process of imperial conquest and genocide. The process inevitably involved the dispossession of the peasantry, and the conversion of land to cash crops produced for export. The result was the physical degradation of the peasantry, and the destruction of the village community.

At issue here is far more than the displacement of millions of human beings and the destruction of their traditional way of life. The village community was for millennia one of the leading forces for the development of human social capacities. It was the village community which made human beings part not only of a particular nuclear family, but of a larger community of workers, which was capable of building and exercising power. It was through participation in the village community that human beings grasped the underlying unity and order of the cosmos and understood the universe as an organized totality. And it was through the medium of the village community that human beings came to understand themselves as participants in that organized totality.

The great tributary empires had often subordinated the village community to the imperatives of predatory warlords, but they had never really attempted to destroy it. And yet this precisely is what the marketplace did in short order as it penetrated the countrysides of the planet.

The destruction of the village community all but wiped out one of the most ancient and powerful centers for the development of human social capacities. It is a loss from which we have yet to recover.

2. The Development of Market Relations in the Cities

The emergence of a market economy did not only destroy centuries of horticultural and communitarian civilization. It also destroyed the incipiently industrial economies of the great cities of Asia and the Mediterranean basin.

It is one of the most serious prejudices of European capitalist historiography to assume that the industrial revolution was a European achievement, and was the result of the emergence of capitalist relations of production. In fact, the early stages industrialization were already under way throughout Asia when Europe was still just a rural backwater. China was, up until the eighteenth century, the technologically and economically most advanced country on the planet, surpassing Europe even in the production of such advanced products as iron and iron implements. India and the Islamic world were the home of thriving advanced tributary societies which, while still predominantly agrarian, had witnessed the development of thriving crafts and manufacturing sectors, centered primarily in the production of such consumer goods as textiles. The European conquest of the planet, and the incorporation of these countries into the emerging world market almost universally led to the destruction of Asian crafts and manufacturing. This process is particularly well documented in the case of India (Akin 1980: 91, Moore 1966: 370, Gough).

By the eighteenth century India had attained a high degree of development in pre-industrial terms. Agriculture was sufficiently developed to support a relatively large number of non-agricultural workers; there were highly skilled craftsmen in iron, steel, textiles, shipbuilding and metalwork. India produced manufactured goods not only for home consumption but for export. India's economic wealth had for centuries been controlled by merchant bankers and princes who siphoned off the surplus of production over consumption in the form of idle hoards of gold and silver bullion; hence this wealth was sufficiently concentrated to represent a potential source of investment funds. India's resources of good quality coal and iron were located in convenient proximity to each other ...

Why did not this combination of apparently propitious circumstances produce a type of economic development capable of generating real momentum? Despite the many complexities and anomalies of the situation, basically the answer is simple. The colonial relationship subordinated India to British political and economic interests; it stimulated Indian economic development in some ways and inhibited it in others (Lamb in Mandel 1968: 442).

The classical Marxist claim that European industry triumphed over Asian purely through the cheapness of its products is not really accurate. European conquest, and the imposition of protectionist measures, played by far the larger role.

Though the beginning of the industrial revolution is to be placed at 1760, another half century passed during which India and China continued to be the world's chief providers of textile products. As late as 1815 India was exporting Ł 1.3 Million worth of cotton goods to Britain, while Indian imports of British cottons amounted to only Ł 26,000. China exported in 1819 nearly 3.5 million pieces of cotton goods, while its imports were infinitesimal. Like calico, nankeen was known and in demand through the world.

British industry succeeded in dominating the world market only by carrying on an extreme protectionist policy.

In 1813 Indian cotton and silk goods were 50 to 60 percent cheaper than British; therefore they were subjected for a long period to an import duty of 70 to 80 percent, after all importation of Indian cottons had several times been prohibited altogether, notably in 1700 and 1720. At the same time, as Britain was following this extreme protectionist policy, it imposed, through the East India Company, the free trade policy upon India (it did the same to China later, through the Opium Wars). At a time when India silk goods were paying 20 percent import duty in Britain, British silk goods were paying only 3.5 percent in India!

... Age old industrial centers died. Dacca was partly overgrown with jungle. The craftsmen, reduced to idleness, spilt over into agriculture. The vicious circle closed when, after 1833, Britain decided to develop on a large scale in India the production of agricultural raw materials, especially cotton plantations. A people who formerly had exported cotton goods to all parts of the world now exported only raw cotton, to be worked up in Britain and sent back to India as textile goods! (Mandel 1968: 446-447).

A similar process took place in those regions of Europe which were conquered by emerging metropolitan bourgeoisies, and transformed into platforms for the primitive accumulation of capital. The forces that came to power after the unification of Italy, for example, were interested first and foremost in creating the conditions for the capitalist development and industrialization of the North. Lacking foreign colonies, they were forced to transform their own countryside into a base for primitive accumulation of the capital necessary for the industrialization. They also needed to undertake formation of a unified national market capable of creating sufficient internal demand to sustain industrialization, free trade policies favorable to the development of a commercial export agriculture, and the capitalization of landed property, which had the result of supplying cheap food, prerequisite for the low wage policies of Northern industry, while at the same time driving the peasantry off the land, creating a large supply of "free" wage labor for Italian, and as it turned out, North American industry.

Transformation of the South and the countryside into a base for primitive accumulation of capital was accomplished largely through the mechanism of fiscal policy. State financed rail construction and other public works projects (in the North) and the mounting costs of repression (in the South) led to creation of a massive national debt. In 1859 the combined expenditures of the Italian states were Ł 570 million, and the combined debt, owed mostly by the Northern states, was Ł 1.5 billion. By 1879, the annual expenses of the unified state had risen to Ł 1022 million, and the combined debt, now being serviced with the help of the Southern states, which had not been responsible for incurring it, to Ł 8.3 billion (Sereni 1968:59). Neither taxation nor expenditures were distributed equitably among the regions. The North controlled 56% of Italy's wealth, and paid only 47% of the taxes. The Center controlled 17% of the wealth and paid 16% of the taxes. The South, which controlled only 27% of the wealth carried 37% of the total tax burden! (Sereni 1968:78) Per capita wealth was, further, lower in the South (Ł 1372) than in the North (Ł 2411), making this taxation especially onerous.

Distribution of state expenditures was even more inequitable than was the distribution of the tax burden. Expenditure per capita in the North was Ł 25.27, in the Center Ł 29, and in the South and the Islands, it was Ł 0.38 (Zitara 1971:74). While expenditures in the North supported land reclamation, railroad construction, and service of the national debt --and thus the accumulation of banking capital-- expenditure in the South covered only the costs of the "brigand war" (i.e. repression of the Southern popular movements) and the generous patronage which secured the continued support of the Southern middle classes (Zitara 1971:69-70). Interest on the debt, paid largely by the peasants and largely by the South, subsidized the creation of banking capital, which in turn financed Northern industrialization.

An outlet for Northern manufactured goods was assured by creation of a unified national market. All internal tariffs were eliminated immediately, and external tariffs were cut by 80% (Sereni 1968:8). The unification of the national market was further assured by the construction of railroads. In 1865 Italy as a whole had only 4420km of railroads; by 1900 it had 15,880 km (Sereni 1968:8). This policy had, not surprisingly, a devastating impact on Southern industry. The removal of internal tariffs opened Southern markets to Northern manufactured goods for the first time, while international competition nearly wiped out, within the space of a decade, the Sicilian textile and sulfur industries. The North, meanwhile, profited from the regime of free trade.

The development of industrial forces of production everywhere proceeded not through the fusion of new technologies with the existing knowledge of the master craftsmen, but rather by means of the wholesale destruction of the artisanate and its way of life. This process was particularly dramatic in Europe, where the absence of a centralized tributary state had permitted the development of a wide variety of relatively autonomous craft guilds which played an important role, alongside the peasantry, in the revolutions of the late feudal period, and which continued to be an important political force up through the middle of the nineteenth century. The guild system provided a complex system of rules designed both to guarantee the production of high-quality products, and to protect the most skilled craftsmen from the vicissitudes of the marketplace and the encroachments of feudal lords. Guild rules regulated training for and admission to the trade, production processes and working conditions, wages, prices, etc.

The guild, however, was not simply an agency for economic regulation. It was also a moral and cultic community which gave the artisans a collective identity. Each guild had its own rites of initiation, it own patron saint, its own feast day and it own special role in the feasts of the city as a whole. The lodge of the guild was a social and religious center——in fact the principal focus of the artisan's life. Those who had not been admitted to the status of master formed *compagnonages*, or journeymens' associations, which served much the same function as the guilds, except that they acted on behalf of the journeymen rather than the masters.

In effect the guild or journeymen's association was the urban equivalent of the peasant community. The skills associated with a particular trade were regarded as the common patrimony of the guild, allocated to individual members for their use, but subject to regulation intended to guarantee that each artisan's share of the patrimony was used for the common good.

The guilds and journeymens' associations played a leading role in the French revolution of 1789, their members swelling the ranks of the *sans culottes*, demanding that the new revolutionary state take up the tasks of economic regulation——particularly the regulation of the price of food——which were beyond the traditional powers and jurisdiction of the workers associations.

In 1793, after the defeat of the *Jacobins*, the guilds and compagnonages were repressed. All combinations of workers to regulate training, the organization of production, the work process, wages, prices, etc. were prohibited as "conspiracies to obstruct commerce" and workers were increasingly subject to market norms (Sewell 1980).

In the United States there were no real guilds. It appears, however, that masonic lodges may have played a similar role, at least for the upper strata of the artisanate. For the lower strata it was above all the tavern which became the public gathering place and the center for the development of an autonomous working class culture. Indeed, the word "pub" itself, undoubtedly comes from the word "public," indicating the vital role of this institution in providing public space for the working class.   P a u l Johnson (1978) has shown that it was precisely the industrial revolution and the emergence of an autonomous working class culture which made the pub a target of bourgeois "temperance movements."

> The drinking problem of the late 1820s stemmed directly from the new relationship between master and wage earner. Alcohol had been a builder of morale in household workshops, a subtle and pleasant bond between men. But in the 1820s proprietors turned their workshops into little factories, moved their families away from their places of business and devised standards of discipline, self-control and domesticity that banned liquor (Johnson 1978: 60).

The temperance movement was, on the one hand, an attempt to secure the level of factory discipline necessary for industrial production, but also an attempt to undermine the emergence of an autonomous working class culture. At more or less the same time, a strong Anti-masonic movement emerged, which sought to prevent the emergence within the United States of the kind of left-Jacobin current which contributed so strongly to the French revolution and which gave French democracy a strongly socialist coloration.

The development of capitalism in the cities, as in the countryside, was therefore, a profoundly destructive process, wiping out in a matter of a few decades the ancient crafts and manufacturing economies of Asia, destroying the guild structures of Europe, and everywhere transforming the skilled artisan into merely an unskilled cog in a more complex machine, performing "fine tuned" motions which the machines could not, as yet, be designed to carry out. Workers were deprived of both the technical skills which they had built up over the course of centuries, and of their traditional forms of social and political organization. The marketplace acted, in effect, as a solvent, which dissolved every type of relationship between humanity and the natural world, and between human beings, and prepared human beings for reorganization according to market imperatives, under the discipline of the factory system. And where the marketplace could not do the job by itself, the capitalist state intervened, to wipe out traditional forms of manufacturing, and to destroy traditional forms of peasant and artisan organization and prevent new forms from emerging.

B. The Marketplace

As peasants and artisans were forcibly separated from the means of production, and the wealth of the great agrarian civilizations plundered, the conditions for a fully developed market economy gradually emerged. On the one hand, the great mass of humanity was forced to sell its labor power in return for access to the means of production and subsistence. At the same time, a relatively small number of individuals had accumulated the wealth necessary to purchase their labor power and employ it mobilizing the new industrial forces of production. Even these entrepreneurs, however, were not really free to deploy their resources in accord with rational judgements regarding the best way to develop human social capacities, but were constrained to invest in those activities which promised the highest rate of return. The result was a fundamentally new economic structure, i.e. a new way of centralizing and allocating the resources of society. We need now to consider this structure in some detail.[1]

---

[1]. For a complete discussion of the operation of the marketplace as an economic structure, cf. Mandel 1968 and Amin 1978. Our analysis of the marketplace as an economic does not differ substantially from that developed by Marx and further elaborated by the dialectical materialist tradition. We do, however, place considerably more emphasis tendency of market relations to undermine the integrity of the ecosystem and the social fabric, and, in general, to hold back the development of human social capacities in a way which, we will see, makes the whole project of socialist revolution

In the centralized redistributional systems which characterized archaic, communitarian civilization, goods and services were centralized and allocated on the basis of their social use values, i.e. in such a manner as to provide for the community's reproduction, and where possible, its qualitative growth and development. The emergence of the warlord state distorted the way in which surplus was extracted and allocated, and often led to oppressive levels of taxation, but in principle at least the structure was rational and transparent. In market societies the central authorities which might make such decisions have either broken down, or have had their authority restricted by the emerging bourgeoisie. Goods and services are now exchanged more or less freely, with individuals and firms seeking the highest return possible for the assets they bring to the marketplace, and competing with each other to procure scarce goods and services at the lowest possible cost. Goods and services, including labor power, have been transformed into commodities. They are, that is, produced for exchange, in order to realize individual consumption interests, rather than for use by the community in its own self-development.

In such a system, it is possible to define two distinct but related forms of equivalence between commodities. The first of these is value. Value is, essentially, a quantitative expression of the complexity or level of organization of the commodity in question, and is equal to the average socially necessary labor time the commodity contains, i.e the average time necessary to produce it under existing technological conditions.[2] Value provides us with a way, however limited, of measuring the substantive contribution of a particular activity to the complexity of the cosmos in general.

Commodities are not, however, exchanged in proportion to their value, but rather in proportion to their price (Amin 1978). Price is determined by the complex interaction of supply and demand, as buyers and sellers bargain among themselves in an effort to secure the highest possible return on their assets. Price is, in this sense, a quantitative expression of the marginal relative individual use value of a commodity, with the one

---

and socialist construction extraordinarily difficult.

[2]. We have already noted above, in Chapter One, the similarity between the Marxist concept of value, and certain contemporary definitions of complex organization, especially the definition advanced by IBM scientist Charles Bennett (1988), which centers on "logical depth," "the work required to derive a" message "from a hypothetical cause involving no unnecessary ad hoc assumptions," or "the time required to compute this message from" its "minimal description."

qualification that the preferences of consumers are weighted in accord with their relative buying power.

In a pure market system, in which an average rate of profit has emerged, it is possible in principle to define an equivalence between value V and price p:

$$V = e+t+l+s,$$

$$r = s/(e+t+l), \text{ and}$$

$$p = (e+t+l)(1+r),$$

where e is capital invested in ecosystem inputs (raw materials), t is the capital invested in technological inputs (tools), l is variable capital invested in labor (wages), s is the surplus, r is the average rate of profit, and p is the price of production.

Capitalists are able to extract surplus value from wage laborers because labor, unlike other commodities, is capable of producing more value than it consumes in its own reproduction. As the productive forces develop, the worker spends less and less of his time reimbursing the capitalist for the means of subsistence and production which the capitalist has advanced, and more and more time producing new value, which was not there before.

Capitalism, furthermore, would seem, unlike the tributary mode of production, to have a built in mechanism which all but guarantees reinvestment of most if not all of the surplus extracted. The capitalist could, of course, consume the surplus entirely in capricious luxury consumption. And a significant portion of surplus has, historically, been consumed in this way. The marketplace, however, tends to impose a very specific kind of discipline on the capitalist. If he fails to reinvest his surplus in capitalist production, then he will eventually be out competed by other capitalists who do. To this extent, the marketplace appears to have a rationalizing influence, requiring that a significant part of capital be reinvested in expanding production.

A difficulty, however, emerges when we begin to examine the precise criterion by which investment decisions are made. The marketplace has no access to information regarding the impact of various investment decisions on the qualitative complexity of the ecosystem, the development of human social capacities, or on the level of organization of the cosmos in general. Even if a capitalist wanted to invest in accord with these criteria, the marketplace would provide him with little guidance. And goods and services produced in a market economy must be sold. This means there must be people who want to purchase them. This places a strict limit on the complexity of the goods and services which can be produced. It is impossible to produce, on a capitalistic basis, goods and services an interest in which presupposes a level of development higher than that of the consuming public, even if it can be shown rationally that the production of such goods and services promotes the overall development of human social capacities and the larger process of cosmohistorical evolution. Indeed, when a wealthy person decides to invest a portion of the surplus which he has accumulated in activities chosen for substantively rational, rather than market driven criteria, this is not regarded as investment at all, but rather as "philanthropy."

The problem is further exacerbated by the emergence of a market in capital. As industry becomes increasingly capital intensive, it is increasingly difficult for individual capitalists to finance new ventures out of their own resources. They must, in effect, borrow capital from others in order to create a pool of resources large enough to sustain the enterprise. This usually means selling stocks or bonds, or else procuring loans from banks——and thus paying dividends or interest. This further increases the pressure to generate high, short term profits. Indeed, other things being equal, capital will tend to flow towards those ventures which promise the highest rate of return, leaving out in the cold not only the profligate consumer and the would be philanthropist, but also the entrepreneur whose venture appears likely to yield a lower than average rate of return, or promises significant gains only in the very long term. In general, this means that capital will flow towards activities for which there is a high level of effective demand——i.e. in producing goods, or performing services, which correspond to the spontaneous consumer interests of those who have the money to pay for them.

This criterion is profoundly irrational. The spontaneous interest of the consumer says nothing about the objective value of an activity. The most valuable activities are those which contribute most to the emergence of complex organization in the universe. Such activities are, by definition, likely to be so complex as to be incomprehensible to most human beings who could not, therefore, have any conceivable interest in them. Market mechanisms allocate resources away from such complex activities, towards simpler activities which consumers can comprehend, and thus desire, and thereby hold back the development of human social capacities and the cosmohistorical evolutionary process.

But matters get still worse. In a market system, the rate of return is always higher, other things being equal, in low wage, low technology activities than in high wage, high technology activities. Assuming

$$r = s/(e+t+l)$$

as e, t, and/or l increase, r will decline. In other words, the more we have to invest in scarce ecosystem inputs, high technology machines, and skilled labor, the lower the rate of return on our investment. This means that in a pure market system, there will be a tendency for capital to be redeployed to low wage, low technology activities. This is the basis of the whole phenomenon of imperialism, which we will document and analyze in some detail later in this chapter. As a result, the marketplace tends to hold back the development of human social capacities, by favoring low technology, low wage production.

The destructive impact of the marketplace is not limited to the process of production. As capital is increasingly redeployed to low wage activities, the price of labor power is driven below its value so that the productive capacities of the work force are gradually eroded. Concretely, this means that workers are forced to work so long and so hard to feed themselves and their families, that they lack sufficient time to maintain the social fabric. Gradually families begin to disintegrate, the socialization process breaks down, and society begins to come apart at the seams.

As the family disintegrates and the social fabric unravels, the society's capacity for self-organization is gradually undermined. Building power means building relationships. The most powerful individuals are those with a broad range of interests, who can, therefore, tap into the interests and talents of other people, and move them to action. Where human interests have been so narrowed, however, that people think only of maximizing consumption, it becomes increasingly difficult to organize people for other purposes. Nonprofit civic, educational, scientific, charitable and religious institutions decline, or else are transformed into nonprofit "businesses" which are forced to operate according to market imperatives. Their interests narrowed to realization of the purpose for which the existing social system is organized, the people lose any interest in the reorganization of society. The result is social stagnation.

At the theoretical level, the impact of the marketplace is reflected in an entire complex of ideological deformations. The marketplace transforms the complex interdependence characteristic of a developing industrial society into a system of relationships between things (commodities), each possessing their own properties (use values), related to each other in a purely external fashion (price). This phenomenon, known as reification, has been studied in some detail by historical materialist sociology (Lukacs 1971). Mediation of human organizing activity through the marketplace conceals the organizing nature of this activity and makes it appear simply as a relationship of exchange. This, in turn undermines recognition of the mutually determining character of the organizing process, in which both organizer and organized are transformed into something other than they previous were. Instead, each appears as a self-subsisting individual, affecting and affected by others, without however being fundamentally transformed in the process.

Now, organization in a system of only-externally related particulars by definition comes from the outside. On the one hand, this means that the relations of holism, self-organization, and teleology, which we have seen characterize matter at every level of organization, are obscured. At the same time, people come to believe that they are at the mercy of mysterious organizing powers which are beyond their comprehension or control. The Chinese, African, or Aztec peasant knew why he was poor: he worked one half of his time to support his warlord. The medieval craftsmen saw some relationship between the quality of his work and his progress from apprentice, to journeyman, to master. The worker in a capitalist society, however, is paid the full value of his labor power, but grows poor as the capitalist grows rich. And the marketplace rewards even capitalists not in accord with their skill in organizing productive activity which develops human capacities or increases the qualititative complexity of the ecosystem, but rather in accord with how well they serve the spontaneous interests of consumers with the money to purchase their products. As a result, laws of development of the capitalist system are obscured and organization appears to be something external to the activities of real social individuals (Fromm 1941).

Mediation of social relations by the marketplace, furthermore, tends to produce a kind of moral alienation. When all activity is simply a means to realizing individual consumer interests, a contradiction opens up between the interest of the individual and that of the community, humanity generally, and the cosmos as a whole. My own interest appears to me simply as what I *want*, one consumer option among may others, and society generally simply as a means of realizing this interest. Rather than understanding their particular interests as partial participation in the common good, humans begin to regard themselves as fundamentally selfish and individualistic. Thus the feeling of "radical depravity" characteristic of individuals in bourgeois society. Any attempt to constitute a properly public good, therefore, tends to take on a negative form: action on behalf of the common good is action for something other than what I want and therefore contrary to my own interest: it is self-sacrifice. This dynamic almost inevitably blocks development of an actual interest in the development of human social capacities, or in the self-organizing activity of the cosmos.

These two dynamics in turn flow together to generate the profound spiritual alienation characteristic of bourgeois societies. Full participation in the self-organizing activity of the cosmos seems increasingly to be possible only by negating one's own immediate interests: i.e. through submission to the inscrutable imperatives of the marketplace. This is the basis in experience for the emergence of evangelical spirituality, and ultimately for the authoritarian movements which have developed in advanced market societies. These evangelical and authoritarian movements at once celebrated the leading role of humanity in the self-organizing activity of the cosmos, while at the same time reflecting a gradually weakening grasp of that self-organizing activity itself and a devaluing the human capacity for virtue. Like the warlord state, capitalism leads people to seek the source of creativity in self-sacrifice, with the one exception that rather than being literally burned on an altar or devoured by the priests, human lives and the products of human labor are sacrificed on the figurative altar of the marketplace, and consumed by the great lords of capital.

III. Capitalist Civilization

The development of capitalism was therefore, from the very beginning, a contradictory process. On the one hand, the industrial, democratic, and scientific revolutions, even with their limitations, represented a qualitative leap forward in human organizing capacities. At the same time, the creation of a market economy brought in its wake incredible destruction. Genocide, slavery, forced labor, the dispossession of the peasantry, and the conversion of grain lands to cash crops very nearly wiped out several thousands of years of communitarian civilization. Discriminatory tariffs and repression of traditional guild structures, similarly, destroyed the productive capacities of the millions of craftsmen in the great preindustrial cities of Asia and the Mediterranean Basin. And the marketplace itself, as we have seen, is an obstacle to the centralization of the resources necessary for investment in infrastructure, education, research, and development.

In spite of this record of many centuries of destruction and stagnation, a broad consensus has emerged in the present period that the marketplace provides the most favorable context for social progress. Defenders of the marketplace can, of course, point to the not inconsiderable achievements of the advanced capitalist countries——and to what they perceive as the failure of socialism. In this section we will analyze the actual development of capitalist civilization. We will focus in particular on the task of isolating the real basis of social progress in the advanced capitalist countries. We will argue that in point of fact, where progress has occurred, it has been due largely to the operation of nonmarket social institutions: either local religious congregations that catalyzed the development of a strong sense of commitment to human technical and social progress, or else the state apparatus.

We will begin by examining the development of capitalism, and show that in no case did a vibrant, industrial capitalist economy emerge exclusively or even primarily on the basis of the spontaneous operation of market forces. We will then examine the way the internal contradictions of capitalism began to undermine social progress by the end of the nineteenth century, and the diverse, and ultimately inadequate strategies developed by bourgeois reformers to once again unleash human productive forces within the context of a still fundamentally capitalist economy.

A. Capitalist Development

There are, fundamentally, two distinct roads to capitalist development: the liberal road characteristic of the Anglo-American experience, and the statist road, which has, in fact been dominant throughout most of the planet. Among societies which have followed the statist road, it is, in turn, possible to identify two variants. Where agrarian, tributary bureaucracies were not able to rationalize themselves sufficiently to become real catalysts for development, democratic revolutions took place which carried out the necessary rationalization. The French experience is typical of this "revolutionary-democratic" variant. Where tributary bureaucracies were able to rationalize themselves, as in Germany and Japan, capitalist development followed what we will call an "authoritarian" road.

In this section we will examine each of these paths of development in some detail. We will focus most of our attention on the Anglo-American, and especially the U.S. experience, because it is here that the strongest case can be made for the spontaneous, market driven, emergence of industrial capitalism——a claim we want to refute. We will analyze the French and Prusso-Japanese experiences only in sufficient detail to demonstrate their statist character and round out our analysis of the process of capitalist development.

1. The Anglo-American or Liberal Road

The development of capitalism in England and North America

proceeded along what, from a global perspective, must be regarded as very unusual lines. Throughout most parts of Europe the peasant revolts of the late tributary period were put down, and a long period of feudal reaction ensued. Much the same was true throughout most of Asia, Africa, and the Americas, where European conquerors put down local revolts and strengthened indigenous tributary structures as a mechanism of surplus extraction and social control (Anderson 1974). In England, on the other hand, the peasant revolts, particularly the revolt of 1381, led to a partial victory, and by the end of the fourteenth century, the peasants had succeeded in eliminating most strictly feudal obligations, and in significantly increasing their total share of the social product. In response to this situation landowners began a long struggle to rationalize agricultural production. Many converted from grain production, which was relatively labor intensive, to wool production, which required fewer hands. Others began to implement new and more sophisticated techniques for cultivating grain. Implementation of these techniques required a smaller and more disciplined work force, with the result that in grain growing areas as well peasants were run off their land. Over a period which lasted nearly four hundred years the greater part of the English peasantry was gradually driven from the land, and transformed into a massive agricultural and industrial proletariat.

When the English conquered North America, they continued this process. The indigenous peoples of the continent actively resisted the imposition of forced labor, and so they were driven off their land, pressed westward, and eventually exterminated, opening up the continent to capitalist development. In the Southern part of the continent they imported African slaves. The northern part of the continent, on the other hand, served as an outlet for English, and later European peasants, displaced by the penetration of capitalist relations into the countryside.

The result was the famous "triangle trade," a powerful engine for the primitive accumulation of capital. English traders captured or purchased slaves from Africa, which they sold to the sugar, tobacco, and later the cotton planters of the Caribbean basin. These planters in turn sold them agricultural raw materials, which the traders then carried to the industrial centers of England, or later the American northeast. The finished goods——rum, tobacco products, textiles——as well as guns and other manufactured goods were then resold to African slave traders or Caribbean planters. Grain farmers in New England, and later in the Northwest Territories, kept workers and slaves alike supplied with cheap food, and provided a secondary but growing market for manufactured goods. Capital was accumulated primarily by the industrialists, plantation agriculture being resistant to rationalization, and the planter culture oriented towards high levels of luxury consumption.

It is in the context of this distinctively Anglo-American experience that liberal theorists are most likely to look for evidence to support their claim that capitalism was a spontaneous development. And it is true that the Anglo-American road created a powerful constituency for liberal economic policies. It was the "progressive gentry" of England during the later seventeenth and eighteenth century who provided the principal constituency for liberal theories, such as that of Locke, who argued that the state existed only to protect the "natural rights" of life, liberty, and property, and that a right to property emerged spontaneously when human beings mixed their labor (or that of agricultural laborers or African slaves under their direction) with the land. And it was the planter elites of Virginia, the Carolinas, and Georgia who provided the single most consistent constituency for limited government during the first century of U.S. independence, arguing for low levels of taxation and state expenditure, low tariffs, and a minimum of state regulation of the economy.

Two points are, however, in order. First, even though a capitalist agriculture may indeed have developed more or less spontaneously on the soil cleared for it by the enclosure movement and by the conquest of North America and the extermination of its indigenous inhabitants, the state in fact played a very active role in creating the conditions for this "spontaneous" development. Enclosure of common lands required approval by Parliament, and often sustained support by the repressive state apparatus. And the conquest and settlement of North America was, of course carried out primarily by joint stock companies granted special royal privileges and monopolies.

More important, however, this "spontaneous" capitalism was not industrial, but agricultural. While it did reflect a certain rationalization of agricultural techniques, technical innovation was in fact rather limited, especially in the plantation areas, for the simple reason that slave labor could not be disciplined to carry out complex methods of cultivation. When we begin to examine regions characterized by high rates of technical innovation in agriculture (as opposed to mere superexploitation), and especially by the emergence of manufactures and eventually of industry, we discover that another, very potent, and clearly nonmarket social force was at work: Reformed Protestantism. It is this force which we need to examine in detail if we hope to grasp the reasons for the incredible dynamism of Anglo-American capitalism. In what follows we will focus especially on the North American experience, as it was in North America that the Reformed dynamic found its purest and most intense expression. In order to industrialize North America, the emerging bourgeoisie had to carry out two distinct but interrelated tasks. First, they had to centralize the resources necessary for capitalist development and insure that these resources would be allocated to productive investment. This meant taxing to raise funds for infrastructure development (roads, canals, railroads). It meant building and staffing schools and universities in order to train a skilled work force. It meant subsidizing research and development. And, finally, it meant centralizing the capital necessary to build factories——not only factories producing final consumer goods, but also factories producing capital goods such as steel and machine tools.

Second, they had to create a home market for manufactured goods. Concretely this meant protecting young industries from foreign competition. But it also meant putting an end to coercive forms of labor organization, such as slavery, which withdraw large sectors of the population from the marketplace and thus restricted the demand for wage foods, clothing, simple household items, and farm implements, and thus threatened to leave North American industry without a sufficient outlet for its products.

In order to carry out these tasks, it was necessary, furthermore, to build a broad constituency not only among the people, but also within the bourgeoisie itself. It was necessary to draw the bourgeoisie away from the numerous opportunities for profiteering available to them in the slave trade or export agriculture, and draw them into the disciplined universe of industrial capitalist production. And it was necessary to break the traditional communitarianism of newly proletarianized workers, and force them to submit to factory discipline, without, however, losing their political support, on which the industrial bourgeoisie would need to call in its struggle with planter and rentier elements.

The Reformed tradition was ideally suited to carry out these tasks. While Puritanism first emerged among the progressive gentry and yeomen of the English countrysides, it soon found support among the merchants and nascent capitalists of the cities (Hill: 1972: 154). And it mobilized the energies of these sectors in a uniquely progressive manner.

Reformed doctrine is characterized by an unusually strong sense both of divine sovereignty, and of humanity's place in God's cosmic plan. God is the absolute ruler of the cosmos, which he governs in accord with a complex, inscrutable plan. Human beings are, essentially, God's instruments in carrying out this plan. Thus the Reformed conviction that all useful work has authentic religious significance. Because God is an infinite, or general system, however, and we are merely finite, we can neither grasp, nor really love God in his essence. Rather we love ourselves, and God and neighbor only in so far as they help us to realize our own finite interests. This is the significance of the doctrine of total depravity. In order to realize his plan for the cosmos, however, God requires that some, but not all humans, know and love him in his essence, for his own sake. He thus elects a limited number of humans to be "saved," the election being made entirely apart from any consideration of merit, in the moral sense, but rather entirely in accord with the imperatives of God's own plan. One almost gets the sense of God requiring so many peasants, so many bakers and butchers, so many merchants and bankers, a fixed number of magistrates and ministers, etc. These are washed clean from sin by the blood of Jesus, irresistibly sanctified (transformed morally and spiritually), and thus predestined to become conscious, committed participants, in God's work. The rest of humanity, is, in virtue of its very nature, doomed to serve God without ever willing to do so, or even really knowing that they are.

It should be apparent from the very beginning that Reformed theology is a reflex of the process of capitalist development: i.e. industrialization under the conditions of a market economy. On the one hand, Reformed theology recognizes human labor as a real participation in the creative life of God. On the other hand, Reformed theology also reflects and in fact embodies the alienating dynamic generated by the marketplace. In a market system, the enormous cooperation of millions of interdependent individuals takes on the form of competition——indeed of a war of all against all——concealing the essentially social character of human nature and creating the appearance of radical depravity. Participants in the marketplace, furthermore, feel themselves to be at the mercy of forces beyond their control, which reward their labor not in proportion to merit, but rather in accord with the inscrutable imperatives of supply and demand (unconditional election). Indeed, the whole purpose served by human productive activity is obscured, undermining the rational teleology of Catholic ethics and setting in its place a demand for blind submission. Resolution of this alienation takes the form of radical self-negation. It is not my labor itself, but rather its negation and sublation by the marketplace which serves humanity. Similarly, in Reformed doctrine, the individual achieves salvation not by coming rationally to understand his or her own will as a real, albeit limited and finite participation in the will of God, but rather by denying that will and "receiving" the will of God in its place.

It is possible to identify three distinct stages in the development of the Reformed program:

a) an agrarian stage, the stage of Puritanism and the Holy

Commonwealths,

b) a mercantile stage, the stage of nascent liberalism and the First Great Awakening, and

c) an industrial stage, the stage of the Second Great Awakening, of the Ante-Bellum Reform Movements, and of the rise of the Federalist, Whig and eventually of the Republican Parties.

We will consider the first two stages only briefly, since they affect our argument only indirectly, and will focus instead on the third.

In its initial, agrarian form, the Reformed program centered on the establishment of a Holy Commonwealth. Here divine sovereignty is represented by the authority structure of the town community——the pastor and the chief magistrate——which acts outside the marketplace in order to insure that the community's resources are used in a way which serves the common good. Within the Holy Commonwealths of Massachusetts and Connecticut land was allocated on the basis of family size and "usefulness to the community," and political participation was restricted to regenerate church members——i.e. those who could give a convincing narrative of God's work in their lives. Both the "regenerate" and the "unregenerate" population remained under the careful supervision of the ecclesiastical authorities and sins of "covetousness," such as profiteering or luxury consumption were punished as rigorously as those of sexual misconduct.

The systematic and detailed regulation of social life by the religious and civil authorities of the Holy Commonwealths, as well as the intense introspection required by Reformed theology, mobilized an unprecedented portion of the Commonwealths' resources for development. The result was a highly skilled, highly motivated work force deployed in an emerging commercial agriculture and a rapidly expanding handicrafts and manufacturing economy centered initially at least on shipbuilding and the provision of naval stores, as well as rapid accumulation of capital.

Inserted as they were in a developing world market, however, the Holy Commonwealths were subjected to intense centrifugal forces. Accumulation of capital meant emergence of a bourgeoisie, which resented its subordination to the ministers and magistrates who governed the Commonwealths, and which soon began to "liberate" itself from church discipline. First the Restoration and later the Glorious Revolution of 1688 undermined the constitutional status of the standing order, so that by the 1730s most of the structures which had made the Commonwealths such a dynamic social force had disintegrated, leaving in place a string of small commercial colonies on the periphery of the world market.

Gradually, however, the Reformed tradition adapted to the penetration of market relations into the towns and the countryside. This second, mercantile phase in the development of the Reformed program is represented by the twin forces of nascent liberalism and the evangelicalism of the New Divinity movement, as well as a centrist trend known as "benevolism" , influential in the foundation of Georgia, before its transformation into just another plantation colony. Here God's plan for the cosmos is conceived in a more immanent fashion, as something present in the forces of nature——in gravity and chemical interactions——and in the expanding web of interdependence created by the marketplace itself.

Where evangelicals, benevolists, and liberals differed was around the character of and conditions for authentic interdependence. Liberals argued that while human beings are fundamentally egoistic, their egoism motivates productive activity which ultimately contributes to the common good. Evangelicals, on the other hand, regarded as virtuous only activity which was actually motivated by what they called "disinterested benevolence"——a loving submission to the will of God and a desire to serve the common good even if this meant acting contrary to one's own interests. Where liberals celebrated the capacity of the marketplace to transform the selfish actions of egoistic individuals into a force for social progress, evangelicals demanded that individuals actually internalize the inscrutable demands of the marketplace——that they actually want what society requires of them, even if this did violence to them and everything which mattered to them. Benevolists took a middle road, arguing that the "invisible hand" of the marketplace needed to be supplemented by "moral sentiments" of benevolence towards one's fellow human beings.

Much as they differed around theological questions, however, liberals, benevolists, and evangelicals were able to unite in the struggle to extend the operation of the marketplace in North American society. They joined in the struggle against the semifeudal monarchy in France, against resurgent feudal elements in England, and eventually against England herself when those feudal elements gained the upper hand. Monarchic power, Catholic or Anglican, royal or episcopal, seemed to evangelical and liberal alike to give dangerous scope to natural human egoism, by permitting an individual or a small group of individuals to work their will on human society, rather than subjecting that will to the discipline of the marketplace, which forced the individual to serve the common good.

Soon after independence, however, it became obvious that the free operation of market forces was not going to be sufficient to transform the United States into an industrial power. On the contrary, it was only the southern landed elite which consistently supported a liberal economic policy, centered on free trade and limited government. The emerging northern industrial bourgeoisie, on the other hand, and to lesser extent the mercantile and farming interests allied with them, recognized the need for stronger state intervention.

It was, once again, Reformed Protestantism, which made it possible for the industrial bourgeoisie to gradually built up, over the course of the first half of the nineteenth century, the kind of constituency it required in order to carry out this intervention.

This did, to be sure, require further modification in Reformed doctrine. On the one hand, as technology advanced human beings found it increasingly difficult to regard themselves as purely and simply at the mercy of forces beyond their control. Reformed doctrines of double predestination gave way to a more complex understanding of divine sovereignty which stressed the importance of human initiative and cooperation. On the other hand, it became increasing clear that in the course of their productive activity human beings were doing more than merely serving their own consumer interests, or those of others. They were participating in a divine, cosmohistorical organizing process. And yet it was also apparent that there were sectors of society——such as the Southern planters——whose unlimited pursuit of a narrow self-interest was in fact holding back the development of North American society. The opposition between advocates of disinterested benevolence on the one hand, and the "invisible hand" on the other, gave way to a kind of "cosmic utilitarianism" which encouraged on active intervention by both the state and the church to promote activity which was useful to society. It was human organizing activity——work——and not human interdependence, or community, which represented the highest level of participation in the life of God.

Concretely this new political theology found expression in support for an interventionist state which taxed vigorously to support investment in infrastructure, education, research, and development, support for high tariffs to protect North American industry, and opposition to the extension, and eventually the very existence of, slavery.

Cosmic utilitarianism meant a rejection of the marketplace as the exclusive or even the principal means of organizing human talents and creativity. Particularly interesting in this regard is the work of economist Henry Carey. For Carey the laws of political economy revealed the wisdom and benevolence of the creator. What labor did was to increase the level of organization of the material world. This, in turn, required progressively more complex forms of social organization, or "association."

The development of human society toward higher levels of association at once reflected and realized the divine love which ordered the cosmos as a whole.

Against the individualism of the Manchester school, Carey argued that human societies were governed by a law of association, which played the same role in organizing human affairs that the law of gravitation played in the organization matter generally.

> The more [man's] power of association, the greater is the tendency toward development of his various faculties; the greater becomes his control of the forces of nature, and the more perfect his own power for self-direction; mental force thus more and more obtaining control over that which is material, the labors of the present and the accumulations of the past (Howe 1979: 114-15).

Because of this

> economic development was a means to human redemption. Addressing himself to women in particular, he argued that economic diversification held out the promise to them of liberation from the role of farmer's wife
> ...

By bringing people into ever higher degrees of association, economic development helped to realize the creator's plan for the cosmos.

Carey was firmly committed to capitalism, but not to the doctrine of *laissez faire* promoted by the Manchester school.

> The economic system that Carey felt offered the best prospect for economic development was, of course capitalism, but he was quite willing to call on government to help the process along. Technological innovation and population increase, fueled by unrestricted immigration, Carey acknowledged important. Government policy would add to these dynamic influences what is today called social overhead——a transportation network and an educational system. To keep the economy expanding the burden of taxation should not become oppressive ... Carey laid the most stress on a plentiful money supply and a protective tariff the prevent this money from being drained off in payment for imports. Taken together these policies would secure investment capital, increase productivity, and raise wages (Howe 1979: 111-112).

The marketplace could be catalyst to development, but it could also become an obstacle. This was particularly true in the arena of international trade.

Within a community, commerce was simply the economic dimension of association, the mutually beneficial interaction between human beings. But trade between two different communities was typically exploitative ... Trade develops, according to Carey, out of a distortion in the natural process of association, like a cancer on an organism. Specialization of function is carried too far, and raw materials are brought to a metropolitan center from distant points. Parasitic middlemen appear. Armies are raised to conquer new lands to supply the metropolis, and armies require taxation. The population of the colonies is reduced to poverty, and often to actual serfdom or slavery ....

Colonialism, furthermore, destroys the social fabric and undermines the traditions which are the precondition for the economic development of any people. In this sense Carey was a

... a now forgotten spokesman for peoples who felt their communal identity threatened by economic colonialism...

He

found that his message was welcomed wherever nations were industrializing and British economic hegemony resented (Howe 1979: 118-19).

Trade also undermined the welfare of the population of the trading metropolis as well.

The same Carey who praised the small to medium scale capitalism of the town deplored the large scale capitalism of the metropolis ... Within these cities ... a submerged "proletariat" ... appeared, just as truly exploited as the distant laborers who produced the staples. Carey's analysis of trade bore a remarkable resemblance to the analysis of capitalism by his great contemporary Marx. The Trading class lived by "appropriation of wealth created by others. The impoverished urban populations of trading empires would eventually cease to be able to consume the finished products of their own labor ... Wars would ensue as the trading nations competed for markets..." (Howe 1979: 118-19).

This kind of thinking had definite political implications. First of all, the tendency of capital to redeploy itself towards low wage, low technology activities such as plantation agriculture, and the whole dynamic of colonial expansion which grew out of this tendency had to be restricted. Slavery had to be abolished. Second, the state had to act as an agent for centralizing and investing resources in the development of the kind of physical and social infrastructure required by an advanced industrial society. Not only roads, canals, bridges, and railroads, telecommunications lines, and the like, but also schools, libraries, museums, and universities were necessary if human beings were to fully realize their potential. Finally trade had to be regulated——encouraged when it opened up new opportunities for human development, and restricted when it undermined those opportunities.

Formation of a political alliance capable of implementing these sorts of policy was a protracted process. Neither of the two main political parties which emerged after independence were really parties of the industrial bourgeoisie. The Federalist party had its principal base among the merchants of the great port cities, while the Democratic Party, while it incorporated many radical farmers and artisans, was first and foremost the party of the Southern planter elite. Alexander Hamilton was able to win Federalist support for a sophisticated strategy of state led industrialization, but his plan offered too little to the workers and farmers to make it politically credible. Aaron Burr, similarly, made a bid to transform the Democrats into a revolutionary democratic party, but was beaten back by the slave owners, who were able to keep the masses in line with promises of cheap land stolen from the Indians.

By the mid-1820s, however, a new party, the Whigs, began to link together the emerging industrial bourgeoisie, the evangelical intelligentsia, and the more commercially oriented farmers. The Whig program included a commitment to a national bank, internal improvements, and high tariffs. The Whig party had close ties to the evangelical united front which was leading the Second Great Awakening. While less overtly hostile to the working class than the Federalist program, the Whigs were still unable to unite around the need to halt the expansion of, and eventually abolish slavery, which had become an increasingly serious brake on development. And the Whigs still offered too little to the working class and the farmers to make themselves a strong majority party. Specifically, there was nothing in their program which pointed towards higher wage levels, and no promise of access to cheap or free land in the West.

It was not until the 1850s, with the formation of the Republican party, that the industrial bourgeoisie was able to advance a program which united all of the progressive forces in U.S. society. To industry the Republicans promised a reform of the national financial system, a commitment to federal investment in infrastructure and high tariffs. But they joined to this program, which they had taken from the Whigs, a firm position against the further expansion of slavery, a promise of free land for settlers anxious to move West, and a commitment the development of human capital. The

Morrill Land Grant College Act of 1862, for example, set aside some 30,000 acres of federal land for each state for each senator or congressman that represented it in Congress. This land was to support the establishment of agricultural colleges, which eventually became the cornerstone of the state university system.

It was this alliance which was finally able to defeat the southern planter elite and to create the conditions necessary for the emergence of a strong, industrial capitalist system in North America. The achievements of the Republican Party during the 1860s are impressive indeed, and rank alongside those of other revolutionary parties of the period. The Republicans

a) abolished slavery, and thus slowed the flow of capital into unproductive low technology activities,

b) raised taxes and state expenditures in order to create the infrastructure necessary for rapid industrial development, including federally subsidized railroads, and the land grant college system which became the cornerstone of the country's land grant state university system, and

c) raised tariffs in order to protect U.S. industry from foreign (primarily British) competition.

It is the victories of 1860-1865, rather than those of 1776-1800, which made possible the emergence of and advanced industrial economy in the U.S.

It should be clear by now that even during the early stages of development, the marketplace was not a catalyst but rather an obstacle to the technological, economic, political, social, and cultural development of North America. It was the marketplace which permitted the emergence of plantation slavery in the South, and which allowed millions of dollars to be siphoned off into unproductive luxury consumption. It was above all nonmarket institutions——particularly Puritan and other Reformed congregations, and eventually the state apparatus——which provided the dynamism necessary to carry out the industrialization of the United States. We will see that in other parts of the world that nonmarket institutions played an even more important role.

2. The Statist Roads

As we noted above, in most parts of the European continent, the peasant uprisings of the late feudal period ended in defeat and were followed by a long period of feudal reaction. When commercial agriculture finally began to develop it took a very different form. In France for example, commercial agriculture was centered in the production of wine for export. According to Barrington Moore

> Viticulture, particularly in the days before artificial fertilizers was ... a labor intensive variety of agriculture, requiring large amounts of fairly skilled peasant labor, and relatively small amounts of capital either in the form of land or equipment ...

Rather than driving the peasants off the land,

> the French aristocrat ... used feudal levers to extract more produce. Then the nobleman sold the produce on the market (Moore 1966: 48).

Similarly, in the grain growing areas around Toulouse,

> the nobles used the prevailing social and political framework to squeeze more grain out the peasants and sell it. Unless the nobles had been able to do this and overcome the peasants' reluctance to part with his grain, the townsfolk would have had nothing to eat (1966: 53-54).

Much the same pattern prevailed in Germany and Japan. The Japanese used to say of their peasants that they "are like sesame seeds; the more you press the more comes out (Moore 1966: 241)." The relations between lord and peasant remained essentially the same. What changed was the growing insertion of the lord in market relations, with the result that his appetite for grain increased enormously. In order to take advantage of the growing market for grain created by the industrialization of Northwestern Europe "the Prussian nobility expanded its holdings at the expense of the peasantry ... and reduced them to serfdom (Moore 1966:435)."

In this context, the industrial bourgeoisie remains very weak, and dependent on state subsidies and state monopolies.

> Under the seventeenth-century monarchy, the French bourgeoisie was not the spearhead of modernization taking the countryside along with it towards the still invisible world of industrial capitalism that their English counterpart had already become. Instead it was heavily dependent on

royal favor, subject to royal regulation, and oriented towards the
production of arms and luxuries for a restricted clientele (1966: 56-57).

The situation was much the same in late Tokugawa Japan (Moore 1966:
56-57). In Germany,

> the state ... served as an engine of primary capitalist accumulation,
> gathering resources and directing them toward the building of an
> industrial plant .. taming ... the labor force ... Armaments served as an
> important stimulus for industry. So did protectionist tariff policies
> (Moore 1966: 440).

Indeed, it would be safe to say that the Anglo-American case is in every
way an exception.

a) The French, or Revolutionary Democratic Road

What separates the French, or revolutionary democratic road to
capitalist development, from the authoritarian road pursued by Germany
and Japan, is the relative inability of the French monarchic state to
rationalize itself, and transform itself into an effective force for
industrialization. Moore describes the problem in these terms:

> In preindustrial societies it was practically impossible to generate and
> extract enough of an economic surplus to pay the members of the
> bureaucracy a salary that would ensure their real dependence on the
> crown ... The French monarchy tried to solve this problem by selling
> positions in the bureaucracy (Moore 1966: 57-58).

The result of this practice was, of course, to undermine the formation of
a rational bureaucratic apparatus which could lead the process of
industrialization. Indeed, while at first this measure may have won the
monarchy allies within the bourgeoisie in its struggle against feudalism,
eventually it tended to feudalize the bourgeoisie itself (Moore 1966: 60).

At the same time, the increasing weight of feudal burdens radicalized
the peasantry.

> Though limited capitalist penetration failed during the eighteenth century
> to revolutionize agriculture or eliminate the peasantry, it came in such
> a way as to increase very sharply peasant hostility to the *ancien regime*.
> Peasants resented the increase in feudal dues and revival of old ones by
> clever lawyers (Moore 1966: 65).

There were, however, sharp divisions within the peasantry. The poorer peasants seem to have been interested primarily in gaining access to a plot of land, and to preserving those customs of the village community, such as gleaning rights, and the right to let their animals graze on harvested fields (Moore 1966: 71-72). Since they produced little or no surplus above what they needed for their own consumption, the poorer peasants had little interest in high grain prices. In short, the poorer peasants vacillated between a struggle for partition and a struggle to restore the traditional rights of the village community, free from the burdens imposed by the nobility, and from the influence of the rich peasants.

The richer peasants, on the other hand, looked forward primarily to the abolition of feudal burdens and the creation of a free market in grain, which would allow them to make the most of the surpluses which they produced. For the most part they rejected attempts to divide up land among the poorer peasants, whether this land was part of the commons, or land seized from emigres after the Revolution (Moore 1966: 71-73).

The urban masses were similarly divided. The great mass of journeymen, organized in the *compagnonages* were deeply committed to an anticapitalist and collectivist vision. They looked to the state, on the one hand, to provide employment and serve as a stimulus to economic growth, and, on the other hand, to regulate wages and prices just as their own organizations had done for centuries on a smaller scale. A minority of wealthier masters, however, together with the middle layers of shopkeepers, lawyers, etc., were seeking primarily to be free of feudal burdens and to seek their fortune in the marketplace.

The resulting upheaval can hardly be called a "bourgeois revolution." On the contrary, it reflected a complex and contradictory complex of bourgeois and socialist elements, with the moderate leaders of the Gironde seeking to abolish feudal privilege and restrictions, and put in place a liberal economic policy, while the poor peasants and the *sans culotes* demanded the establishment of communal granaries, and strict regulations on the price of food.

That the revolution failed to take on a fully socialist character was due primarily to the ideological and organizational weakness of the Jacobin leadership, which vacillated between these two extremes. The intelligentsia which the Jacobins represented had not yet clearly grasped its interest in transcending, or rather in preventing the emergence of, a fully developed market society. Nor was it clearly committed to liberalism. This mid-point between liberalism and socialism might well be called the "Rousseauean moment" of the revolution. Unlike Locke and most of the English theorists, Rousseau recognized no "natural rights" to life or liberty, much less to property, which the state was obliged to respect. On the contrary, prior to the establishment of the state, there was no right or morality. It was participation in the political process itself (one might even be tempted to say "participation in the revolution") which lifted human beings out of their animal state and made them fully social human beings. In becoming part of society, human beings alienated all of their rights to the state. Still, at the end of the process, the state establishes a "right" to property, or, rather, bestows property on the individual to be managed in such a way as to best serve the common good.

Despite its reputation as a highly disciplined, conspiratorial organization ready and willing to employ violence, the Jacobin party was in fact fragmented and undisciplined by comparison with later revolutionary parties. It had no common analysis of the conditions of French society, no grasp of the "line of march" of the revolution, and above all no strategy for uniting the diverse interests of the poor peasants, the emerging proletariat, and the intelligentsia. And it was unwilling to engage in the degree of violence against the French population, and especially the rich peasants, which later revolutions proved to be necessary if villages which entered the revolution to defend their traditional communal rights were to be turned over night into engines for industrialization. In the end the Jacobins went too far in challenging the right to private property to maintain the support of the rich peasants, imposing price controls and sending "revolutionary armies" into the countryside to requisition grain, but not far enough to sustain the enthusiasm of the workers and peasants (Moore 1966: 89-91). As a result the radical democratic revolution collapsed, and the bourgeoisie regained the upper hand.

The "capitalism" which emerged from the Revolution was, however, very different from that of England or the United States. The intelligentsia played from the very beginning an unusually powerful role in the state apparatus, which, with Napolean's support, it remade into a rational, bureaucratic structure. The French state supported a complex of schools and universities which created and inculcated a rational, secular, humanistic civic culture. And the French state provided unusually strong protections for its peasantry, so that as feudal restrictions were swept away, and capitalist relations gradually penetrated the countryside, large sections of the peasantry were able to survive and even prosper. This insured a home market for manufactured goods, just as the settlement of the Midwest and Great Plains by grain farmers had in the U.S.

In short, it was the organized, at least partially conscious, action of the masses, and not the spontaneous workings of the marketplace, which created the conditions for the industrialization of France.

b) The Prusso-Japanese Road

In other parts of the world, the state *was* able to rationalize itself, and transform itself into an instrument for industrialization. The two most important examples of this path of development are Germany and Japan. There are a number of reasons why these states in particular were able to carry out the process of industrialization in the context of a still primarily authoritarian state apparatus. The first, and most important, has to do with the ability of these stages to develop an effective taxing mechanism. Even before the Meiji Restoration, the Tokugawa had developed an effective mechanism for extracting an agrarian surplus from the peasantry, without cutting so deeply into peasant incomes as to undermine the incentives for increasing agrarian productivity (Moore 1966: 232-233). They were also able to restrict the independence of the great nobles (the *daimyo*), requiring them to reside for part of the year at the capital at Edo, and severely restricting their political initiative.

After the Meiji Restoration this system was further rationalized.

The new government needed a regular and dependable source of revenues. The Land Tax, adopted in 1873 was ... the only economically and politically feasible one under the circumstances. The peasants provided most of the revenue for the government. Since the government undertook most of the first steps in industrialization ... the peasant did pay for the beginning stages of industrial growth. On the other hand ... the Meiji land tax constituted no increase over the Tokugawa levy. The new government merely redirected it into new channels, thereby achieving modernization without reducing rural living standards. It was possible to do this because agricultural productivity continued to rise (Moore 1966: 270-271).

Similarly, in Prussia, "The Hohenzollern rulers managed to destroy the independence of the nobility and crush the Estates, playing nobles and townsmen off against one another (Moore 1966: 435-436)," and permitting the state to establish a rational taxing mechanism.

Second, Germany and Japan were able to create a rationalized bureaucratic structure which could lead the process of industrialization. In both cases this structure emerged out of the transformation of elements of the old warrior aristocracy. In Japan, it was the *samurai*, the lower echelon of the old warrior aristocracy, which had been marginalized as it lost its military function during the enforced peace of the Tokugawa period, which provided the principal base of support for the Meiji Restoration (Moore 1966: 244). These *samurai* transformed themselves into government and corporate bureaucrats, arguing that if Japan was to defend itself from the threat of European conquest, it had, first and foremost, to industrialize. In Prussia, as the Hohenzollern monarchy gradually brought the Prussian nobility to heal, it simultaneously gave them a central place in the highly militarized and authoritarian state apparatus (Moore 1966: 435-436).

Finally, in Germany and Japan, it proved possible to transform traditional religious ideologies, which served to legitimate the authoritarian state, into ideological supports for capitalist industrialization, without undermining their effectiveness as underpinnings for authoritarian rule. In Japan, the old warrior ethic, *bushido* which centered on unconditional loyalty to and service of the Emperor, and to Ameratesu, the Sun Goddess of whom he was the descendant, was simply transformed into an ethic of service to the state through promotion of economic growth. In Germany, the process was somewhat more complex. Lutheranism provided religious legitimation for economic activity, which it regarded as a service to one's neighbor, and thus a type of God's love for humanity. At the same time, it regarded human motives as radically depraved. Human beings served each other only for selfish motives, and not out of an actual interest in the good of the community. This, in turn, meant that it was necessary, if order was to be maintained, to establish a strong central authority. Lutheranism thus created an ideological basis for state action to restrict the liberty of the individual in order to protect the common good

The capitalism which resulted from these essentially authoritarian processes of industrialization differed, once again, from the liberal, or Anglo-American model. In both Germany and Japan the state continued to be the principal agent of economic development, centralizing capital and providing a market for heavy industrial goods——especially armaments. Capital remained dependent on and subordinate to the state apparatus until well into the twentieth century. In Japan complex corporate networks, the so called *zaibatsu*, developed. These networks were characterized by interpenetrating relationships of ownership and control, and well established supply networks. Japanese corporations owned each other, and negotiated stable supply relationships (Moore 1966: 287-288, 301-303). This meant that a free market in capital never really developed in Japan, nor did a market in capital goods. Industrialization proceeded, rather, through a rationalization of the old feudal structure. This pattern prevails to this day in Japan, where the economy continues to be dominated by great corporate families, now called *kereitsu*. Less than 20% of Japanese capital is publicly held and traded. In Germany, corporate networks were less formalized, and operated largely through the dominance of finance capital, but served much the same function, facilitating the centralization of capital for new investments and permitting the subsidization of new ventures over a long period of growth and gestation.

It should be apparent by now that industrialization was nowhere the product of the spontaneous operation of the marketplace. Even in England and North America, which come closest to the classical liberal model of capitalist development, industrialization was the work of men and women motivated by their participation in a powerful nonmarket social movement (Reformed Protestantism) to contribute through productive activity to God's inscrutable plan for the cosmos. And even in the U.S. the progressive, industrial sectors of capital eventually had to have recourse to the state to create the conditions for industrialization, both by abolishing slavery and by using the state's taxing authority to centralizing the resources necessary to create the infrastructure of roads and canals, railroads and bridges, schools, libraries, and universities necessary for an advanced industrial economy. Elsewhere, the state played an even more significant role in industrialization. In France, industrialization took place under the auspices of a highly rationalized state apparatus, product of a revolution which stopped just short of socialism. In Germany and Japan, the old tributary state structures were able to reorganize themselves, and become effective mechanisms for centralizing and allocating the social surplus product in such a way as to build powerful industrial machines.

B. Capitalist Underdevelopment

The second half of the nineteenth century, the period following the consolidation of power by the industrial bourgeoisie, witnessed significant social progress. These were the years of the second stage of the industrial revolution: the development of electricity and of the internal combustion engine which harnessed, for humanity the incredible power of electromagnetism. This was the period during which the railroad network was completed and industry was electrified. It was the period during which the telephone, the automobile, and the airplane were invented. It was the period during which the modern corporation, which as Engels himself pointed out reflected a movement towards the socialization of capital, first emerged. The late nineteenth century was the golden age of the realist novel, which chronicled the struggles of the emancipated individual to find meaning in an increasingly complex society. It was a period of great progress in physics, chemistry, biology, and sociology, as capitalists invested a portion of their profits in the development of great research universities, such as Johns Hopkins and the University of Chicago.

But realization of the full potential of these developments was held back by the operation of market forces. We noted above that as the economy becomes more technologically developed and more capital intensive, there is a tendency for the rate of profit to decline. There is a considerable body of evidence to indicate that this has in fact been the case. Mandel notes that in the U.S., the average rate of profit in manufacturing sectors fell from 26.6% in 1889 to 20.5% in 1899, to 18.1% in 1909, to 16.2% in 1919 to 13.8 in 1948 (Mandel 1968: 166, 454). Profits in the low technology, low wage economies of the Third World, however, were much higher. While the rage of profit for manufacturing companies operating in the U.S. averaged 10.6% between 1945 and 1948, the rate of profit for U.S. companies operating in the Third World averaged 15.7%. The figures for Belgian industry are even more dramatic. The rate of profit for manufacturing companies operating in Belgium between 1951 and 1957 was 8.56%, while the rate of profit for Belgian companies operating in the Congo was 20.74%.

As a result of this difference in the rate of return, there was a tendency for capital to be redeployed from Europe, North America, and eventually Japan, to the underdeveloped countries of Asia, Africa, and Latin America. Mandel gives the following figures for the principal capitalist countries, in billions of gold francs of 1913 (Mandel 1968: 450).

Level of Foreign Investment

	Britain	France	Germany	USA	Japan
1862	3.6	-	-	-	-
1870	20.0	10.0	-	-	-
1885	30.0	15.0	6.5	-	-
1902	62.0	30.0	12.5	3.0	-
1914	87.0	40.0	30.0	15.0	1.0
1930	90.0	20.0	5.6	75.0	4.5
1938	85.0	15.0	-	48.0	9.0
1948	40.0	3.0	-	69.0	-
1957	46.0	6.0	2.0	120.0	-
1960	60.0	?	4.0	150.0	1.0

The impact of this pattern of economic development on the countries of Asia, Africa, and Latin America has long been a matter of controversy, with liberal developmentalists, as well as some Marxists, especially those from the Trotskyist tradition, maintaining that foreign investment contributed to economic growth, and most Leninist and dependency theorists arguing for different reasons that, on the contrary, imperialism has held back the development of the Third World. In reality both positions have merit.

Foreign investment did, in a certain sense, accelerate the process of economic growth in these countries, both by financing the development of agricultural and later industrial export activities, and by catalyzing the formation of an agricultural and industrial proletariat, so that growing numbers of people now had to purchase goods they had formerly produced for themselves. But it did little to actually catalyze the development of human social capacities. This is because the bulk of foreign investment has been in relatively low technology, low wage activities which do very little to promote technological innovation, and which leave the masses even more impoverished than they were before the penetration of market relations.

At the same time, the impact on the advanced industrial countries has been devastating. Formation of new business capital has showed a consistent tendency to decline. Mandel gives the following figures from Seindl (Mandel 1968: 166).

1869-1878	3.75%
1879-1888	4.65%
1889-1898	3.20%
1899-1908	3.75%
1909-1918	2.76%
1919-1928	2.18%
1929-1938	0.38%

Similarly, the percentage of new equipment destined to extend, rather than merely to replace, existing capacity, declined precipitously (Mandel 1968: 167).

1879-1888	52.2%
1889-1898	57.9%
1899-1908	54.1%
1909-1918	43.1%
1919-1928	36.6%

Imperialism, by opening up profitable, but low wage, low technology, outlets for capital in the Third World, effectively held back the process of technological innovation in the advanced industrial countries, without doing anything to develop the Third World, and effectively putting to an end the technological utopia of the nineteenth century.

Imperialism also had a profound impact on the political and cultural dynamics of the advanced industrial countries. Nowhere was this more apparent than in the United States, where the progressive alliance between industrial capital, the farmers and peasants, Euro-American and African-American, and the working class which had defeated the southern landed elite collapsed within two decades, and was replaced by a reactionary alliance between industrial capital and the old landed elite. Parallel alliances formed in Italy, Germany, and parts of the Slavic east. The state did, to be sure, continue to play a progressive role in certain respects. It subsidized development of the country's railroad network. Through its taxing mechanism, the state played a leading role in the development of schools and universities. And it continued to regulate trade to prevent competition which was destructive to U.S. industry.

But as the rate of profit declined, capital began increasingly to look to the United States' internal colony in the South, and eventually to Asia and Latin America. Capitalists who were themselves withdrawing capital from industry to buy banana plantations in Central America could hardly serve as the vanguard of a liberation struggle on their own continent. After the Civil War the Republican Party soon divided over the question of Reconstruction. Radical Republicans, representing the most advanced sections of the bourgeoisie (steel, railroads), the intelligentsia, and the industrial working class, argued that the lands of the slave owners should be expropriated and each freed slave granted 40 acres and a mule, as well as the right to vote. Moderates, representing the textile industry and some midwestern farmers, fearful that cotton production would drop as black farmers shifted land to subsistence agriculture, opposed land reform, while supporting black suffrage, which they hoped would guarantee a permanent Republican majority. Bourgeois enthusiasm for African American suffrage waned rapidly, however, as capitalists realized that African American suffrage could give industrial workers and African American peasants a majority within the Republican Party. The Northern bourgeoisie soon reached an accommodation with the Southern landed elite. When Union troops were withdrawn from the South after 1876, African Americans were quickly deprived of any effective means of participation in the political process (Moore 1966: 141-155).

Similar developments affected the other advanced capitalist countries. Even as England moved towards universal suffrage at home, it was building a vast colonial empire, subjugating the peoples of much of Africa and of the Indian subcontinent, carrying kings into captivity (Sarkisyanz 1965) and brutally suppressing popular uprisings, without giving any serious thoughts to the "natural rights" of the colonized peoples. In France, the revolutionary democratic spirit of the Third Republic soon gave way to the manifest antisemitism of the Dreyfuss affair. The new Italian state, as we have seen, spent the second half of the nineteenth century trying to put down a series of peasant revolts and transforming Sicily and the *Mezzogiorno* into an internal colony on the back of which it could undertake the tasks of a rather belated primitive accumulation of capital. Indeed, it is only in the countries which were on an essentially authoritarian road to capitalism——Germany and Japan——that the last half of the nineteenth century can be regarded as a period of relative political enlightenment. These countries had only limited colonial empires, and were not yet strong enough to begin the aggressive process of imperial expansion which would characterize their politics in the next century.

The reactionary turn of this period was reflected in the dominant forms of bourgeois ideology. Where earlier capitalist theory had recognized that it was labor itself which created value, and regarded the entrepreneur first and foremost as himself a worker, who organized the general conditions for the productive process, bourgeois economics increasingly understood the marketplace itself as the creator of value and the organizer of production. According to the theory of "marginal utility" developed during this period, forerunner of the contemporary neoliberal Austrian school, value is an entirely subjective reality determined by the play of supply and demand in the marketplace, which allocates land, labor, and capital in such a way as to best satisfy the desires of consumers.

> From Petty to Ricardo and Marx, every theory of value was *objective*, that is, its ultimate starting point was *production* ...

> The neo-classical school, however, approached the problem in an altogether different way ... The separation of exchange value from use value, the starting point of the classical school, was questioned. It was declared on the contrary, that exchange value is essentially a function of use value, of the utility of the given commodity ...

> They laid down individual scales of needs; this is why this school has correctly been described as being *subjectivist* ...

> ... neo-classical theory states: it is not the intensity of the need in itself, but the intensity of *the last fragment of need not satisfied* (of the *marginal utility*) that determines value.

> Starting from this general idea, the neo-classical school worked out a series of curves the intersection points of which are supposed to show conditions of equilibrium ... (Mandel 1968: 713-714).

According to one writer, in a situation of equilibrium

the satisfactions obtained by consumers are at their highest because any
transfer of a factor of production would result in a reduction of the
"value" created by this factor (Denis in Mandel 1968: 715)

Where classical economics had as its aim, implicit or explicit, increasing
the quantitative complexity of human society by maximizing the value
added, and differed from Marx principally in its assessment of the
adequacy of the marketplace as a means of achieving this aim, neo-classical
theory——the first fully capitalist economic theory——seeks to maximize
consumer satisfaction. Not surprisingly, this tendency had considerable
difficulty grasping the "marginal utility" of capital goods (Mandel 1968:
715). If classical economics represented the self-consciousness of the
industrial bourgeoisie, neo-classical economics represents the self-
consciousness of the rentier strata generated by imperialism.

The claim that it is competition itself, and not work, which creates
complex organization is reflected in the Darwinist theory of evolution
which ascribes the evolution of the species entirely to the process of natural
selection, without ever explaining what it is that produces the adaptations
for which the environment selects. This approach becomes still more
problematic when applied to the evolution of complex forms of *social
organization*——so-called Social Darwinism.

One of the facts difficult to reconcile with current theories of the
Universe, is that high organizations throughout the animal kingdom
habitually serve to aid destruction or to aid escape from destruction. If
we hold to the ancient view, we must say that high organization has
been deliberately devised for such purposes. If we accept the modern
view, we must say high organization has been evolved by the exercise
of destructive activities during immeasurable period of the past. Here we
choose the latter alternative ...

Warfare among men, like warfare among animals had a large share in
raising their organizations to a higher stage (Spencer 1873/1972: 167-
168).

Eventually, however, as human society becomes more complex, the
destructive impact of warfare begins to exceed its benefits. This does not,
mean, however, that social evolution begins to take the shape of peaceful
collaboration. On the contrary

After this stage has been reached, the purifying process, still an
important one, remains to be carried on by industrial war——by a

competition of societies during which the best, physically, emotionally, and intellectually spread most, and leave the least capable to disappear gradually, from failing to leave a sufficiently-numerous posterity (Spencer 1873/1972: 173-174).

By the end of the Civil War the principal voices of the Reformed tradition were themselves sounding increasingly like Social Darwinists. Often it is possible to chart this transformation in the biography of a single figure. Before the Civil War Henry Ward Beecher had supplied arms to John Brown in his guerilla war against planters in Kansas. He once remarked that his rifles (shipped in crates marked "bibles") had done more to spread the gospel in Kansas than had the bible itself: thus the term "Beecher's Bibles." After the Civil War, he could write from his comfortable Brooklyn manse that

Looking comprehensively through city and town and village and country, the general truth will stand, that no man in this land suffers from poverty unless it be more than his fault——unless it be his sin . . . There is enough and to spare thrice over; and if men have not enough, it is owning to the want of provident care, and foresight, and industry, and frugality, and wise saving. This is the general truth (in Ahlstrom 1972: 789).

By roughly 1914 the contradictions of the imperialist phase of capitalist development were becoming apparent. Redeployment of capital to labor intensive, low wage activities precipitates a secondary economic contradiction: the tendency towards overproduction or underconsumption. In a market system all production must find an outlet either in final consumer demand, or else in demand for capital goods used to produce goods for final consumption. But if capital flows by preference into low wage activities, then real wages, and therefore effective demand, will remain low and eventually become an obstacle to economic growth. Historically this became a critical problem beginning towards the end of the nineteenth and the beginning of the twentieth centuries, when corporations began to develop durable consumer goods such as automobiles, various home appliances, and eventually radios and televisions, which were too expensive for the average worker. The result was the period of protracted stagnation which lasted from 1914-1945, and more specifically, the Great

Depression which began in 1929 (Marx 1863/1978: 465, Davis 1986).[3]
Clearly it would be necessary for the state to intervene actively in the
management of the economy if "equilibrium" was to be restored.

C. Bourgeois Reformism

The most important theoretician of bourgeois reformism was John
Maynard Keynes. Mandel explains Keynesian theory.

... households and firms may take two decisions regarding the sums of
money they possess: a decision to spend (consume) them, or a decision
to hoard them. And since it is the volume of demand that determines the
level of economic activity, the latter will fluctuate with the propensity
to consume, that is, according to whether incomes as a whole are spent
or not. As household usually spend the bulk of their income, it is
fluctuations in the expenditures of firms, fluctuations in investment, that
ultimately determine the volume of demand, employment, and
production

The Keynesian theory is an income theory, since it makes the level of
employment dependent in the last analysis on the allocation income. And
since a certain allocation of income (of demand) is essential if full
employment is to be achieved, Keynes proposes that public expenditure
be brought in to make up for the inadequacy of investment when there
is a full in income and excessive unemployment (Mandel 1966: 718-
719).

In short, the state must do what the market will not: make investments
which are (relatively) unprofitable, but which are necessary for the
development of human social capacities.

Historically, bourgeois state intervention has taken two distinct forms,
one authoritarian and associated especially with fascism, and the second
liberal and associated with the so called "welfare state."

1. Fascism

Fascism and social liberalism in fact have a great deal in common. In

---

[3]. The precise timing of this crisis varied from one country to the next. In the U.S.
the full impact was not felt until 1929 because postwar tax cuts unleashed a period of
high levels of luxury spending which propped up the economy for almost a decade.

both systems, high levels of military spending subsidize high technology development not supported by the marketplace (electronics, information processing, rocket, and nuclear technologies, among others), while various types of income maintenance programs shore up effective demand and secure the support of the working classes, temporarily relieving the pressure for socialist revolution.

The two tendencies differ primarily in the way in which they mobilize working class support. Fascism was strongest in countries in the middle ranks of industrial development——Germany, Japan, Italy——which lacked a colonial empire, and where revolutionary socialist movements represented (or had at one point represented) a real threat to the bourgeoisie.[4] The bourgeoisies of these countries found themselves doubly constrained. On the one hand, the rate of profit was declining rapidly in high technology, high wage sectors, creating a crisis of capital formation. On the other hand, they lacked an outlet for capital in low wage, low technology activities of the kind provided to the U.S., England, and France by their growing colonial empires. Their solution was creation of a permanent war economy, which would permit them to build the kind of military machine necessary in order to win colonies, while providing reliable state subsidies for high technology development.

In order to do this, of course, they needed not only the acquiescence, but in fact the support of the masses, who were more or less profoundly attracted by a socialist solution to the crisis. They were able, in this regard, to tap into a profound authoritarian undercurrent characteristic of advanced capitalist societies. We have already noted that in an industrial society, human beings become real participants in the self-organizing activity of the cosmos, in a way which was never before possible. At the same time, mediation of human organizing activity by the marketplace makes this activity simply a means of realizing individual consumer interests. The social significance of work lies not so much in its performance, but rather in its negation: its sublation by the marketplace and thus its transformation into a means of realizing the needs of others. But human beings experience the marketplace as a mysterious force beyond their control. There is no direct relationship between their contribution towards meeting the needs of others, and their own economic situation. Because of this, any desire to serve universal purposes, human or cosmic, is regarded as a negation of individual interest——an act of submission and self-sacrifice, rather than the realization of the individual human's deepest nature. In a capitalist

---

4. Japan, of course, is an exception in this regard.

society human beings serve the community through submission to the will of the collective, whether that collective is represented symbolically as God, Nation, or both. This tendency was already implicit in Protestantism and affects even the most progressive movements in bourgeois society. But when it becomes explicit, when the masses begin to seek order, meaning, and purpose in submission and self-sacrifice, and thus in the negation of their own capacities, then the conditions for fascism are ripe (cf. among other studies, Fromm 1941).

This is precisely what happened as market relations marginalized growing numbers of people in the advanced capitalist countries during the late nineteenth and early twentieth centuries. This process of marginalization not only deprived the people of the resources they needed in order to develop as workers; it deprived them as well of the complex social networks which had made life meaningful in precapitalist societies. People saw all around them the tremendous power of industrial society. They lacked the organized networks through which they could access and participate in developing that power. And so they sought to participate through submission.

Fascist ideology served as a link between the spontaneous authoritarianism of the masses, and the bourgeoisie's program of militarization and social discipline. Often fascism drew on the religious traditions of the masses, i.e. their spontaneous experience of the self-organizing activity of the cosmos. This was case on the Iberian peninsula and in Japan, where fascism mobilized elements in the Catholic and State Shinto traditions respectively. In other places it drew on the symbolism of the nation state, of even of the democratic revolution. Mussolini, for example, appropriated the symbolism of the *Terza Italia* from the radical democratic ideology of Mazzini. The term "fascism" itself derives from a root which means "bound together," and derives from the revolutionary socialist *fasci contadini*, the peasant leagues which led the struggle against the penetration of market relations in to the Sicilian countryside during the 1890s. Nazism, of course tapped into the German nationalist tradition, and surrounded itself with symbols drawn from a largely fictitious Aryan past. Fascist ideology mobilizes the national, religious, and democratic traditions of the people in such a way as to galvanize support for the political program of the fascist bourgeoisie (Laclau 1977 passim).

Once in possession of state power, the fascists set in motion a process of rapid militarization and rearmament. This military apparatus at once promised to win for their countries the colonial empires which they lacked, and at the same time, provided a focus for state financed investment in research and development activities which the marketplace itself would not support———e.g. the development of rocket technology, nuclear energy, etc. (Mandel 1968: 536). In the short run this strategy proved quite successful. What fascism neglected to consider, however, were the real conditions for advanced industrial development. The authoritarian socialization and legitimation strategies on which fascism depended in fact undermined the conditions for scientific and technological creativity. While the fascist countries were certainly able to make some real advances in certain technological fields (rocket and nuclear technology), fascism was unable to mobilize either the talent and creativity of the best minds or the energies of the masses. Much of Germany's intelligentsia, particularly its Jewish intelligentsia, fled to the U.S. or the U.S.S.R., while the other fascist countries, which were less developed technologically and scientifically at the beginning of the war, were never able to develop a real cadre of scientists and engineers. As a result most of the fascist systems had collapsed within 15-20 years.

2. Social Liberalism

Social liberalism,[5] on the other hand, developed in the advanced capitalist countries (Northwestern Europe, North America), which already had large colonial empires, and where revolutionary socialism was not really a threat. Here too, the strategy of the bourgeoisie centered at least in part on the development of large national security bureaucracies which used state expenditures to finance the development of technologically advanced defense systems which were directed first against fascism, and later against the socialist countries and the national liberation movements. State expenditures were, in effect, used to defend the marketplace against the threat of movements advocating statist development strategies———and simultaneously to finance research and development in the electronics, telecommunications, and aerospace industries which the marketplace itself could not support.

What differed was the way in which the bourgeoisie related to the working classes. Rather than simply disciplining the working class (or rather mobilizing the spontaneous authoritarianism of the workers as a

---

[5]. In this section we will focus on the unfolding of the social liberal dynamic in the United States, as it represents the purest expression of bourgeois reformism. European "welfare states" have been profoundly influenced by Social Christian and Social Democratic parties which at least claim to look forward to a more profound break with the marketplace and will be considered in the next chapter, when we discuss the socialist experiment.

means of promoting self-discipline) social liberalism attempted to make the workers active partners in the process of development, "sharing" with them part of the surplus which they produced, so that they could develop themselves as human beings, and creating political forms which guaranteed the representation of working class interests in the decision making process. At the economic level, this meant creation of a legal framework for collective bargaining which permitted workers organizations to drive up real wages and thus effective demand (but not to mount a real contest for power), and set in place a system of transfer payments designed to mitigate the effects of cyclical crises on real wages and effective demand (Mandel 1968: 536; Davis 1986). At the political level it meant the development of an alliance between the progressive bourgeoisie, the intelligentsia, and the working classes. In the United States, where market relations were more developed and liberalism was stronger, this alliance took the form of a social liberal bloc centered in the Democratic Party. In Europe, where both traditional communitarian and socialist tendencies were stronger, it meant the formation of an alliance with the working class through the medium of reformist Social Christian or Social Democratic parties.

Ideologically, social liberalism found expression in a whole cluster of attempts at "holistic" thinking, which attempt to grasp "whole systems" while remaining within the framework of the dominant positivistic problematic. The most important and interesting of these attempts is the "general systems theory." It is possible to distinguish within this trend between a more or less rigidly positivistic right, an incipiently dialectical or synergistic left, and an eclectic center. The positivist right is characterized by a tendency to understand systems in atomistic and mechanistic terms as aggregates of discrete particles of matter or information which are related to each other in only a purely external way. Organization is either something imposed on the system from the outside, or else something which emerges spontaneously through the operation of natural selection, the marketplace, etc. But ultimately all organizing processes are reducible to mechanics. I would include in this "positivistic" tendency the information-theoretical and game-theoretical work developed by such cold-war theoreticians as Claude Shannon, Warren Weaver and John von Neumann. The dialectical trend, on the other hand, has increasingly stressed the relational, self-organizing, and teleological character of matter, in a way which converges with the most advanced elements in the dialectical idealist and dialectical materialist traditions. Alexander Bogdanov (1928/1980) was an early exponent of this trend, as was R. Buckminster Fuller (1975-1979, 1981, 1992). More recently this trend has won support from the emerging discipline of complex systems theory (Pines 1988, Campbell 1990, Zurek 1990), postdarwinian evolutionary biology (Waddington 1957, Lenat 1980, Denton 1985, Sheldrake 1981, 1989, Margulis 1991, and Wesson 1991), and "participatory" as opposed to "many worlds" interpreters of the anthropic cosmological principle. The "General Systems Theory" of Ludwig von Bertalanffy (1968) and his collaborators, as well as the structural linguistics of Ferdinand Saussure (1973), the structural anthropology of Claude Levi-Strauss (1963), and the structural psychology of Jean Piaget (1957) stand roughly in between these two trends. On the one hand it shares with the semidialectical left a concern to understand such "system" properties as holism, self-regulation, and self-organization. At the same time, it conserves a positivistic concept of system as a network of discrete elements which are mutually determining in their behavior, but not in their essence or existence.

General systems theory in all its varieties clearly represents an advance over the historic atomism of bourgeois ideology, which was utterly unable to comprehend the teleological, self-organizing, holism of human societies. At the same time, at least in the "rightist" and "centrist" forms dominant in the U.S. it remained unable to theorize adequately the contradictions of a market economy, and was thus unable to generate authentic solutions. The relationships between individuals within a social system determines the very nature of those individuals——their social character (Fromm 1941). This is true in a market society as in any other type of social system. In a market society, we have seen, it tends to determine them first and foremost as *consumers*. But general systems theory does not comprehend this level of system interdependence. Interventionist economic and social policies based on systems theory generally assumed that changes in "system parameters" such as the levels of taxation and state expenditure could regulate the behavior of both the system and its constituent elements (individuals) without any need to alter the basic structure of the system. Not surprising, while such policies often achieved valuable goals, such as expanding the higher education system or beginning the exploration of the solar system, they were not by themselves able to counter the disintegrating impact of the marketplace on the ecosystem and the social fabric, nor were they able to do anything more than mitigate tendency of the rate of profit to decline in high technology, high wage sectors, the tendency of capital to be redeployed to the periphery, or the tendency towards overconsumptionism which began to affect the system during the postwar period.

This is apparent from the actual evolution of the welfare state.[6] Social liberalism could have developed in either of two directions. On the one hand, it was possible for the surplus centralized by the state to be invested in infrastructure, education, research, and development, setting in motion a process of human capital formation and of working class development. Politically this would have required strengthening progressive democratic party structures, so that the spontaneous interest group organizing of both the working class and the progressive bourgeoisie was gradually transformed into organizing on behalf of the common good. On the other hand, part of the surplus centralized by the state could be redistributed to the workers, and used to create an expanding market for consumer durables (and in particular for suburban houses and automobiles) which did little to advance human creative capacities, while the remainder was invested in a military industrial complex designed to defend the off shore investments of the imperialist bourgeoisie.

While both dynamics did in fact operate under the social liberal regimes of the postwar period, it was the latter which eventually became dominant. There were a number of reasons for this. The most important, of course, was the corrosive impact of the marketplace itself, which constituted workers as consumers, who organized not in order to increase their productive capacity, but rather to augment their consumption levels. Second, particularly in the United States, it was the commercial banks, multinationals, and the consumer durables industries which exercised hegemony within the ruling bloc. Producers of high technology capital goods were incorporated into the bloc largely as arms manufacturers and thus as agents of the cold war, so that their progressive potential was lost and, if anything, they and their employees became an important constituency for the right.

The implementation of social liberal policies did, to be sure, at least temporarily unleash the development of human social capacities. Throughout the postwar period, scientific research and technological development proceeded at a rapid pace. U.S. scientists completed the work begun by the Nazis in the fields of nuclear physics and rocket technology. Perhaps more important, the development of microelectronics laid the groundwork for extraordinary advances in information processing and has made possible a comprehensive automation of industry, putting us on the

---

6. In what follows we focus primarily on the United States, because it is there that the disintegrating dynamics of the marketplace are most apparent. The same argument holds, albeit with somewhat less stringency, for the social democratic and "organized capitalist" countries of Europe and Japan.

threshold of an era in which routine manual labor, as we have understood it since the late eighteenth century, is all but obsolete. Research in biophysics and molecular biology laid the groundwork for understanding the genome, and for the biotechnological industries which emerged in the 1980s and 1990s. Other research, less fully developed, has laid the groundwork for development of clean, effectively limitless sources of energy (solar, wind), and an extraordinary prolongation of human life. We have begun our long journey towards the stars. There is a broad consensus that the developments of the postwar years constitute at the very least a Third Industrial Revolution (Mandel 1978)———and perhaps the first signs of a new, postindustrial society in which human labor will no longer be absorbed in direct material production, but rather in the design, monitoring, and organization of the production process itself.

The product of North American popular culture which best expresses the possibilities latent in this period is the television series *Star Trek*, which has captured the imagination of the U.S. public for over a quarter century. In the 24th century society depicted in the series, every worker is at least a trained technician, contradictions between nations and to some extent even between intelligent species have been overcome. Indeed, the market economy itself appears to have been transcended.

The social liberal strategy was, however, marked from the beginning by several profound contradictions. It did nothing to reverse the dynamics of imperialism and of unproductive consumption which are intrinsic to capitalism itself[7]. The export of capital to the Third World, was, first of all, a response to the tendency of the rate of profit to fall as industry became more technologically developed and thus more capital intensive. The banana plantations of Central America———and, for that matter the textile mills and electronic assembly plants of Korea and Thailand———provided investors with a higher rate of return on their investment than they would have received from investment in the

---

[7]. There has been much debate about the reasons for the success of the Japanese and European economies by comparison with that of the United States. Corporate structures which encourage reinvestment rather than payment of high dividends, and state investment in the civilian rather than military arena are certainly important factors. It must be remembered, however, that Germany and Japan lost the Second World War, and that the other European countries lost much of their colonial empires even though they were formally on the winning side. This meant that European and Japanese capital had less access to outlets in the Third World than U.S. capital. These countries were forced to invest at home, in high technology, high skill, high wage activities rather than redeploying capital to technologically and economically backward activities on the periphery of the world system. The result is hardly surprising.

automation of the steel industry, the development of mass transit systems, or research and development of commercially viable solar power systems. Investment in the Third World, in other words, held back the development of the productive forces.

Second, the export of capital to the Third World (and to Third World regions within the imperialist metropoles) was predicated on the low wage rates characteristic of these regions. This limited demand, creating underconsumptionist tendencies in the system as a whole, as consumer durables markets became saturated in the late 1960's (Davis 1984:14). Neither the Alliance for Progress, with its strategy of import- substitution industrialization of the Third World, or the War on Poverty was able to resolve this contradiction. The first of these initiatives failed largely because the principal incentive for attracting foreign investment to the Third World --low labor costs-- simultaneously blocked formation of internal demand, limiting production for domestic consumption (Jonas 1981:xi)——and thus the market for U.S. produced capital goods. The second initiative failed because unlike the New Deal, which actually transferred income from the bourgeoisie to the working class, increasing the final demand for consumer goods, most Great Society programs depended on transfers *within* the working class. The War on Poverty, therefore, created no new demand (Davis 1984:18), and exacerbated considerably contradictions within the ruling liberal democratic bloc. In other words, existing markets for mass consumer durables (automobiles, refrigerators) derived from the technologies of the second industrial revolution were saturated, and development and penetration of new markets in the Third World were held back by the low level of effective demand.

As wages rose in the advanced capitalist countries, and capital flowed to low technology, low wage activities in the Third World, or to unproductive, speculative investments, the advanced industrial countries generally, and the United States in particular, experienced a decline in the rate of growth of productivity, and ultimately a decline in the rate of profit. While the rate of profit is notoriously difficult to measure, most estimates suggest a rather dramatic decline during the postwar period: from 8.1% in 1958 to 5.8% in the mid 1970's (de Janvry 1981:57) or, according to a different measure, from 16% in 1952 to 9% in 1980 (Dumenil 1984: 149). Most of the revenue redirected towards the working class under the pressure of the labor movement was used for unproductive consumption ——i.e. for consumer goods such as larger homes, automobiles, fancier clothes, increased meat consumption, etc., which did not increase the productivity of this class, rather than for education and organizing initiatives which might have increased the productivity of workers in the advanced industrial countries. This was especially true in the U.S. High levels of deficit spending drove up interest rates, blocking capital formation and further restricting productive investment. This was, once again, particularly true in the U.S. where investment in technological and human development competed for state funding with a massive military apparatus.

The crisis of social liberalism is thus best understood as a crisis of *overconsumption*——the overconsumption of luxuries by the bourgeoisie, the new middle strata, and to some extent even the working classes of the advanced industrialized countries——which restricted the revenues available to finance both basic human development on the periphery of the system, and advanced research and development in the centers. This is, of course, precisely the result we would expect given the limitations of the "general systems theory" which provided the theoretical rational for social liberal polities.

Meanwhile, the growth of the consumer-industrial complex had a devastating effect on the ecology and social fabric of the planet. Resources which might have been invested in research and development, in ecologically sound infrastructure systems, and in education were squandered on a war of attrition against socialism and on ecologically and economically wasteful consumption in which a globally very small fraction of the working class shared. The penetration of market relations into the countrysides of Asia, Africa, and Latin America effectively completed the destruction of the peasant community and impoverished millions. Most of the real growth and development which took place during this period was the result either of the persistence of older dynamics (state investment in civilian research and development, infrastructure, and education) or of spin offs from technology developed by the military industrial complex. The working class became increasingly mired in the hegemonic consumerism, and abandoned its historic vocation to lead humanity's struggle towards self-development, while the technologically most advanced sectors of the bourgeoisie (electronics, telecommunications, aerospace) and much of the intelligentsia was absorbed into the anticommunist military industrial complex and the national security state. Those intellectuals who were not tied to the military industrial complex, meanwhile, faced rapid proletarianization.

From a purely economic standpoint, the crisis could have been resolved rather easily. Sharply increased, highly progressive taxation, coupled with reduced military outlays, would have made it possible to maintain and increase public investment in technological and human development, increasing productivity, while bringing down state deficits and thus holding down interest rates. Radical land and wage reform in the Third World (and among "Third World" sectors in the advanced capitalist countries), would have increased effective demand and opened up new markets. Redirection of capital in the advanced industrial countries into high technology capital goods production would have permitted capture of lower technology industries by developing countries while maintaining a healthy market for exports. Payments to labor, meanwhile, in both the advanced industrial and the developing countries, could have been channeled into collective expenditures on housing, health care, mass transportation, education, and culture in order to promote the all-sided development of the productive capacities of the working class.

While these measures represented in many ways simply an extension of existing policies, the political conditions for carrying them out did not exist in the United States, and existed only imperfectly in the other advanced industrial countries. Support for these measures among the U.S. bourgeoisie was weakened by the fact that higher rates of return were available from investment in low technology, low wage production in the Third World, from speculation of various kinds, and, for what should have been the cutting edge, advanced industrial sectors (electronics, aerospace), from defense contracts completely shielded from market pressures. European and Asian capitalists, on the other hand, were able to profit from the abandonment of intermediate technology production (steel, auto) and increasingly of non-military high technology activities, by the North Americans, reducing the pressure to create new markets for high technology capital goods in the Third World. High rates of taxation, and high rates of expenditure on physical and social infrastructure created the conditions for continued economic development in Europe and Japan, while complex corporate and state structures guaranteed reinvestment of a high percentage of surplus in strategically important industries.

The penetration of market relations into every sector of society, meanwhile, undermined the political organization and solidarity of the working class. This process is most visible in the decline of the so-called "mediating institutions," institutions intermediate between the individual and the state, which historically served to organize the complex interests which characterize industrial society. Among the most important mediating institutions, we would note the trade union, the neighborhood organization, the mutual benefit society, local political clubs, and local religious congregations. As capitalism has developed, the functions once performed by these intermediate institutions are increasingly taken over by the marketplace. As the network of mediating institutions has eroded, people have forgotten how to form public relationships. And already by the end of the 1950s consumerism was gaining control over the hearts and minds of the North American people. This made it increasingly difficult to organize the working class, and deprived the progressive forces within the bourgeoisie of a critical strategic reserve.

It would be a mistake, however, to conclude that there were no forces, even within the bourgeoisie, which did not grasp the progressive potential of social liberalism. John Kennedy's "New Frontier," and his brother's "Newer World" programs were fundamentally attempts to reform the hegemonic social liberal bloc from within. The fate of these programs demonstrates the difficulty of profound structural reform at the imperialist center. At the international level, they attempted to reign in the military industrial complex while setting in motion a process of detente with the Soviet Union, and supporting reformist development strategies in the Third World, with a view to creating an expanding market for capital goods produced in the U.S. On the domestic front, they attempted to shift the thrust of domestic social policy from increasing consumption to catalyzing development of human social capacities, with growing investment in scientific space exploration serving as both a symbolic focus and substantive leading factor. Domestic volunteer programs and expanding state expenditures promised to rescue the nontechnical intelligentsia from proletarianization and enlist them in a campaign to comprehensively develop human social capacities. Their attack on the hegemonic consumerism was summed up in John Kennedy's inaugural call for Americans to ask not what their country could do for them——but rather what they could do for their country.

This effort at reform was cut short by the bullets of assassins who may well have been in the employ of threatened elements of the U.S. ruling class. The result was economic crisis and the rapid disintegration of the hegemonic social liberal bloc. Continued U.S. support for reactionary landed elites and dependent bourgeois elements in the Third World, symbolized by escalation of the war in Vietnam, blocked the process of agrarian reform and industrialization which would have created the expanding Latin American, Asian, and African market which U.S. capital goods industries required. The market for consumer durables became saturated, and European and Japanese products began to compete with those produced in the U.S. The less than visionary programs of the Great Society did little more than redistribute income within the working class, failing to increase the total surplus available for investment in either new technologies, or human development. The rate of growth of productivity declined, and with it the rate of profit. Redeployment of capital to low technology, low wage, activities in the Third World only deepened the crisis (Davis 1986).

Income transfers within the working class, meanwhile, created contradictions between lower strata workers, drawn largely from oppressed nationalities, and the newly suburbanized workers in the consumer-and military-industrial complexes whose unions had won them a remarkably high standard of living. Movements of resistance to proletarianization among younger members of the nontechnical intelligentsia, meanwhile led to cultural contradictions with the rest of the working class (Davis 1986).

The result was a period of economic stagnation which began around 1968 and which has persisted up to the present period, the speculative boom of the 1980s notwithstanding. The first attempt to address this crisis was rather contradictory in character. On the one hand, the Nixon administration led the military industrial complex in what must be regarded as a temporary, tactical retreat, coupled with a second effort at detente with the Soviet Union——this time balanced with a new strategic alliance with China to restrict Soviet influence in the Third World, and thus protect the interests of U.S. multinationals with direct investments overseas. Meanwhile, Nixon moved to protect the interests of industries increasingly under the pressure of foreign competition, instituting wage and price controls, going off the gold standard, and effectively devaluing the U.S. dollar. Most of the income transfer programs of the Great Society were left intact, but new initiatives, designed to develop human social capacities rather than merely buy off poor communities, were blocked. A modest but unsuccessful effort was made to rationalize income transfer programs with proposals for a "negative income tax." These proposals were especially aimed at elements in the social development bureaucracy, which capital increasingly perceived as a drain on its resources and a strategic reserve for the left. The rising tide of oppressed nationality movements and the emerging new left were crushed in a new wave of domestic repression (Davis 1986).

Nixon was forced from office after overstepping himself in the Watergate conspiracy. A new coalition centered in the Trilateral Commission, led by the large commercial banks, with elements of high technology export oriented capital as junior partners, came to power under Ford and Carter. In order to understand the character of this bloc it is necessary to make a distinction between transnational and multinational capital. By transnational capital we mean corporations the ownership of which is increasingly international in character. Transnational corporations are above all located in the sphere of finance (investment banking) and in certain high technology sectors. By multinational corporations we mean U.S. corporations with direct investments in (mostly low technology, low wage activities) abroad.

Trilateralism reflected the interests of the transnational as opposed to the multinational sector of the ruling class. It was first and foremost an attempt to rationalize production on an international scale, permitting redeployment of low and intermediate technology activities to Europe, Asia, Latin America, and Africa, while catalyzing the development of a high technology export oriented economy in the U.S. The Trilateralist strategy centered on increasing loans to countries in the Third World, to finance development of low and intermediate technology industries. This would simultaneously initiate a downward pressure on wages in low and intermediate technology sectors in the U.S., "rationalizing" production, while creating an expanding market for high technology capital goods produced in the U.S. The Trilateralist strategy assumed that the discipline of a growing debt burden would bring the national liberation movements which were coming to power throughout the Third World more or less to heel, while rapid industrialization undercut the predominantly rural social basis of these movements and laid the ground work for capitalist modernization——the strength of populist, socialist, and communist ideologies notwithstanding. At the same time, Trilateralists mounted a diplomatic and ideological offensive against socialism, slowing the pace of detente, responding sharply to the Soviet invasion of Afghanistan, and attacking alleged violations of "human rights." For some within the Trilateralist coalition (e.g. Brzezinski) this was to be the beginning of a new offensive against the Soviet bloc, for others (Vance, and probably Carter himself), it represented an effort to set in motion internal reforms within the Soviet bloc which might lay the groundwork for a new period of peaceful coexistence and collaboration.

The pressure toward economic rationalization in the U.S. was intensified by a program of fiscal austerity which tended to reduce transfer payments and weaken the bargaining position of workers in the U.S. Reduction in spending for entitlement programs was to be balanced by new investments in infrastructure, education, research, and development, particularly development of alternative energy sources and community development programs which empowered working class communities. Carter's first government included, at the subcabinet level, a new generation of community organizers such as Gail Cincotta and Gino Barone who saw their role less as advocates for the consumer interests of the working class than as conservers of the social fabric and leaders in the struggle to develop human social capacities.

Up until 1978 the more progressive elements in the Trilateralist coalition predominated, and rationalizing market pressure was accompanied by new initiatives, especially in the field of energy and community development. But ultimately the Trilateralist coalition was too weak to resist growing pressures from the right. On the one hand the forward march of victorious national liberation movements put multinationals with investments in the Third World in a panic. These were the years of victory for Vietnam, Laos, and Cambodia, Guinea Bissau, Angola, and Mozambique, Ethiopia, Iran, Grenada, and finally Nicaragua. The Soviet Union increased its aid to national liberation movements, sending troops to Afghanistan in 1979. And even the globalist elements within the Trilateralist camp found it increasingly more difficult to resist the temptation of imperialist superprofits. Growing debt burdens soon began to undermine capital formation in the Third World in any case, making rising demand for U.S. capital goods seem increasingly unlikely. Powerful reactionary interests in the U.S. meanwhile, were profoundly threatened by the policies of the Carter administration. The petroleum industry feared the impact of a successful alternative energy strategy on its monopoly superprofits. Rentier elements were hurt by high levels of inflation which undermined the return on their investments. Intermediate technology sectors joined the clothing and textile industries in demanding protectionist measures to dampen the effects of foreign competition.

Antitax sentiment among the working class, meanwhile, was growing. Trilateralist elements used this to garner support for reductions in transfer payments and entitlement programs. This was a serious error, as "fiscal austerity" measures soon extended beyond reduction in spending for entitlement to attacks on spending for infrastructure, education, and nonmilitary research and development. The new community organizations which were subaltern partners in the Trilateralist alliance were not strong enough to stem this growing antitax sentiment among the workers, much less to affect the administration's policies (Davis 1986).

By the end of the 1970s the progressive alliance which had governed the United States since 1932 had collapsed.

3. Neoliberalism and The End of Bourgeois Progressivism

Towards the end of the Trilateralist period a new hegemonic bloc took shape within the Republican Party, composed of:

a) the large commercial banks and the multinational corporations, increasingly concerned about the rising tide of the national liberation movements,

b) the military industrial complex and the national security bureaucracy, and

c) broad strata of small investors, speculators, entrepreneurs, etc. who had little interest in investment in infrastructure, education, research, and development.

The new bloc was united first and foremost around a commitment to global free trade. The dominant sectors within the bloc——the large commercial banks and the multinational corporations, have been more or less committed to a neoliberal analysis of the global economy, according to which taxation, state expenditure, and state regulation all constitute obstacles to growth, though clearly the military industrial complex and the national security bureaucracy can only embrace this sort of analysis in a very conditional way.

This bloc was able to secure ideological and programmatic hegemony over broad sectors of the working class and traditional petty bourgeoisie along two distinct lines. As market relations penetrate every sphere of social life, all activity is transformed into merely a means of realizing individual consumer interest. This undermines interest in the public sphere and in promoting the development of human social capacities generally. Thus the rise of antitax movements throughout the 1970s.

At the same time, life in an advanced industrial society *is* characterized by an extraordinarily high level of interdependence. Mediation of this interdependence through market relations undermines humanity's grasp of its own underlying sociality, while at the same time subjecting individuals to powerful forces beyond their comprehension and control. Human interdependence receives alienated expression in ideologies in which the self-organizing activity of the cosmos appears as an alien power, beyond human knowledge or control. This is most apparent in evangelical Protestantism, which has historically stressed human sinfulness and the need to submit to the will of a divine sovereign whose inscrutable cosmic plan is forever beyond our comprehension (Fromm 1941: 56-122). Similar tendencies, however, have manifested themselves in the Catholic Church, specifically in the *Communio*-theology which is currently dominant in the Vatican (von Balthasar 1968), and in the popular twelve step programs, which, like the evangelical churches, teach that human beings cannot control their drives and must turn their lives over to what in this case is an unnamed "higher power."

Ideologically the link between neoliberalism and cultural conservatism was forged with the help of Austrian school economists who stressed the superiority of spontaneous social forms (the family, the marketplace, traditional religious institutions) to the rationalizing constructs of bureaucratic planners and redistributors. According to thinkers such as Frederick Hayek (1988: 70) morality arises spontaneously as a result of human adaptation to the environment. Traditional morals may not be rationally justifiable according to the canons established by rationalist thinkers from Plato and Aristotle, through Aquinas and Spinoza, to Hegel and Marx, but they have nonetheless demonstrated that they have survival value.

The new ruling bloc thus appealed both to a growing anti-tax sentiment rooted in the hegemonic consumerism, and to an emerging cultural conservative trend centered in the evangelical churches. For this reason it is appropriate to refer to it as the neoliberal-conservative bloc.

Once in power, the neoliberal bloc pursued a two pronged strategy. On the one hand they set in motion a largely successful counteroffensive against the socialist countries and the national liberation movements, which resulted by the end of the 1980s in the effective defeat of the Soviet bloc and its Third World allies——though not of China. This greatly weakened the position of the peasantry, the proletariat, and the intelligentsia around the world, and undermined the credibility of even moderate progressives proposing modest programs of increased taxation, state expenditure, and regulation.

High levels of military spending secured the alliance between multinational finance capital and the military-industrial complex, the national security bureaucracy, and the peculiar cadre of "defense intellectuals" and "rocket scientists" created by the cold war. At the same time they set in motion an assault on the national bureaucratic redistributional and regulatory apparatus, attacking funding for infrastructure development, education, social services, and nonmilitary research and development, and interfering systematically with the activity of critical planning and regulatory agencies, while drastically reducing taxes on the bourgeoisie and the upper middle strata. The result was a speculative boom based on high levels of military spending and luxury consumption, while the country's physical and social infrastructure literally rotted away (Davis 1986).

Already by the middle 1980s the limits of this strategy were becoming apparent. As defense spending peaked, and rising federal deficits prevented further reductions in levels of taxation, the economic engine which had driven the boom ground slowly to a halt, first in the farm-belt and the industrial midwest, then in Sunbelt metropoles heavily dependent on military spending, and finally on the coasts. The collapse of the Soviet bloc, and the resulting prospect of even lower levels of military spending, in the context of the prevailing neoliberal strategy, has meant not increased funds for the civilian sector, but a further contraction of state expenditures, and a deepening crisis throughout all sections of the country.

The impact of these policies on the income of the working class and the middle layers of society has been nothing short of disastrous. Real wages have fallen every year since 1973! The changes since Reagan took office have been particularly dramatic.

According to one study utilized by the AFL-CIO, 55% of the American population fell between the boundaries of $17,000 and $41,000 (in 1983 dollars) in 1978, By 1983 only 42% of the population remained within these middle income parameters. One quarter of the missing 13% moved upward, into the ranks of the affluent, but three quarters fell downward into the limbo of the working poor (Davis 1985:48)

The after tax incomes of the lower 20% of the population fell by 7.6% between 1980 and 1984; the income of the upper 20% rose by 8.4% during the same period.

The redeployment of much of the country's industrial base to Europe, Japan, and the periphery has been equally dramatic. Between 1960 and 1984 the portion of the domestic market controlled by imports increased from 4.2% to 25.4% for steel, from 4.1% to 22 % for automobiles, and from 3.2% to 42% for machine tools (Davis 1985:51). Beginning in the 1970s consumer electronics as well began to feel the pinch of competition, as imports, primarily Japanese in this case, began to flood U.S. markets.[8] And in the 1980s even relatively high level intellectual labor, such as computer programming and software engineering began to be outsourced to Third World markets. Bangalore, India is now one of the most important software centers on the planet. Indian programmers work for roughly 10% of their U.S. counterparts (Longworth 1994)

The 1992 election brought to power a new government which campaigned on a platform centered on defense of the ecosystem and the social fabric and investment in infrastructure, education, research and development. There is growing recognition, not only among the masses,

---

[8] . The newly industrialized countries of the Third World are not, of course, themselves immune to the pressures of the world market. The collapse of the Soviet bloc has opened up "super-low" wage labor markets such as Vietnam to capitalist exploitation, so that even countries such as Thailand and Indonesia are beginning to lose industry to "foreign competition."

who are largely unorganized, but also among certain sections of the ruling class, that we cannot continue on our present path. The record of bourgeois reformism, however, is sobering. One need not question the sincere commitment of bourgeois progressives to human social progress in order to understand that they will be able to accomplish very little so long as the larger framework of market relations remains intact. And the record of the first two years of the Clinton government is not encouraging. The only significant victory of the government has been passage of the North American Free Trade Act, which simply extends further the pressures of the world marketplace. The government has, on the other hand, been defeated outright or forced to retreat on economic policy, health care reform, social policy, and a whole series of other initiatives. Education budgets remain stagnant and research budgets have actually declined, as several important initiatives have been scrapped or scaled down. The record thus far is not encouraging. And the midterm elections of 1994 seem to have returned to power, under new and more extremist leadership, the neoliberal-conservative alliance, which governed throughout the 1980s.

It should be clear by this point that the marketplace is a profoundly destructive social force, and that reformist movements which operate within a still fundamentally capitalist framework are doomed to failure. The penetration of market relations into every sphere of life, together with the collapse of bureaucratic forms of regulation and social integration, has set in motion an ever deepening crisis of the ecosystem and of the social fabric. The marketplace, as we noted above, has no way to know how the activities it organizes affect the integrity of the ecosystem. So long as customary or legal restraints are in place, the damage to the ecosystem may well be gradual and even imperceptible. But once these restraints have been stripped away, the delicate balance of life itself is in danger.

Similarly, at the social level, the marketplace functions like a solvent, which dissolves the bonds of family, clan, village, region, and nationality so that individuals can be combined in new ways by industrial modes of production. In this sense, market societies have always depended on a reservoir of nonmarket institutions——a literal or figurative countryside——from which to draw individuals who have been socialized by stable families, trained in quality schools, and in general formed into productive, powerful, rational human beings capable of forming stable bonds with others. The destruction of this nonmarket hinterland means that such individuals are no longer being socialized, with results that are apparent from the ghettos of South Central Los Angeles, to the prosperous suburbs such as Plano, Texas——teenage suicide capital of the country (Lasch 1979, Bellah 1985, 1991).

The marketplace, furthermore, has no way to know how the activities it promotes affect the development of human social capacities. Once again, so long as the marketplace was surrounded by a complex of (private and public) nonmarket political, social, cultural, scientific, and religious institutions, it was possible to centralize the resources necessary to develop the infrastructure, educate workers and citizens, promote artistic and cultural expression and scientific discovery, and cultivate human spirituality, even where these activities did not themselves lead to the accumulation of capital. As traditional communitarian institutions declined, or where they were unable to carry out the organizing tasks required, bureaucratic institutions took their place. Twelve years of neoliberal hegemony, however, have completed in short order the destruction of these nonmarket institutions, something which has been taking place gradually for several hundred years, with the result that social progress, understood in terms of qualitative advances in technology, organization, art, science, and spirituality, is slowly grinding to a halt. Nonmilitary state expenditures have failed to keep pace with the need for investment in infrastructure, education, research, and development, and in many cases have declined, while generous tax cuts for the very wealthy have undermined the incentives both for private investment in new technologies, and for contributions to nonprofit institutions. Thus the current "crisis of capital formation."

Finally, as market relations undermine the social fabric, and the communitarian traditions of the working classes, it becomes increasingly difficult to organize workers to support the levels of taxation necessary to centralize resources in order to defend the ecosystem and the social fabric, or to promote the development of human social capacities. Consumerism appears to have triumphed. The result has been an enormous waste of potential. As Soviet Cosmonaut Volodya Manarov says to a North American counterpart in William Barton and Michael Capobianco's novel *Fellow Traveller*.

In 1970 you had a twenty-year technological lead on the rest of the world, but you pissed it swiftly away. You know, Marx was right, in a sense, but for reasons he could never have anticipated. Your capitalism, whatever else it may be, denies the future. It has no underlying philosophical unity, no goal other than immediate personal gratification. A half-century ago, under inspired leadership, you had one "brief, shining moment," of triumph, then you exchanged it all for military adventurism and backward looking economics. ... Money doesn't dream, you see, and now we will bury you after all (Barton and Capobianco 1991:381).

CHAPTER SIX:
The Mandate of History

The industrial, democratic, and scientific revolutions, by enabling humanity to grasp the underlying, organizing principle of the cosmos, and to use that knowledge to create new forms of organization, tapping into the potential latent in the physical, chemical, biological, and social forms of matter, marked an enormous step forward in the cosmohistorical evolutionary process. At the same time, regulation of this organizing activity by market mechanisms has held back full realization of the potential of these great revolutionary movements. This has been particularly true on the periphery of the world system, where penetration of market relations destroyed millennia of communitarian civilization, and where "development" has been largely restricted to low technology, low wage activities to which investment has flowed as the organic composition of capital increases and the rate of profit declines in the advanced capitalist countries. But this same dynamic has also held back development in the capitalist centers, as the export of capital has made it increasingly difficult to centralize the resources necessary for investment in infrastructure, education, research, and development. And the penetration of market relations has everywhere undermined the integrity of the ecosystem and the social fabric, which form the vital matrix out of which advanced social capacities emerge.

As the contradictions of the market system mounted, mass movements emerged which attempted to realize the full potential of the industrial, democratic, and scientific revolutions by restricting, and eventually overcoming entirely, the operation of market forces and constructing a rationalized redistributional system capable of centralizing and allocating the resources of society in such a manner as to best promote the development of human social capacities. In accord with the customary usage, and for want of a better word, we call these movements socialist, though it must be understood that we mean by socialism not only movements inspired by dialectical materialism, but also movements with an idealist or religious ideological orientation which aim at transcending, or at least significantly restricting the operation of, market forces. In this sense socialism represents just another stage in humanity's ongoing struggle to "realize the mandate of heaven," with the difference that, having been mediated through the experience of capitalist development, this mandate is now understood not simply as a demand to bring human society into harmony with the larger divine, cosmic order, but rather as a call to participate actively in the creative, self-organizing activity of the cosmos. Thus understood, socialism necessarily involved radically new developments in the theoretical arena. The dialectical philosophy first developed by Hegel, and elaborated in very different ways by the objective or religious idealists on the one hand, and the dialectical materialists on the other hand, marked for the first time humanity's understanding of the universe as a holistic, self-organizing, and teleological system within which it played a vitally important creative role.

From the vantage point of the mid-nineteenth century it appeared that, whatever struggles humanity faced in the transition to socialism, and in completing the difficult tasks of socialist construction, the path at least was clear. Humanity was well on the way to completing its prehistory and taking its place as a mature participant in the life of the universe. Today of course, things look very different. The planet's first socialist state, the Soviet Union, has been destroyed, and the socialist movements are everywhere in crisis. This is true not only of materialistic socialism, but also of movements grounded in objective or religious idealism, including Labor Zionism, African, Buddhist, and Islamic socialism, and Social Christianity. The forces of religious fundamentalism, neoliberal stasis, and postmodern nihilism are everywhere triumphant.

In this chapter we will analyze the origins, development, and crisis of the socialist project. We will begin by discussing the larger vision of dialectical philosophy which was first developed in idealist form by Hegel. We will explore the reasons for the collapse of Hegel's synthesis and the fragmentation of the dialectical tradition into competing trends, none of which was able to theorize adequately either the larger process of cosmohistorical evolution or the specific tasks facing humanity as it struggles to come to terms with the contradictions of industrial capitalist society. Specifically, we will argue that the basic insights of dialectical theory found inadequate support in the atomistic science of the nineteenth century. Some theoreticians (the objective idealists) responded to this situation by seeking a ground for organization outside of matter. Others attempted to derive complex organization from an essentially mechanistic pattern of interactions within matter itself:thus dialectical materialism. Neither doctrine was able to grasp the complex, self-organizing dynamic latent within matter, and thus neither was able to develop an adequate strategy for reorganizing human society and transcending the limitations of the marketplace. We will analyze the limitations of the strategies advanced by various trends and tendencies in some detail. Finally, we will explore some of the creative trends which have emerged out of the crisis of socialism, and which point beyond objective idealism and dialectical materialism towards a new, synergistic interpretation of the dialectical tradition.

I. Idealist Dialectics

A. Hegel's Synthesis

We have already analyzed in earlier chapters both the enormous contributions, and the very real limitations of the philosophical traditions which emerged out of the salvation religions generally and the Hellenic and Judaic traditions in particular. On the one hand these traditions grasped the universe as an organized totality, and were even beginning to understand the creative potential of humanity within the cosmic system. On the other hand, in so far as they regarded the cosmic order as something fixed by God or by the prime mover from all eternity, authentic development towards higher levels of organization remained essentially impossible.

Hegel was the first thinker to transcend these limits and, while conserving the ancient philosophical conviction that the cosmos is an organized totality, to theorize development as a integral, even constitutive dimension of cosmic organization. Hegel recognized that the cosmos was a complex, self-organizing system. The organizing principle, which he called the Idea, was implicit in even the most elementary forms of the natural world, and realized itself through the development of increasingly more complex forms of organization. In human society (what Hegel called the "objective Spirit") the Idea becomes conscious of itself. It does this first of all through the emerging thought processes of the human mind. But these processes can work themselves out only in the context of human social institutions: family, civil society, and the state. It is only on the basis of the historical experience of humanity that what Hegel calls "Absolute Spirit," the realm of art, religion, and philosophy, can emerge.

Hegel understands human history as at once a process of *differentiation*, i.e. of the gradual emergence of an increasing diversity of interests and capacities, and of *integration*, i.e. of the emergence of structures capable of comprehending these diverse interests and capacities. The emergence of a market economy, for Hegel, represented an essential moment in this process. The marketplace creates a complex web of social interdependence which is at least a partial realization of the underlying relationality of the Idea itself. To this extent he upholds the liberal doctrine.

> In the course of the actual attainment of selfish ends——an attainment conditioned in this way by universality——there is formed a system of complete interdependence, wherein the livelihood, happiness, and legal

status of one man is interwoven with the livelihood, happiness, and rights of all (Hegel, *Philosophy of Right*, section 183).

The division of labor makes necessary everywhere the dependence of men on one another and their reciprocal relation in the satisfaction of their other needs. Further, the abstraction of one man's production from another's makes work more and more mechanical, until finally man is able to step aside and install machines in his place (Hegel, *Philosophy of Right*, section 198).

The fact that I must direct my conduct by reference to others introduces here the form of universality. It is from others that I acquire the means of satisfaction and I must accordingly accept their views. At the same time, however, I am compelled to produce means for the satisfaction of others. We play into each other's hands and so hang together. To this extent everything private becomes social (Hegel, *Philosophy of Right*, section 192).

The operation of the marketplace, however, is inherently contradictory.

When civil society is in a state of unimpeded activity, it is engaged in expanding internally in population and industry. The amassing of wealth is intensified by generalizing a) the linkage of men by their needs and b) the methods of preparing and distributing the means to satisfy these needs... That is one side of the picture. The other is the subdivision and restriction of particular jobs. This results in the dependence and distress of the class tied to the work of that sort, and these again entail inability to feel and enjoy the broader freedoms and especially the intellectual benefits of society (Hegel, *Philosophy of Right*, section 243).

The poor still have the needs common to civil society, and yet since society has withdrawn from them the natural means of acquisition and broken the bond of the family ... their poverty leaves them more or less deprived of all the advantages of society, of the opportunity of acquiring skill or education of any kind, as well as of the administration of justice, the public heath services, and often even of the consolations of religion, and so forth(Hegel, *Philosophy of Right*, section 241).

Hegel even recognized the tendency of capital to seek outlets in foreign markets.

This inner dialectic of civil society thus drives it ... to push beyond its own limits and seek markets, and so the necessary means of subsistence, in other lands, which are either deficient in the goods it has overproduced, or else generally backward in industry, etc. ... (Hegel, *Philosophy of Right*, section 246).

This far flung connecting link affords the means for the colonizing activity——sporadic or systematic——to which the mature civil society is driven and by which it supplies to a part of its population a return to life on the family basis in a new land and so also supplies itself with a new demand and field for its industry(Hegel, *Philosophy of Right*, section 248).

Hegel proposes two possible was of resolving these contradictions. The first is the restriction of the operation of market forces by nonmarket social institutions. Hegel argues against the abolition, but rather for the reorganization and revitalization of the corporate structures (guilds, professional associations, etc.) which characterized precapitalist European societies. These corporations, he argues, provide a link between the self-interested activity of the individual and the universality of the state.

Under modern political conditions the citizens have only a restricted share in the public business of the state, yet it is essential to provide men——ethical entities——with work of a public character, over and above their private business. This work of a public character, which the modern state does not always provide, is found in the corporation (Hegel, *Philosophy of Right*, section 255).

It is vitally important that the corporations become authentic organizations of the working class.

For some time past ... the lower classes, the mass of the population, have been left more or less unorganized. And yet it is of the utmost importance that the masses should be organized, because only so do they become mighty and powerful. Otherwise they are nothing but a heap, an aggregate of atomic unites. Only when the particular associations are organized members of the state are they possessed of legitimate power(Hegel, *Philosophy of Right*, section 290).

Ultimately, however, the common good can be guaranteed only by the direct action of the state, in which Reason, the organizing principle which has been implicit in nature and society all along, becomes a fully developed, conscious actor, able to organize and direct all particular entities in such a way as to serve the universal common good. The state itself must be in the hands of what Hegel calls the universal class: the civil service or the bureaucracy.

> The universal class has for its task the universal interests of the community. It must therefore be relieved from direct labour to supply its needs, either by having private means or by receiving an allowance from the state which claims its industry, with the result that private interest finds its satisfaction in the work for the universal (Hegel, *Philosophy of Right*, section 205).

The state acts in two ways to promote the universal common good. First, the state regulates economic activity.

> ... public care and direction are most of all necessary in the case of larger branches of industry, because these are dependent on conditions abroad and on combinations of distant circumstances which cannot be grasped as a whole by the individuals tied to these industries for their living(Hegel, *Philosophy of Right*, section 236).

Hegel seems to suggest as well that the state must provide in some way for those impoverished by the operation of the marketplace

> When the masses begin to decline into poverty a) the burden of maintaining them at their ordinary standard of living might be directly laid on the wealthier classes ... b) As an alternative they might be given subsistence directly through being given work ... (Hegel, *Philosophy of Right*, section 245).

Second, the state serves the universal common good negatively, as it were, by making war. In peace time the operation of the marketplace tends to enclose individuals within the narrow sphere of their immediate interests.

> In peace, civil life continually expands; all its departments wall themselves in and in the long run men stagnate ... for health the unity of the body is required, and if its part harden themselves into exclusiveness, that is death (Hegel, *Philosophy of Right*, section 324).

War subjects the particular and finite elements of the system (individuals, firms, etc.) to the imperatives of the universal and infinite.

> In times of peace the particular spheres and functions pursue their paths of satisfying their particular sims and minding their own business, and it is in part only by way of the unconscious necessity of the thing that their self-

seeking is turned into a contribution to the reciprocal support and to the support of the whole ...

In a situation of exigency, however, whether in home or foreign affairs, the organism of which these particular spheres are members fuses into the concept of sovereignty. The sovereignty is entrusted with the salvation of the state at the sacrifice of these particular authorities whose powers are valid at other times, and it is then that that ideality comes into its proper actuality (Hegel, *Philosophy of Right*, section 278).

Ultimately for Hegel, the contradictions of human social existence could not be resolved by any political-economic system, because even the State in all its grandeur was inadequate to the Concept——the self-organizing systematic unity which constitutes the universe. These contradictions thus are resolved only beyond the state and beyond history in the realm of Absolute Spirit——the realm of art, religion and philosophy in which Spirit knows itself as it really is.

Hegel's vision thus remains profoundly ambivalent. On the one hand, he provides an adequate ontological ground for the emergence of complex organization, and understands that the cosmohistorical evolutionary process points beyond human society to something which we humans can grasp only in art, religion, and philosophy. At the sociological level, Hegel understands that the marketplace reflects authentic social progress——the emergence of a more complex ensemble of talents and interests than was possible in precapitalist, tributary societies. He also recognizes that the marketplace is unable to actually tap into those interests and talents in a manner which, to use his language, which is fully adequate to the idea——which, that is, comprehends fully the underlying relationality of which human society is at once agent and expression. Thus his concern to actually *transcend* market structures, rather than merely repressing them and returning to the "substantial unity" of premarket societies. It is no doubt for this reason that rejects any attempt to simply abolish the market in favor of centralized state planning.

At the same time, Hegel's actual solutions reflect a residual idealism. It is one thing to recognize that the Idea cannot be adequately realized within the limits of finite human social existence. It is quite another to argue that it is impossible to definitively transcend market relations and build a new, more complex mode of social organization, or to argue that human beings overcome their egoism only in the context of war.

Ultimately this failure reflects the low level of development of human organizing capacities generally and of the special sciences in particular, and the ideological alienation generated by the very market system which Hegel criticized. Even as Hegel was grasping, as no previous philosopher really had, the holistic, self-organizing, and teleological character of the universe, the atomistic paradigm dominant in the sciences and generated by the market system was gradually undercutting the scientific foundations of his vision. Of particular importance in this regard was the development of thermodynamics and the "discovery" of the second "law" which states that any closed system of particles will, over time, tend towards ever increasing disorder, or entropy. While Newtonian atomism had restricted humanity's ability to *theorize* complex organization, it did not, by itself, rule out either the emergence of such organization or the gradual evolution of the universe generally in the direction of ever higher levels of organization. The Second "Law" of Thermodynamics, on the other hand, suggested that the emergence of organized complexity is, at best, a counterpoint to a larger process of cosmic decay and disintegration.

We have already outlined, in the first chapter, the principal philosophical reactions to this development. Many thinkers abandoned entirely the belief in the ultimate meaningfulness of the universe, opting for one or another brand of nihilism——positivistic, existentialist, etc. Those who remained committed to the cosmic order faced the difficult task of finding some way to ground this order. There were two main alternatives. Some, the dialectical materialists, attempted to ground complex organization in the fundamentally mechanistic interactions of the natural world. We will examine their approach to the problem in the next section. Others, the objective or religious idealists, sought this ground in an organizing principle outside of matter.

It is not possible in this context to trace the complex philosophical itinerary of objective idealism. We have already suggested the main outlines of this itinerary in Chapter One. Our principal interest here is in the sociohistorical manifestations of the trend. Generally speaking it is possible to identify two principal tendencies. The first is essentially a prolongation of the ancient millenarian tradition of the peasantry. The second, associated with progressive elements within the religious hierarchies, aims at a more comprehensive rationalization of the salvation religions, and in particular at assimilating into the religious tradition some notion of historical progress and a more positive valuation of humanity's role in cosmohistorical evolution. We refer to this latter tendency as religious socialism. We will examine each tendency in turn, and the explore some of the reasons for the collapse of objective idealism and the rise of religious fundamentalism in the present period.

B. Objective Idealist Politics

1. Millenarian Movements

As the penetration of market relations began to undermine the integrity of the ecosystem and the social fabric on one continent after another, a new wave of peasant movements emerged which continued the ancient tradition of autonomous struggle against the forces of cosmic disorder, but with two important differences. First, the peasant movements now directed their main blow not against the warlord state, but rather against the forces of the world marketplace, and the invading European powers which attempted to make the world safe for capitalism. In the process they often entered into alliance with their old adversaries in the military and priestly aristocracies, which now seemed benign by comparison. Second, unlike earlier movements which simply attempted to restore human society to harmony with the divine order, these new movements now attempted to actually raise human society to a qualitatively higher level of participation in the self-organizing activity of the cosmos.

In speaking of millenarianism as a form of objective idealist politics, we are not, of course, implying that the peasant movements self-consciously appropriated and attempted to apply the philosophy of Hegel or other objective idealists. Rather, we are suggesting that the ideologies which did guide these movements reflected in popular form the same ideological dynamics which characterize the "high tradition" of objective idealist philosophy: the idea that the universe represents the gradual unfolding of some extramaterial organizing principle, and that human history is, in some sense, a progressive working out of that principle as it struggles towards realization in the millennial future. This said, it should be noted that if the peasant leaders were not reading Hegel they were, very often, reading Joachim of Fiore or similar theorists working in the Chinese or Indian traditions, or at least assimilating their ideas through popular tradition. And Joachism, we have seen, prefigured many of the characteristics of idealist dialectics.

We need now to examine a few of these movements in some detail. It is not possible in this context even to survey the tremendous variety of anticolonial millenarian movements. We have chosen, instead, to discuss a few movements in greater detail in order to document the characteristics we have outlined above.

a). Movements in the Americas

The main features of anticolonial millenarian movements are particularly well illustrated by the *Tupac Amaru* revolt in Peru in the eighteenth century. William Stein argues that the revolt originated as

> a challenge to the colonial system of *repartiamiento*, the forced sale of commodities by *corregidores* in their rural domains, designed ... to break the self-subsistence of peasant producers who had to accept the goods distributed and were necessarily obliged to sell their produce or their labor power in order to pay for the 'commodities' which had been distributed to them. The *repartos* contained some goods, such as cloth or mules ... which were useful; however, they also contained useless items, such as books in foreign languages, out of fashion apparel, overstocks for which there was no demand, and other assorted commodified garbage, sold to rural people at prices significantly above their rates in the market (Stein 1987: 102).

This system hurt both the peasantry, which was forcibly drawn into the market economy, and the *kuraka* class——descendants of the Inca nobility, who had been reduced to the function of collecting taxes and maintaining order for the Spanish overlords, but who had also developed into a stratum of wealthy merchants. *Tupac Amaru* himself was a member of this class. Born Jose Gabriel Condorcanqui in 1738, he took the name *Tupac Amaru* after the last Inca ruler, who had been executed at Cuzco in 1572. He owned coca fields, mines and 350 mules. He had attended a Jesuit school, spoke Spanish, read Latin and was fluent in Quechua (Stein 1987: 100-102).

The diversity of the interests aligned against the Spanish meant that the revolt put forward somewhat contradictory goals. On the one hand the movement demanded "the abolition of forced labor and forced sales, cancellation of duties and fees," and the restoration of the Inca monarchy in one form or another (Stein 1987: 105). This "minimum program" appealed to both the peasantry and the indigenous gentry who led the revolt. At the same time, the symbolism which developed around the revolt reflected deeper, more profoundly revolutionary aspirations. According to Andean cosmology, the cosmos has gone through several cycles of destruction, in which it has been transformed into stone, water and fire, and rebirth, and will go through several more until the perfect pattern of creation is at long last achieved. After the conquest this cosmology merged with the old Joachite prophecies which we discussed above, which look forward to a Third Age of the Holy Spirit in which all things will be held in common. The peasant masses who formed the main force behind the revolt thus saw it as the beginning of a great cycle of destruction and reorganization which would ultimately issue in the reestablishment of *Tahuantinsuyo* (the perfect Inca state), in which "people will no longer suffer, the high altitudes will produce subtropical fruits, valleys will feed flocks from the heights, and all people will live in peace and harmony. All people will have wings (Stein 1987: 106-107)."

Instrumental in the transition to this new order was *Inkarri*, the Inca king, identified variously with *Atahualpa* or *Tupac Amaru* I, "who has been beheaded, but is growing a new head, and who, when he is complete, will reestablish order in a new world (Stein 1987: 107)." *Tupac Amaru II* took on the role of *Inkarri* in the popular imagination, further fueling millenarian expectation.

Tupac Amaru was able to build an army of several thousand, to liberate much of the *altiplano*, and to hold it against the Spanish for several months. Excommunication by the church, dissension and vacillation within the *kuraka* class, and the arrival of colonial reinforcements, however, prevented him from taking Cuzco, and led ultimately to his defeat, capture, and execution.

The story of the Tupac Amaru uprising illustrates both the potential power, and the limitations of this type of movement. On the one hand the network of solidarities and the millenarian aspirations conserved by the peasant community make it possible to mobilize a large army relatively quickly and to deploy them in pursuit of the revolutionary reorganization of the social formation. The inability of the peasantry to develop its own leadership cadre, however, vacillation on the part of the nobles, bourgeoisie, or intellectuals who do lead the movement, and the technological superiority of European colonial forces generally undermine the movement in the long run. As market relations penetrate the countryside ever more deeply, gradually undermining the organization of the peasant communities, these weaknesses become more and more serious, until it is no longer really possible for the peasants to resist the capitalist offensive.

b) Asian Movements

Remarkably similar to the *Tupac Amaru* revolt in the way it integrates indigenous and European elements, and in the way it integrated an aspiration for an objectively synergistic social order with a reactionary neotributary program, is the Chinese *Taiping* revolt of the nineteenth century. Once again the revolt was based in resistance to the penetration of capitalist relations into the countryside.

The Taiping Rebellion began in the southern provinces of Kwangtung and Kwangsi, a natural unit ... oriented towards the port city of Canton. It was at Canton that foreign traders had first set foot on Chinese soil, and it was through Canton that foreign influence drove its main entering wedge after the opening of China to foreign trade (Wolf 1968: 119).

We have already noted the impact of the world market on Chinese industry.

After the opening of the ports of Shanghai, Nigpo, Amoy, Foochow and Canton to foreign trade under the terms of the Treaty of Nanking (1842) foreign cotton textiles and other consumer goods poured in. The total value of the merchandise imported from Britain in 1842 was L 969,381, by 1845 it was already L 2,394,827, of which L 99,958 was spent on cotton yarn and L 1,635,183 on cotton goods ... By the mid 1840s there were no longer good sales for cotton goods from Chekiang or Kiangsu, and stocks of native-made cloth were piling up in Fukien. The regions of Singkiang and Taitsang in Kiangsu province, once the most prosperous centers of native textiles, had only half the former volume of trade ... China's old social economic order was destroyed, handicraft manufacture went into decline and the path to industrial development was blocked. Large numbers of peasants and handicraft workers went bankrupt or became unemployed (Compilation Group for the History of Modern China Series 1976: 3-4).

The *Taiping* ideology, like the ideology of the *Tupac Amaru* revolt, integrated indigenous elements with millenarian motifs imported from Christianity. The movement had important antecedents in smaller revolts led by various Buddhist and Taoist secret societies.

... uprisings were initiated and organized by the popular, antidynastic secret societies such as the *Pai Lien Chiao* (White Lotus Sect) and the *Tien Li Chiao* (Heavenly Reason Sect) in north China, the *Nien* in Honan, Anhwe and Shantung, and the *Chia Chiao* (Society of Vegetarians) in Szechuan, Hunan, Kiangsi, Chekiang, Fukien, and other provinces. The biggest of these secret societies and the one which led most uprisings was the *Tien Ti Hui*, which had spread all along the Yangtze and through the southern provinces (Compilation Group 1976: 9)

These traditional influences were coupled with a heavy dose of fundamentalist Protestantism.

The *Taiping* leader, Hung Hsiu-Ch'uan, (1814-1864) had been a poor Hakka peasant from Kwantung. His family had sacrificed itself to pay for his studies which were to make him a school teacher, but he failed the examinations which would have allowed him to enter the bureaucracy. A Protestant missionary tract served as the catalyst in setting him upon an alternative career as a religious leader In the course of a vision, he came to seem himself as the younger brother of Jesus Christ and hence as the second son of God, chosen to destroy the demons on earth in order to create a new Kingdom of God. He also received two months of training in an American mission at Canton under the tutelage of the fundamentalist Rev. Issachar Roberts from Sumner County, Tennessee. The Bible took its place among the sacred books of the new religion (Wolf 1968: 120).

The resulting synthesis

combined the Western religious idea of "equality" with the Great Harmony concept of ancient China. Hung Hsui-Chuan exposed the darkness of his society, saying: "all love and hatred stem from selfishness," and expressed a longing for the society of great harmony as envisaged in ancient China. Contrasting the ideal society of mutual love and understanding with the contemporary world filled with rivalry and brutality, he wrote of his detestation of the social order of the time and his aim of establishing an ideal kingdom on earth in its stead. Hung Hsiu-chuan further clarified his opposition to the feudal system of ranks, feudal class oppression and autocracy in the lines: "All under Heaven have the same Heavenly Father and so are of one family," "What reason is there for the Emperor to grasp everything in his own hands?" (Compilation Group 1976: 20).

Hung founded the *Pai Shang Ti Hui* (Society for the Worship of God) in 1843, and began to build a revolutionary army. By 1853 he had taken the city of Nanking, which he renamed *Tienching*, the Heavenly Capital, establishing a "peasant state power ... in opposition to the landlord power which the Ching dynasty represented (Compilation Group 1976: 38).

The new authorities established a rigorously egalitarian centralized redistributional system.

Most of the men who were fit went into the army, while the rest were sent to workshops or took up other types of productive employment ... All the women were organized in Women's Camps ...

... not only were all levies and war booty to go to the "Sacred Treasury," it was also laid down that the capital of all the merchants, since it belonged to the Heavenly Father, should be contributed to the "Sacred Treasury." Subsequently all houses, precious metals, grain, merchandise, and so forth were turned over the public ownership ...

... no one, from the Heavenly King down to the ordinary soldiers should receive wages ... no one, soldier or civilian was allowed to own private property, and the needs of all were supplied form the public purse ...

The *Taipings* adopted a system of centralized management of the various handcraft industries ... (Compilation Group 1976: 43-44).

The Taiping program had equally revolutionary implications for the countryside.

... the Heavenly Land system ... contained two measures of paramount importance. Firstly, it abolished the feudal landlord system, made the land public property and distributed it to the peasants to till, and secondly it laid out a new basis for state power, the "System of Local Officials" ...

... land should be divided into nine grades depending on yields at the two annual harvests and distributed evenly among the people. A mixture of land of the nine grades was to be distributed to every family in conformity with the number of its members, regardless of sex ...

... local political administration was based on the Taiping military structure. The *chun* (army) *shih* (division) leaders, and so on down to the *liang* (platoon) leaders function as local officials. There were 13,256 families under an "army" leader, each of which was supposed to supply one soldier. In every "army" unit there were officers in charge of land distribution, the criminal court, cash and grain, revenue, and expenditures, each headed by an official and his deputy. The basic unit, the "platoon" of 25 families, was approximately the size of a village. Each "platoon" had a hall for worship where religious instruction was given and children were educated. Law suits, punishments, and awards were also administered at this basic level ... Thus a design was formed for a closely integrated ideal society (Compilation Group 1976: 46-51).

The new regime "insisted on the equality between men and women" and guaranteed this by ensuring equal rations and land allotments to all, regardless of gender, and by abolishing marriage by purchase (Compilation Group 1976: 52). It also issued edicts against foot binding and prostitution (Wolf 1968: 123).

Later these policies were supplemented by a strategy for economic development laid out by one of the movements' leaders, Hung Jen-kan in his *New Guide for Government*. Having spend time in Hong Kong, and elsewhere, Hung Jen-kan was impressed with the achievements of European civilization, and argued that the Heavenly Kingdom should establish trade and cultural relations with Britain and the United States in particular, though only on the basis of strict equal exchange. This presupposed development of a physical and social infrastructure of roads, railroads, and waterways, banks, currency, reformed public administration, hospitals, schools, and social welfare institutions (Compilation Group 1976: 88-92).

Holding much of China for over 10 years, the revolution must be regarded as far more successful than most previous peasant revolts. It was put down only after it was weakened significantly by internal dissension, and only after the Ching government was bolstered by foreign intervention (Compilation Group 1976: 73-81, 126-159).

Maoist scholarship has traditionally understood the revolt as a peasant uprising with radically egalitarian aims, which, due to the low level of development of the productive forces, could only promote the development of a capitalist society. "Whatever the subjective desire of the Taiping leaders, objectively the *Taiping* Revolution was bound to pave the way for the development of capitalism (Compilation Group 1976: 48)." Maoist analysts point in particular to the emphasis in the *New Guide to Government* on promoting better relations with Europe and the U.S. and on development of the productive forces.

I would like to argue that both dimensions of this assessment are inaccurate. First of all, while radically egalitarian in the sense of eschewing any inequality in the distribution of goods, the aim of *Taiping* system seems to me to have been directly less at guaranteeing rigorous equality than at eliminating all luxury consumption and mobilizing all of the resources of the society for the common good. Indeed the Heavenly land system, and the System of Local Officials, both seem to owe a great deal to the ancient *Chou-li*, which we analyzed in an earlier chapter. The difference between communitarian and tributary forms has less to do with the rate of surplus extraction than it does with the way in which the surplus is extracted (through persuasion or coercion) and the way in which it is invested (productively or in military/luxury expenditure).

Similarly, the principal economic measures of the *Taiping* regime seem to have been directed more at restricting the development of commodity production than at suppressing tributary forms of exploitation and promoting capitalist development. A highly centralized domestic economy was supposed to provide the basis for rapid development of the productive forces, and for the assimilation of European technology and development of international trade on the basis of equal exchange. In this sense the *Taiping* anticipated much of the economic development strategy of the socialist states of the twentieth century. If pressure eventually mounted for reintegration into the world market, then this is only because like later socialist systems, the Taiping state was effectively surrounded by aggressive capitalist powers.

Similar movements developed in India and Burma, where the penetration of capitalist relations of production was accompanied by the presence of direct British political domination, which gave the revolts in question a more immediate target. Popular Hinduism had given birth, after around 200 C.E. to a cyclical cosmology according to which the cosmos underwent successive cycles of birth and destruction (called *yuga*), each cycle relatively worse than its predecessor. This process had culminated in the present *kaliyuga*, or evil age, which is dominated by disorder and violence. At the end of this period one or another of the principal Hindu deities (generally *Siva* or *Vishnu*) will become incarnate and establish *dharmraj*, the kingdom of law, virtue, or order. Some sects looked forward to the last of the 26 or 28 incarnations of *Siva*, others to *Kalki*, the last incarnation of Vishnu. Hindu forms merged with Jain, Buddhist, and Islamic expectation regarding a coming period of 63 saints, or the advent of Maitreya, or the Mahdi. (Gough 1973). Most of these millenarian sects drew their members from the lower castes.

This millenarian milieu gave birth during the seventeenth century to a unique form of social banditry: the so called *Thuggee*, which emerged in Delhi and Agra, and later spread to Oudh, Bengal, Orissa, Rajaputana, the Punjab, Mysore and Karnatak, in response to Mogul and British domination. Displaced peasants and disbanded soldiers dedicated themselves to the Hindu goddess *Kali,* or occasionally to the Islamic *Fatima* (daughter of the prophet) who, they believed, had created their order to root out evil beings and save humanity from destruction. They would rob and strangle, with their characteristic yellow scarves, merchants, soldiers, money carriers, and employees of the British East India Company. They were forbidden to kill women, children, youths, holy men, carpenters, the poor beggars, bards, water carriers, oil vendors, dancers, sweepers, launderers, musicians, and cripples. They shared what they stole with the peasants, killing at least 1 million people and plundering several million rupees in the hope that by rooting out the evil doers they would contribute to the advent of the *dharmraj* (Gough 1973: 104ff).

Eventually this pattern of social banditry led to several full scale insurrections, the best known of which is the North Indian Revolt of 1857——the so called Sepoy Mutiny. Harvey Alper summarizes the events

> The revolt of the native troops in the Company's army began in the Winter of 1857. Rumors concerning the use of animal fat——repugnant to both Hindus and Muslims——to grease the cartridges for the army's new

Enfield rifles began circulating as early as January. Disturbances in March led to the disbanding of several native regiments. On May tenth, the garrison at Meerut revolted. The troops marched on Delhi. They occupied it on the eleventh, and proclaimed the last of the Mughals, Bahadur Shah, once again Emperor. As news of the fall of Delhi spread, other mutinies followed. By June, British authority had disappeared from vast areas of central India. The rebel triumph was short lived. British forces reoccupied Delhi in September. In March 1858 they retook the cities of Lucknow, Jhansi, Bareilly and Gwalior. Although fighting continued for another year, by June 1858 it was clear that the British had reasserted their paramountcy over India (Alper 1987: 37).

It is interesting to note that Hindus, Sikhs, and Moslems fought side by side in this revolt.

Similar revolts occurred in Burma as early as 1838, centered on popular expectations of the coming of *Mettya* (*Maitreya*) Buddha, the last world teacher who was to restore the rule of *dhamma* (*dharma*), and *Setkya-Min* or *Cakkavati*, the last world monarch who was to enforce his decrees. The tradition of millenarian peasant revolution continued in Burma up into the twentieth century, culminating in the peasant revolt led by Saya San in 1930-1932.

In the context of rural Burma's folk-Buddhism, Saya San was rumored to be the Setkya-Min, the idea Buddhist ruler of the Four Island Continents, and reputed to be of royal descent. In accordance with such traditions, he had a "Royal City," called "Buddha-Yaza-Myo," pegged out and erected a palace of bamboo (in the context of ancient cosmocentric symbolism) on the jungle mountain Alaun-taung ... (Sarkisyanz 1965: 161-162)

This revolt, like the others, was repressed and Saya San was hung by the British in 1931 (Sarkisyanz 1965: 151-164).

Islamic anticolonial movements deserve special attention, both because of their unique character, and because one of these movements——the Wahabi sect——succeeded in establishing a state power which has endured up to the present period. Islamic revivalism has roots which reach back to the earliest centuries of Islam itself, as the Karijites, and the various Shi'ite sects, resisted the transformation of Islam into a relatively progressive, cosmopolitan religion. The rapid development of European society, however, and the eventual conquest of most of the planet by the European powers created new difficulties for Islamic societies. Islam, like Hindu, Chinese, and other ancient civilizations, experienced a sense of profound humiliation when they were suddenly eclipsed by what could only seem to them to be European barbarians, who had only recently acquired a thin veneer of civilization. Islamic societies, like the rest of the Third World, experienced the disintegration of their social fabric under the pressure of market relations, and the redeployment of capital from relatively high wage, high technology activities in the advanced industrial countries to low wage, low technology activities on the periphery which further undermined traditional ways of life while doing little or nothing to promote the development of human social capacities. In the case of those Islamic countries which possess large petroleum reserves, this meant transformation into dependencies of large British and U.S. petroleum corporations.

But for Islam, the problem was much more profound. Islamic doctrine required that the entire planet be subjected to Islamic rule, and during its first centuries Islam experienced rapid imperial expansion. Now, however, Islam was suffering even more rapid territorial losses, as Islamic rulers were displaced by Europeans. The Islamic response to European imperialism thus incorporated two distinct dimensions. On the one hand, like anticolonial movements globally, the Islamic resistance provided a vehicle for the aspirations of the peasant, and later proletarian masses. Even more so than in the case of the Tupac Amaru or Taiping movements, however, the Islamic resistance was always a movement of imperial restoration, and because of the nature of Islamic empires, a movement for renewed global expansion.

Among the most important Islamic resistance movements are the Wahabi, founders of Saudi Arabia, the Mughal restorationist movement initiated by Shah Wali Allah in India, the Libyan Sanusi and the Sudanese Mahdists. All of these movements centered, on the one hand, on the purification of Islam——specifically the restoration of its radical monotheism——and, on the other hand, on the establishment of an Islamic state in which Shariah would be the only law. In most cases the leader of the movement claimed broad authority to interpret the law (*ijtihad*). In the case of the Sudanese *Mahdiyya*, the leader actually claimed divine inspiration.

The Wahabi movement had its origins in the eighteenth century, in the Arabian peninsula. Its founder, Muhammad ibn Abd al Wahab was educated at Mecca and Medina, where he studied law with teachers of the Hanafi school——the strictest of the Sunni tendencies. Abd al-Wahab rejected the innovations introduced into Islam by both Sufi mystics and rationalizing philosophers and jurists, and argued for a literal interpretation of the Quran and the Sunnah. This interpretation centered on a radical monotheism which rejected all veneration of saints, kings, martyrs, etc. and even of the prophet himself. Indeed, al-Wahab's followers referred to themselves as "Muwahiddun" or unitarians. Abd al-Wahab joined forces with a tribal chief, Muhammad ibn Saud, and together they destroyed all of the sacred tombs of Mecca and Medina, as well as Karbala, a major Shiite shrine in Iraq, setting in motion the movement which led to reunification of most of the Arabian peninsula under Saudi rule (Esposito 1984: 33-35).

The Wahabi movement is interesting because it is the only tributary restorationist movement to succeed in establishing a lasting state power——something which, to be sure, it accomplished only with the aid of colonial authorities. It thus permits us to assess both the possibilities and the limitations of this kind of movement. The results are not encouraging. Even allowing for ideological differences between Wahabism and, for example, the Taiping movement, it is apparent that little was actually achieved in the way of resistance to the penetration of market relations or the restoration of autonomous imperial authority. On the contrary, the Wahabi state in Saudi Arabia merely puts an Islamic cultural veneer on what, from an economic standpoint, is still a backward economy dependent on the export of primary products——albeit primary products which draw significant monopoly rents on the world market. At most it might be said that the Saudi regime is able to secure a share of these rents for itself, and for the Saudi people. But there has been no real break with capitalism. It is only on the basis of a rational scientific understanding of the corrosive impact of market relations that it is possible to develop a strategy for resisting European colonialism. Movements which understand anti-colonialism or anti-imperialism in a primarily political-military or cultural sense will inevitably end by succumbing to imperialist hegemony.

Similar movements developed during roughly the same period in India and Africa, which sought to cleanse Islam of popular accretions, and, at the same time, to restore Islamic rule (Esposito 1984: 35-39). The Indian movement initiated by Shah Wali Allah succeeded for a time in establishing a state in what would eventually become Pakistan, as did Uthman dan Fodio in Northern Nigeria, Sayyid Muhammad ibn Idris in Libya, and Muhammad Ahmad in the Sudan——though all of these regimes were later supplanted by pro-imperialist forces, by Islamic socialism or by new, more radical forms of Islamic fundamentalism.

c) European Movements

We have already noted above that the penetration of capitalist relations of production, while it benefited European capital, had a devastating effect on the European countryside. This was particularly true in those regions of Southern and Eastern Europe which stood in a quasi-colonial relationship to the developing metropoles of the North and West. Once again the Italian case illustrates well the larger pattern. During the early stages of *Risorgimento*, the peasants supported the struggle against the Bourbon monarchy, and for the unification of Italy. Garibaldi, for his part, realized that the bands of brigands which Sicily and the *Mezzogiorno* had harbored for centuries, and which had expanded their ranks in expectation of the coming anti-Bourbon insurrection, could be of no small value to him in his conquest of the island, and he courted the peasants with promises of an end to the tax on flour, and vague promises of a division of the lands of the clergy and those of the nobles who backed the Bourbon dynasty (del Carria 1966:34). But when the peasants demanded more, he responded with fierce repression (del Carria 1866:52), which in turn unleashed a long period of guerilla warfare and social banditry. Indeed, it soon became apparent that the new government, far from carrying out a comprehensive agrarian reform in the interest of the peasantry, planned to implement a series of policies which threatened to undermine the very existence of the rural communities and to destroy their traditional way of life.

This drove many peasants into the arms of the Bourbons and the church. Land invasions and sacking of palaces and town halls, which earlier had taken place under the slogan "Death to the Bourbons! Long Live Garibaldi!" now went forward to cries of "Death to the Liberals! Long Live San Erasmo! Long Live Francesco! (the Bourbon King)" (del Carria 1966:57-92)

The following incident in Northern Basilicata is typical.

On the 7th of April 1861 the brigands descended from the woods of Volture and Lagopesole, met up with the miserable peasants of the latifundia of Prince Doria, and adorned with the red cockade of peasant revolt, and carrying the white banner of the Bourbons, invaded the lands to the cry of "Long Live Francis II! Death to the Usurping Lords!" Encouraged by rumors that Bourbon and Austrian troops had disembarked on the coast, the peasants armed themselves with agricultural implements and invaded the feuds. Two to three hundred peasants from the mud hut settlements of Filani, Frusci, Iscalunga, San Ilario and Lavagnone struck Ripcandila on the night of the 7th, where the lower classes were already in revolt, killing the commander of the National Guard, disarming the militia, and proclaiming a Bourbon government, destroying the coat of arms of the House of Savoy, electing new municipal officers ... devastating the palaces of the landlords, and celebrating their victory with a Te Deum, fireworks, and a feast. Carmine Crocco, known as Donatelli, proclaimed himself a general of King Francis and became the leader of the peasant movement (del Carria 1966:76).

Similar developments took place in the North and Center after the reimposition of the tax on flour on 1 January 1869 (del Carria 1966:124). Gramsci writes

> The tax on flour was unbearable for the small peasants who consumed what little grain they produced; the tax required them to sell grain in order to procure money to pay the millers tax, and was the occasion for usury of the worst kind (1949b:161).

Forced to sell what formerly they had eaten, their lands threatened with expropriation for non-payment of taxes, or for failure to repay the loans they had taken out in order to pay their taxes, the peasants of the North and Center, and especially small holders who were more numerous in these regions, now followed the peasants of the South into the piazze to seek redress. "For the first time from the Alps to Sicily the peasant world was moved by the same struggle: hunger and desperation unified the poor of the North and the poor of the South" (del Carria 1966:91) Their battle cry:

Long live bread without the tax on flour!

Long live the Austrian government, the Pope, and religion!

And in the South:

Long live King Francis!

These movements reflect the persistence of monarchic ideology among the peasantry. The king and the pope were guardians of the moral and even of the cosmic order, and the peasants looked to them to defend their traditional way of life against the forces of capitalist development and modernization. Sereni sees this as evidence of the inherently backward and reactionary character of the peasant movements.

> Apart from the economic, organizational, and moral influence of the clergy, the religious nostalgia which manifested itself during the anti-tax movement among the peasant masses ... finds its explanation in the persistence of the social basis of religious ideology in the Italian countryside. The profound social crisis which sowed destruction among the agricultural population accentuated in those years the motives for religious sentiment ... Clerical and reactionary ideology could all the better adapt itself to the peasant movements in so far as these movements themselves had a reactionary dimension of resistance to the advance of capitalism in

the countryside (Sereni 1968:92-3).

This assumes, however, that resistance to capitalist development is always and everywhere a mark of reaction. It would be more accurate to see the peasants as communitarians attempting to conserve one of humanity's most precious resources: the village community. Weakened by the penetration of market relations, and forced to choose between a liberal revolution which threatened to undermine their entire way of life, and a clerical elite which at least upheld their traditional values, the peasants opted for an alliance with the clergy.

The Church, of course, soon made an uneasy peace with the new state, and peasants were left to resolve their problems on their own. Disillusionment with the clerical-legitimist alliance further radicalized the peasantry, and set in motion a process of rationalization which issued first in the emergence of neo-Joachite tendencies, and ultimately in the emergence of a Christian socialism.

Of particular interest in this regard is the movement of the *giurisdavidici*, which developed in the region of the Monte Amiata. Matters were particularly difficult for the peasantry of this region, where the capitalist transformation of the countryside was in full swing, and small-holders and *mezzadri* were being driven off the land in large numbers. American grain had flooded the world market after 1875, and wheat prices were plunging, making payment of the hated tax on flour ever more difficult. (Hobsbawm 1959:70) By 1878, the peasants were ready to act.

The peasants of the Monte Amiata found their leader in a certain Davide Lazzaretti, a carter from Arcidosso. Lazzaretti's spiritual development reads like a political history of Italy at mid-century. Born in 1834, he experienced a vision in 1848, and underwent a more profound conversion in 1868 --the year when the tax on flour was reimposed (Hobsbawm 1959:68)-- after which he began to elaborate his prophetic teachings systematically.

> He believed himself to be a remote descendent of a French king (France at the time being the principal defender of the papacy) ... he foresaw a prophet, a captain, a legislator, and a reformer of laws, a new pastor from Sinai who was to arise and liberate the peoples now groaning "as slaves under the monster of ambition" . . . Increasingly he left the old orthodoxy behind. . . hitherto had been the Kingdom of Grace, which he identified with the papacy of Pius IX. It would be followed by the Kingdom of Justice and the Reform of the Holy Ghost, the third and last age of the world. Great calamities were to presage the final liberation of men at the hand of God. But he, Lazzaretti, would die (Hobsbawm 1959:69).

Lazzaretti almost certainly stands as the last of the important popular interpreters of the old Joachite prophecies of the Third Age, a genuine heir to the *fraticelli* of the Middle Ages. The alliance with the church and the crown nearing the end of its usefulness, the peasants thus began to reassert the traditional, autonomous form of their Christianity.

Lazzaretti's monarchist pretensions notwithstanding, the *giurisdavidici* were strong Republicans. In 1878 Victor Emmanuel and Pius IX both died. Lazzaretti returned from France in July and proclaimed himself Messiah, whereupon he was excommunicated. On August 18th, during the octave of the Assumption, a traditional time of rest and feasts in Italy, he descended from his holy mountain. Crowds greeted him singing

We go by faith to save our fatherland,
Long live the Republic, God and Liberty!

They ran up their banner, on which were inscribed the words:

The Republic is the Kingdom of God.

Lazzaretti addressed his people:

What do you want of me? I bring you peace and compassion.
Is that what you want?

The crowd responds "Yes!"

Are you willing to pay no more taxes?

Again the crowd responds in the affirmative.

Are you for the Republic?

Again "Yes!"

But don't think it will be the Republic of 1848. It will be the Republic of Christ. Therefore all cry with me: Long live the Republic of God (Hobsbawm 1959:66).

The *carbinieri* fired, killing Lazzaretti and a few others. His "apostles" and "Levites" were tried and sentenced, and the *giurisdavidici* ceased to be a mass movement. . . though a few still hold out in the district of the Monte Amiata.

The peasants, who, only a few years before had been making common cause with the Bourbons and the Hapsburgs, now joined their autonomous popular Christianity to the republican cause. It would be only a little while before they realized the necessity of an alliance with the working class.

The similarity among these movements is striking. All were set in motion to a large extent by the penetration of market relations into the countrysides of peripheral or semiperipheral regions of the emerging world system, a process which destroyed indigenous industry and severely disrupted traditional patterns of agrarian life. They all drew their ideology primarily from millenarian religious traditions which had been conserved by the peasant community, or by religious sects or secret societies, from which they also drew their organizational networks. These indigenous elements, however, were layered over by motifs taken from European Christianity (Joachism, fundamentalist Protestantism) or occasionally from organization theory (in the case of the *Taiping* revolt) or republican ideology (in the case of the *giurisdavidici*). All were directed first and foremost at the radical restriction of market relations. Indeed, it might be said that these revolts turned on indigenous tributary ruling classes only when and to the extent that they appeared to be collaborating with foreign capital or promoting the penetration of capitalist relations of production into the countryside. To the extent that they were able to elaborate a program it was nearly always the same: reconstruction of a highly rationalized centralized state redistributional system which would radically restrict or even eliminate the necessity of commerce, and permit rapid economic development. All failed largely because the structures which they were able to develop were not able to organize sufficient technological, economic, or political-military power to mount an effective contest with the metropolitan bourgeoisies. This was undoubtedly due in part to the fundamental inability of centralized state redistributional systems to organize complex technological, economic, political, and cultural activity. But it was also due in part to the fact that the communitarian and tributary structures from which these revolts drew their organizational models had already begun to disintegrate under the pressure of advancing market relations.

2. Religious Socialism

Political initiatives rooted in an objective idealist theoretical problematic have not been confined to the peasant communities. On the contrary, progressive elements within the religious hierarchies also rose to meet the challenge posed by the contradictions of capitalist development. On the one hand, they recognized the enormous contributions of the industrial, democratic, and scientific revolutions, and attempted to draw out the principal theological implication of these revolutions: that humanity is an active participant in the cosmohistorical evolutionary process. At the same time, they rejected the global process of secularization which was accompanying the penetration of market relations, arguing that secular ideologies provide an inadequate ontological ground for the cosmic order or for principles of value. For this reason they rejected dialectical materialism and secular socialism generally as simply a new manifestation of the same disorder which had produced the marketplace, and began to search for a new road forward for humanity which was faithful both to the traditional values of the religious traditions and to authentic achievements of the industrial, democratic, and scientific revolutions.

Religious socialism differs from peasant millenarianism in its vision and strategy as well as its social basis. Religious socialists have generally been less confident than their peasant comrades in humanity's ability to "realize the will of God" or to build a society "adequate to the idea" and have thus been less audacious in their aims. Even where they have advocated armed struggle, it has generally been in the context of a "just war" rather than a "holy war" theory. The real work of building a new society takes place in the institutions of civil society and in the hearts and minds of the people, not on the battlefield.

Religious socialism is, in many ways, similar to the "populism" which we will analyze later on in this chapter. The principal difference is that most of the movements discussed in this section have specifically affirmed the truth value of the religious claims of their tradition, so that their invocation of this tradition is not purely sociological or strategic as is the case with many populist movements. The line is a fine one however, and readers may dispute specific elements of my classification without calling into question the larger argument of the section.

Generally speaking millenarianism is a phenomenon of the earlier stages of capitalist development, when intact peasant communities face sudden disruption in the face of rapidly penetrating market forces. Religious socialism, on the other hand, is a phenomenon of mature capitalism and of the imperialist era, for it is only then that the religious hierarchies begin to understand that their interests are radically incompatible with those of capital, and that their old alliances with the landed elite can no longer protect them.

a) Asian Socialism

The popular socialism which remained closest to its millenarian roots was

300

Towards Synergism

clearly the Buddhist socialism which developed in the Theravada Buddhist countries during the period around the Second World War, and which became the dominant force in Burma for some time after the war. Once again, the starting point of the Burmese revolution was the marginalization of the peasant masses as a result of colonialism and the penetration of market relations. As we noted in the previous chapter, by the time Burma achieved independence, only 15 % of the land was actually owned by the Burmese peasants, most of it having been expropriated by the British colonial elite (Sarkisyanz 1965: 1980). The economy was based almost exclusively on the export of rice and other agricultural products, industry being all but non-existent.

The Burmese national liberation movement of the 1930s and 1940s developed out of a more or less self-conscious effort on the part of the Burmese intelligentsia to rationalize the traditional peasant millenarianism represented by the Saya San and to forge out of it the basis for the creation of an independent Burmese state. Burmese Marxists used Buddhist language to explain Marxist concepts.

> Thus Thakin Soe attempted to explain the Leninist unity of revolutionary theory and practice in terms of the Buddhist ... "*Priatti, Prapti, Privedi* (preservation of scriptures, possession of knowledge, practice based on scriptural knowledge ...)

He also

> used the characteristically Buddhist concept of Perfection (*Parami*) to describe the special perfections that a "revolutionary leader" must possess ... the Buddhist term for periodical (cyclical) generation and destruction of worlds (*Upathi bin*) was used to designate the eternal flux of Matter in the context of Dialectical Materialism ...

> Of Buddhist origin is the Burmese strike slogan, "turn down, turn down," (the alms bowl) ...

> Of Buddhist origin is also the term often used in Burmese Marxism and radical nationalism for the goal of the revolutionary struggle, the perfect society, *Lokka Nibban* ("The Earthly Nirvana") (Sarkisyanz 1965: 168-170).

At the same time, Buddhist theoreticians began to carry out a systematic rationalization of Buddhist doctrine.

That Nirvana is to mean not absolute non-existence but a life of fellowship in an atmosphere of Truth, Goodness, Freedom, and Enlightenment was claimed already in 1907 by the Buddhist modernist Lakshmi Narasu ... (Sarkisyanz 1965: 170-171).

... within modernistic Buddhism, a book by the Abbot San Kyoug Sayadaw attempted to endow the productive activism of labor with a Buddhist ethos——by reinterpreting the meditation effort (viriya) so as to identify it with the effort of toil (Sarkisyanz 1965: 217-218).

The result of this dual process of rationalization was the emergence, in the 1930s and 1940s, of a Buddhist Socialist movement. This movement looked to the traditional Burmese village community as the basis for creation of socialist society.

Some countries are trying to experiment with that classless society with varying degrees of success and at the cost of much money, labor, blood, and lives. But if you really want to see a classless society which has not been forged by blood and iron but which is the result of natural growth and which stands in peace and harmony, visit a Burmese village and see the society there which is older than Communism ... All people belong to the same class or in other words, there is a classless society (U Ba Yin in Sarkisyanz 1965: 198).

As in the case of other populist movements, Burmese socialism was characterized by sharp differences over both the ultimate goal, and the method of the revolution. These differences reflected contradictions between the peasantry and its "organic intellectuals" on the one hand, and the intelligentsia, bureaucracy, and national bourgeoisie on the other hand. Initially these differences were submerged in the struggle for independence. During the 1950s, however, they began to come to the surface. U Nu, Burma's postwar leader, argued for a strategy which gave priority to agriculture over industry, and to the moral transformation of the Burmese people to the struggle for economic growth and development. One post-independence observer wrote

The Burmese peasant is now tenaciously clinging to his nearly self-sufficient village economy, which resembles more and more the Burmese village under the kings than the partially modernized rural scene of the immediate pre-war period (Pye in Sarkisyanz 1965: 190).

U Nu adopted more or less wholesale the traditional Buddhist vision of human society.

> He explained that, in accordance with Buddhist tradition there had been not
> private ownership before men started taking more than they needed ... He
> narrated the Buddhist tradition on how the fencing off and demarcation of
> property had been followed by theft so that men were induced to elect one
> of themselves to judge and punish ... From the primeval classless society
> emerged the different classes ... Under the Great Charity (Buddhism's
> Maha Dana, identified by U Nu with common ownership) everybody would
> take only what he needs without seeking profit ... By doing Dana Charity
> (in observance of Dhamma) the material welfare of the Cakkavatti's and
> Setkya Min's ideal states could be brought about ... In this sense U Nu
> declared in contradiction of all Marxism that the main difficulty for the
> establishment of a socialist world is low morality. Therefore the main task
> was to be the establishment of the *ethical* basis for it (Sarkisyanz 1966:
> 222-224).

Members of the technocratic intelligentsia, the bureaucracy, and the
national bourgeoisie resisted this focus on restoration of the village
community and moral and spiritual reform, in favor of industrialization.
This did not, however, mean that they rejected either Buddhism or
socialism. Rather, they regarded the ideology which had emerged from the
struggle for independence as a means of legitimating the process of
industrialization in the eyes of the masses (Sarkisyanz 1965: 233).

A military junta led by U Ne Win, overthrew U Nu in 1962. Ne Win
continued to develop Burma in a broadly socialist direction, but began
increasingly to give priority to criteria of efficiency and productivity
(Sarkisyanz 1965: 235ff). The resulting system was characterized by a
strong state sector, the persistence of the village economy, and a strong
resistance to European influence.

The results of the Burmese socialist strategy have been rather more
impressive than most foreign analysts have been willing to acknowledge.
Despite a miserable and not entirely undeserved international reputation for
human rights abuses, by the end of the 1980s Burma's UN Human
Development Index stood well in advance of its GDP index, a sign that
while economic growth has been slow, Burma has effectively exploited its
limited economic capacities to increase literacy and life expectancy (United
Nations Development Program 1990: 128). At the same time, it does not
appear that the Burmese socialists have been able to tap into the creative
energy of the Burmese people to unleash real development of the
productive forces and thus to expand the surplus available for investment.

b) African Socialism

African Socialism reflects much the same pattern as Asian socialism. The most eloquent exponent of this "natural" African welfare state has been Tanzanian President Julius Nyere, who in the 1960s set forth in considerable detail his view that socialism was the original state of the African man. "The foundation and the objective of African socialism ... is the extended family." He literally defined socialism as "familyhood," or "*ujamma*" in the Tanzanian national language of Swahili. This "natural" African tribal man which Nyere set out to revive, lived in basic harmony with his brother and society was free of social conflict. "The idea of class or caste was nonexistent in African society," according to Nyere, because it was only the Agrarian and Industrial Revolutions of Western Europe which gave birth to such social cleavage and thus to such conflict. The task of an African socialist leader, then, was to restore African society to its pristine classless self, and this could be done by ending private ownership of land and other means of production and reestablishing the communal approach to all human activities. It was a question in effect of turning the clock back to Africa's precolonial times (Ottoway and Ottoway 1981: 14).

Similar ideologies emerged at various points in Ghana, Guinea, Sao Tome and Principe, the Seychelle Islands, Mali, and Zambia. Concretely, implementation of the vision differed significantly, in part because of the very different conditions in the various African socialist countries.

In Tanzania, which remained primarily rural and agrarian, development efforts centered on moving families into *ujamma* villages, at first voluntarily and then through coercive means. Concentration in villages gave the peasants access to schools, clinics, clean water, and, eventually it was hoped, to agricultural equipment which would raise agrarian productivity. It would also help to combat individualism (Ottoway and Ottoway 1981: 46).

The results of this strategy have been mixed. Between 1967 and 1975 per capita GDP increased only 1.4 per cent. From 1970 to 1977, the average annual growth rate in the agricultural sector, which received priority under Nyere's rule, was only 3.5%, barely more than the rate of growth in population. The control of most of industrial sector permitted the state to reinvest fully 20 per cent of the GNP between 1964 and 1975, with the result that industrial production increased an average of 7.8% annually (Ottoway and Ottoway 1981: 48-49).

Investments in health and education have produced mixed results. "By the end of the 1970s, Tanzania had achieved universal primary education, several years ahead of the initial target, and had enrolled five million adults in literacy classes ... the number of rural health centers increased from 42 in 1967 to 152 in 1976 (Ottoway and Ottoway 1981: 49-50). As in the case of Burma, Tanzania's UN Human Development Index stood well in advance of its GDP index (United Nations Development Program 1990: 128).

With such low overall growth rates, however, Tanzania has been able to finance improvements in health and education only with very high levels of foreign aid. "By 1976 foreign loans and grants covered 67% of total government spending (Ottoway and Ottoway 1981: 51)." As the advanced industrial countries turned to the Right during the 1980s, assistance levels declined. Grantors and lenders increasingly demanded adherence to market criteria as a condition of aid. As a result, Tanzania has increasingly been forced to abandon its populist road to development in favor of a more market oriented approach.

The Zambian revolution has been guided by principles similar to those which governed Tanzania. Zambian President Kenneth Kaunda has stated that "Our ability to maintain and develop the traditional community based on mutual aid society principles demand that we recognize the village ... as the most important political, economic, social, scientific, and cultural unit for development (Kaunda in Ottoway and Ottoway 1981: 39)." The difficulty, however, is that unlike Tanzania, which is primarily rural and agrarian, Zambia achieved independence as a highly urbanized and already partially industrialized country, in which more than 50% of the population lived in the cities and copper accounted for one third of the Gross Domestic Product (Ottoway and Ottoway 1981: 40). This was reflected in Zambia's early National Development Plans, which allocated roughly 38% of investment to infrastructure and transport, 21% to industry and mining, 18% to social services, and only 11-12% to agriculture. Much of the industrial investment, furthermore has been in the luxury consumer rather than the capital or wage goods sectors (Ottoway and Ottoway 1981: 40-41). The result has been what some analysts call an "upper-class welfare state," in which most of the economy is state controlled, but is managed largely for the benefit of the intelligentsia and the bureaucracy. Even so, Zambia's Human Development Index, like that of Burma and Tanzania, is well in advance of its GDP index.

As David and Marina Ottoway point out, African Socialism, in both Tanzania and Zambia (as indeed throughout the continent) has lacked the dynamism necessary to mobilize the creativity and energy of the peasant masses for the tasks of development. Collectivist structures have ensured that the existing resources of the country are deployed in such a way as to serve the interests of the class(es) which dominate the revolutionary coalition, but no strategy has emerged for actually increasing overall levels of production (Ottoway and Ottoway 1981: 67).

c) Arab Socialism

One exception to this pattern, among the religious socialist movements, is the Arab and Islamic socialist trend. Beginning as a minor literary current in Arab Christian circles during the late Ottoman period, Arab Nationalism was from the beginning characterized by a strong collectivist orientation, even before it took on an explicitly socialist character. Syrian author Constantine Zurayk, for example, located the roots of the crisis of the Arab world in the absence of a collective conscience which could mobilize diverse individual talents and desires for a common purpose——and thus unite Arabs into a single nation. The basis for such a collective conscience could only be in Islam. For Zurayk, who was himself a Christian, however, this did not mean the construction of an Islamic state. Islam was, rather, simply the medium through which Arabs participated in the universal moral truths contained in all religions. Renewal of this collective conscious was to be accompanied by an enthusiastic assimilation of Western science and technology, the introduction of which, he believed, would undermine the residues of tribalism and feudalism and open the way for the emergence of a modern Arab nation. Carrying out these tasks, he argued, would require the formation of a strong intellectual elite (Ismael 1976: 3-6).

Iraqi theorist Abdul Rahman al-Bazzaz, similarly, argued for

a happy mean between the absolute individualism that gave rise to capitalism and Marxist-inspired communism ... Arab Nationalism strives for social justice in every sense of the term, while at the same time it seeks to reinforce the bases of social solidarity between the individuals of the entire community in order to prevent exploitation and class domination (al-Bazzaz in Ismael 1976: 8).

The state played a critical role in mediating between differing private interests and integrating the interests of individuals with those of the body politic. Even more than Zurayk, al-Bazzaz stressed the distinctiveness of Arab culture and the centrality of Islam to the Arab identity. The Arab nation was the core of the Islamic community, and Islam one of the principal defining characteristics of Arab nationality. The role of non-Muslims in the Arab nation was, therefore, problematic from the very beginning.

For both Zurayk and al-Bazzaz, Zionism represented the single most important threat to the Arab nation. For Zurayk this threat was primarily technological and organizational, and could be met only if the Arabs adopted European technology and organizational rationality. For al-Bazzaz the threat was first and foremost a cultural one. Israel was "a center from which dangerous principles are disseminated which conflict with the basic formative elements of Arab culture (al Bazzaz in Ismael 1976: 9)."

Despite their commitment to the liberation, unification, and development of the Arab nation, the first generation of Arab nationalists in fact achieved very little. When the Arab countries achieved independence, it was, for the most part, under the rule of monarchies installed by, and still largely beholden to, the colonial powers. These monarchies had little interest in Arab unity, or in a vigorous struggle to "liberate" "Palestine." The failures of liberal nationalism opened the way for emergence of a more radical, and more explicitly socialist nationalism. Arab nationalists came increasingly to recognize that without fundamental reorganization of the internal structure of Arab societies it would be impossible to unite the Arab nation and develop Arab society technologically, economically, politically, and culturally (Ismael 1976: 12). It was not simply the absence of a "collective conscience" which held back the unification and development of the Arab nation, but rather the complicity of landed elites and of the large commercial bourgeoisie with the European oppressors.

At the same time, the emerging Arab socialist movement explicitly rejected Marxism. Michel Aflaq, for example, one of the founders of the Ba'ath Party, argued that Marxism was a project of eighteenth century European philosophy and wholly foreign to the Arab way of life (Ismael 1976: 44). Nasser, similarly, even as he began increasingly during the 1960s to speak of a "scientific socialism," distinguished his position from Marxism-Leninism by emphasizing that unlike Lenin, he was not a materialist and did not reject religion, that he rejected the notion of a vanguard party in favor of a broad coalition of popular forces, and that he did not reject the existence of small private property. Perhaps most important, for Nasser socialism was a means of uniting and developing the Arab nation. There is no element in his vision which points beyond national development towards global integration in the context of a fully communist social order (Ismael 1976: 89).

Some tendencies within the Arab socialist movement made a more explicit appeal to Islam. Especially interesting in this regard was Muammar al-Qaddafi's attempt to elaborate a comprehensive Arab and Islamic alternative to Liberalism and Communism. Far from representing a sort of fundamentalism, Qaddafi's "Third International Theory" represents a radical reinterpretation of the Islamic tradition in the light of the experience of the national liberation struggles. While Qaddafi upholds the authority of the Quran, he rejects the traditions of the prophet and the entire medieval legal and philosophical heritage, as well as the interpretive authority of the *'ulamma*. In its place he has put the Green Book, which makes socialism the concrete expression of the historic Islamic ideal of social justice (Esposito 1984: 155-160).

During roughly the same period, outside the Arab world, in Iran, Ali Shariati was elaborating a more coherent and sophisticated Islamic Socialism. At the core of Islam, for Shariati, was the belief in the radical unity of God, from which, he believed, followed the underlying unity of all creation ——and thus of human society. Radical monotheism thus necessarily implied a struggle to build a classless and communal social order——or what he called a Tawhidic society. Shariati explicitly rejected the political quietism which grew out of Shiite messianic expectation. The hope and promise of the Mahdi or Imam should be a catalyst for action. God had

> promised the wretched masses they would become the leaders of humanity; he had promised the disinherited they would inherit the earth from the mighty. Belief in the final Savior, in the Shi'i Imams and the Twelfth Imam means that this universal Revolution and final victory is the conclusion of one great continual justice seeking movement of revolt against oppression (Shariati in Esposito 1984: 187).

Shariati's ideas became the principal inspiration for the *Mujahideen-i-Khalq* (The Peoples Warriors), one of the principal groups resisting the old Pahlavi regime, and now the principal opposition to the Iranian fundamentalist regime (Esposito 1984: 186-187).

The Pakistan Peoples' Party of Zulfikar Ali Bhutto and his daughter Benizar Bhutto also belongs to this general trend (Esposito 1084: 162ff).

Arab and Islamic socialist movements drew broad support from the peasantry and working masses, who welcomed the promise of land reform, nationalizations, etc. The principal social basis of these movements, however, was in the intelligentsia, and the emerging bureaucracies——especially the political-military bureaucracies——which found in Arab and Islamic socialism a doctrine which legitimated their claim to be the leading force in Arab and Islamic society (Hinnebusch 1982: 140-150).

Beginning in 1952 in Egypt, 1954 in Syria, 1958 in Iraq, and 1969 in Libya, military coups removed from power the monarchies which had been installed by departing colonial powers to protect their interests. Gradually the struggle for independence in French North Africa gained momentum, achieving complete victory only in the early 1960s. It was during this period as well that Yasir Arafat's *al-Fateh*, which should be seen as part of the Arab socialist trend, really came into its own.

The new regimes which came to power——the Ba'ath movement in Syria and Iraq, Nasser's Arab Socialist Union in Egypt, the Algerian National Liberation Front, Muammar al-Qaddafi's Arab Socialist Jamahirayah, and to a lesser extent the Tunisian Destourian Socialist Party——began to carry out radical land reform and in some cases even nationalization of large industries, especially those owned or controlled by European capitalists. A significant portion of the surplus thus recovered was invested in education and in the development of an industrial infrastructure, decisions which accelerated the process of social development, especially by comparison with those states such as Saudi Arabia which remained in the hands of the old feudal and mercantile elites (Hinnebusch 1982: 149).

This process was both accelerated and held back by the exploitation of petroleum reserves. On the one hand, the ability of the certain progressive Arab states, especially Libya and Iraq, like the reactionary Saudi and Gulf regimes, to extract increasingly significant monopoly rents on petroleum meant that unlike other countries in the Third World, the Arab states had significant resources available for development. These resources were, furthermore, shared to some extent with less well endowed countries such as Egypt and Syria. At the same time, an economic development strategy based on extraction of monopoly rents, or on aid from petroleum producing allies, gave even the progressive Arab states a rentier character. In this sense they were rather more like the reforming monarchies of the tributary period, which expropriated the aristocracy in order to create a mass base among the peasantry, and centralize the resources necessary for social development, than they were like authentic socialist states with a base in industrial production. The availability of petroleum rents, furthermore, made it possible to build a large military apparatus long before the process of industrialization was completed, by buying French or Soviet weapons——or, in the case of Iraq, which took this business most seriously, by investing in developing the expertise necessary to develop an indigenous military-industrial complex.

At the center of the Arab Socialist strategy was the unification of the Arab nation into a single state, and the "liberation" of "Palestine." The Arab socialist parties expended significant political capital throughout the 1960s in unsuccessful efforts at unification——the most important such efforts being the Egyptian-Syrian United Arab Republic proclaimed in 1958 and the Libyan-Egyptian-Syrian Tripartite Union of 1971. The Arab socialist states also expended an unusually large portion of the surplus which they extracted on the development of a large, effective, military machine and on financial support for various organizations dedicated to the "liberation" of "Palestine," as well as other national liberation struggles (Pipes 1983: 301-302). United Nations figures show that in 1986 Syria spent 14.7 percent of its GDP on the military, Libya 12%, and Iraq 32% (the highest of any country on the planet) by comparison with 6.7% for the United States, 2-3% for most of the European countries, and 1% for Japan (United Nations Development Program 1990: 162). This has meant that while unlike Burma, Tanzania, and Zambia, the Arab Socialist countries have generally experienced significant economic growth and industrialization, they have not invested their surplus in a way which promotes the full development of human social capacities. not surprisingly, their U.N. Human Development Index scores are uniformly behind their GNP per capita scores (United Nations Development Program 1990: 128-129).

d) Social Catholicism

One of the most highly developed forms of "objective idealist" politics, from both the theoretical and the strategic standpoint, is to be found in the tradition of Social Catholicism, which has had a profound impact on the development of European and Latin American societies, and which has had considerable influence in North America, and to a certain extent Africa, as well. We have already suggested in an earlier chapter that historic Catholicism must be understood as the ideology of a rapidly developing peripheral tributary social formation. On the one hand, Catholicism, especially in its completed, Thomistic form, provided a basis for understanding human life as a real participation in the life of God. At the same time, based as it was on a static, agrarian view of the world, classical Thomism, at least, provided very few tools for understanding the industrial, democratic, or scientific revolutions, and could, in fact, be used as a weapon against them.

Up through the nineteenth century, it was the reactionary potential of Thomism which the Church exploited. Itself a large feudal landowner, the Church's interests were threatened by the emergence of capitalist relations of production, and particularly by the rise of merchant, financial and industrial capital which sought to restrict and even eliminate the "unproductive consumption" of the clergy. The democratic revolution made human beings the masters of their own destiny, and undercut confidence in the existence of an unchanging natural law which provided a standard against which human laws should be measured. The new sciences threatened the scientific and philosophic underpinnings of Catholic theology ——particularly the teleological conception of the universe—— and called into question even the existence of God and the possibility of supernatural faith.

Within this context Thomism functioned as a conservative and even a reactionary force. Natural law theory was deployed as a bulwark against the contractarian and individualistic natural rights doctrines advanced by the bourgeoisie and the urban masses. Conservative Catholic social theorists such as de Maistre argued that the *ancien regime* was an "organic, corporative" social order which embodied the will of God. This organic corporative ideal was chained politically to the fortunes of the reactionary dynasties of Europe ——the Bourbons, the Hapsburgs, etc., and served as ideological rationale for the restorationist movements of the first half of the 19th century.

By the end of the century, however, it was apparent that this strategy was not going to succeed. The legitimist project was crushed once and for all in 1870 when the Vatican finally lost its temporal authority, and it became clear that if the church was to survive, it would have to seek allies among the popular classes. The result was a sharp change in strategy. The church began to deploy its Thomistic heritage as an instrument for developing a moderate, reformist critique of liberal capitalism. Thus the papal encyclical Rerum Novarum (1891) defended the rights of workers to organize, to secure a living wage, etc. Some forty years later Pius XI called for participation by workers in the ownership and management of capital and decreed that the right of property is subordinate to the common good. Gradually the church developed what amounted to a distinctive social ideology centered on

a) the dignity and social character of the human person,

b) the priority of the common good over private property,

c) the desirability of decentralization or "subsidiarity" in economic and political structures,

d) the rights of workers to organize and even participate in the ownership and management of capital, and

e) a commitment to the creation of effective international political authorities.

By the middle of the 20th century mainstream Catholic political philosophers such as Jacques Maritain (1951/1973) were arguing that Social Christianity was a "third way" distinct from both liberal capitalism and Marxist communism, which would lead to the construction of a communitarian society (Maritain 1951/1973). Maritain even spoke of a "New Christendom," in the context of which the church would play a critical leading role in realizing the full potential of modernity

At first hesitant to give these new teachings political expression, the Church gradually began to support the development of Catholic trade unions, Christian Democratic and Social Christian parties and other parallel organizations as a way to win workers away from the rival socialist and communist movements. Catholic Action groups composed of lay people but under tight clerical supervision, engaged in a process of social analysis, theological reflection and strategic planning (the now famous "see/judge/act" methodology) which situated this social Christian political activity in the context of a fairly conservative interpretation of Thomistic theology.

Christian Democratic parties came to power in several countries during the period after the Second World War. The Christian Democrats were the principal ruling party in Italy, one of two or three dominant parties in Germany, Austria, and the Benelux countries, and were at least influential in France. Christian Democratic parties also came to power at various points in Chile and Venezuela, and had a significant impact on policy in several other countries of Latin America. In the U.S., where the political system made confessional parties unviable, Christian Democratic initiatives were centered on the labor movement, the formation of community organizations, and the ideological struggle against dialectical materialism. Electoral initiatives were channeled through the Democratic party.

For the most part these parties followed a moderate course, protecting the peasantry and the petty bourgeoisie from the penetration of market relations, supporting the establishment of institutionalized collective bargaining arrangements and the workers' demand for a "living wage," and using the state apparatus to centralize and allocate resources necessary for development, without really challenging the hegemony of the bourgeoisie or the larger framework of market relations. In this sense, they differed from social liberal and social democratic parties only in their approach to social policy, supporting church control of education, opposing liberalization of abortion and divorce laws, and generally preferring to use nonstate, nonmarket "mediating institutions" rather than the state bureaucracy to implement social welfare programs. It should be noted, however, that a few Christian Democratic governments, such as that of Eduardo Frei in Chile and Rafael Caldera in Venezuela when much further.

Frei's government carried out a land reform the radicalism of which has been forgotten only because it was followed up by the more comprehensively socialist policy of his successor, Salvador Allende, while Caldera developed highly innovative proposals for workers' comanagement——and has now returned to power on an explicitly antimarket platform more radical than that advocated by most Latin American socialists.

Not surprisingly, Christian Democrats in Latin America who took their vision seriously, met with the same vicious repression as the Marxist left. Indeed, many of the new revolutionary organizations which emerged in Latin America during the 1960's and 1970's had their roots in the left wing of Christian Democracy, including the Ejercito Revolucionario del Pueblo (ERP) in El Salvador, the Movimiento Amplio Popular Unido (MAPU) in Chile, the Movimiento del la Izquierda Revolucionario (MIR) in Bolivia, and others.

Christian Democracy must also be credited with having played a leading role in the development of international organizations. Jacques Maritain made important contributions to the theory of international institutions, arguing for the establishment of an effective international political authority on natural law grounds. Pope John XXIII made this view official Catholic Doctrine when, in his encyclical *Pacem in Terris*, he argued that the absence of such an authority constitutes a "defect in the international system." Christian Democratic politicians such as Robert Schumann, similarly, made important contributions to the formation of the European Union, which they saw as the crucible for establishment of a New Christendom.

It must be remembered, however, that the underlying motive of the Church in supporting Christian Democracy was not so much to carry out a social revolution as it was to regain hegemony over European society, and eventually over the emerging global civilization. Its strategy of alliance with the legitimist monarchies and the aristocracy having failed, and recognizing that it had little common ground with the bourgeoisie, the church turned to the petty bourgeoisie, the peasantry, and the working class. Direct appeals for popular support through progressive policy initiatives were everywhere accompanied by efforts to undermine the establishment of independent working class organizations.

Theologically, Christian Democracy was a mixed bag as well. On the one hand, Christian Democracy mobilized the solidaristic values of the Christian tradition in a way that was at least implicitly critical of capitalism. It affirmed the power of reason to understand the structure of, and determine, without reference to revelation, what was good for human society. Furthermore, it affirmed the capacity of human beings, without the assistance of sanctifying grace, to build a just society, what Jacques Maritain called the "true city of human laws."

At the same time Christian Democratic initiatives were always accompanied by an intense ideological struggle against dialectical materialism. At the highest levels of abstraction this took the form of a sharp polemic against dialectical materialist claims that matter is self-organizing, which were seen as an attempt to deny the need for a prime mover, and thus to evade theism (Wetter 1958). Christian Democratic theory preserved and perhaps even deepened the Thomist distinction between nature and grace and thus between the secular and sacred. The creation of a just social order is the work of lay people, using the power of natural reason and drawing on the natural human capacity for virtuous conduct. This work is theologically meaningful, in the sense that it represents a real, albeit limited, participation in the life of God. But it is not a means of salvation. The task of sanctifying humanity, of the "salvation of souls" is something reserved to the clergy, who, guided by revealed knowledge, mediate to the laity the infused supernatural virtues of faith, hope, and charity which are necessary for salvation. The result of this distinction is to make the struggle for social justice marginal to the institutional life of the church, and thus to render the social teachings of the church relatively ineffectual even when, as is often the case, they are quite radical in their content.

By the end of the 1950's and the beginning of the 1960's it was clear that the Christian Democratic promise of a New Christendom was not being realized. Even as Christian Democratic parties established themselves and not infrequently formed governments, European, North American, and even Latin American society was becoming increasingly secularized. The Church was clearly failing to capture the hearts and minds of the people. Behind this process of secularization was the penetration of commodity relations into every sphere of life, which undermined the direct communitarian relationships which formed the social basis of Catholicism. But to the Church this process remained opaque. The hierarchy sought a solution in theological *aggiornamento* and ecclesiastical reform.

The Second Vatican Council was first and foremost an attempt to carry out this work of reformation. Central to the work of the council was an implicit rejection of the Thomistic framework that had guided Catholic theology for centuries, in favor of a new theological problematic which, for want of a better term, we will call the "conciliar theology."

At the center of the new theological paradigm was a rejection of nature/grace dualism. Thus the council fathers write

God did not create man for life in isolation, but for the formation of social unity. So also it has pleased God to make men whole and save them not merely as individuals, without any mutual bonds, but by making them into a single people...

This communitarian character is developed and consummated in the work of Jesus Christ ... In His preaching He clearly taught the sons of God to treat one another as brothers ...

He founded after his death and resurrection a new brotherly community composed of all those who receive Him in faith and in love.

This solidarity must be constantly increased until that day on which it will be brought to perfection. Then, saved by grace, men will offer flawless glory to God as a family beloved of God and of Christ their Brother (Vatican II: Gaudium et Spes 32).

The vision here is one of a humanity which, created in the image of a triune God, is essentially social in nature, developing its capacity for solidarity throughout the course of one single history, a solidarity which is consummated in the work of Christ, and brought to perfection in the Kingdom of God. In the place of the old dualistic theology with its sharp distinction between the finite human, secular, lay, realm on the one hand, and the divine, sacred, clerical, sacramental realm on the other, we see a unified process of divine—human activity pointing towards an Omega point which transcends history only in the sense of being beyond our finite human comprehension. Everything we do, however, which authentically builds up solidarity, is a real contribution to the building of the Kingdom, and not merely a finite, non-salvific participation in building the true city of human laws.

In the wake of the council, the Church was searching for pastoral methods which would bring this theological reformation to the people, and transform the church into an effective force for the all-sided development of human social capacities. Two principal alternatives emerged. Some argued for what amounted to a strengthening of the Church's alliance with the poor and the working classes——the so-called preferential option for the poor——even if this meant collaboration or alliance with the communist left. The so called "basic Christian community," and "liberation theology" are the products of this strategy. Others——members of the *Communio* group, argued for a strategic alliance with capital in order to combat communism, which they regarded as the Church's principal competitor for the allegiance of the masses, and for a vigorous struggle against secularization. Both strategies, it should be remembered, however, operated within the context of the conciliar theology, and of a larger commitment to the re-establishment of Catholic hegemony.

The "base community" represents first and foremost and attempt to use the "see/judge/act" methodology developed by Catholic Action to strengthen the Church's base among workers, peasants and other "marginalized" elements in society. In rural areas the base community was roughly coterminous with the village. In urban areas it at once drew on residual village solidarities, and re-established the kind of tightly knit community which newly proletarianized workers had left behind. Members gathered to discuss what was happening in their lives, and with the assistance of priests, or more often lay or religious "pastoral agents" to begin to analyze the complex social relationships affecting their interests and to discuss the significance of what they discovered in the light of the their religious traditions.

The *communidades de base* at once tapped into the prepolitical class consciousness conserved by the village community and the popular religion, and began to give it increasingly rational form. The following testimony from Norma Lopez, a Nicaraguan woman interviewed by Roger Lancaster, is worth quoting at some length.

With the youth of the *barrio* we reread and rethought the Bible, from start to finish. Exodus was a choice to move forward, out of slavery and oppression. The major and minor prophets embodied a radical stance against the sin of exploitation. The mission of Christ and the message of the Bible is that of God stooping down to help the poor, of God drawing the poor closer to him, of God intervening in history to save the poor from sin.

And then we realized that politics did not lie outside of religion. We realized that we were victims of Somoza and of exploitation, that the poor are victims of the radical sin of social injustice ...

We developed new songs, protest songs, and eventually, forms of passive resistance against the regime. These brought down massacres upon the community ... They we realized that passive tactics were not enough, and that we had to join or support the armed struggle. Many went away to the mountains, others stayed in the community and did clandestine work ... We were learning that we could only be good Christians by participating int he struggle for freedom (Lancaster 1988: 63-64).

Many, perhaps most participants in base communities developed only to the point of taking greater collective initiative in resolving immediate problems of surviving and of conserving the integrity of the social fabric. But many, enough to make a difference, became militants in peasant organizations, trade unions, community organizations, and other agencies of mass struggle. A few——but again, enough to make a difference——became militants in revolutionary political organizations, while conserving their basic fidelity to the Catholic church and the Catholic tradition.

Participation in popular struggles tended to produce a "leftist" or "liberationist" interpretation of the conciliar theology. Partly this resulted from the application of historical materialist sociology in the social-analytic stage of the "see/judge/act" process. Increasingly, leaders and participants in the base communities alike began to understand that realization of the historic aims of Social Catholicism were impossible within the context of a market driven global economy. The political aims of Catholic organizers began to drift leftward until they were indistinguishable from those of secular socialists.

There were important changes at the more specifically theological level as well. Interpretation of the scriptures and the tradition in the light of the experience of the base communities, sometimes with the assistance of analytic tools derived from historical materialism, produced a new reading of the conciliar theology, one which gave not only the human civilizational project generally, but the struggle for social justice in particular, a central place, even the central place, in the emerging theological problematic. Unlike Christian Democratic theory, liberation theology stresses that there is only one history, in which the salvation proclaimed by the Gospel, and the struggle for a just society are integrally bound up together (Segundo 1985, Boff 1986).

Taken together, the new social analysis and the new theology pointed towards a new strategic direction. Even before the council Social Catholics had been willing to engage in dialogue with socialists and communists, and even to enter in alliance with them in the struggle against fascism. Such alliances had, however, always been tactical in character, and "dialogue" usually took the form of a sharp, if not always fruitless, ideological struggle around key philosophical questions. And Christian Democratic theory had always attempted to sharply distinguish between its own "third way" and the programs of liberal capitalism and atheistic communism, even when the programs put forward by Christian Democratic parties looked rather like a moderate social democratic reformism. Liberation theologians, on the other hand, began increasingly to reject talk of such a "third way" out of hand, and argued for a more or less unambiguous strategic alliance with secular socialists.

The character of this relationship is remarkable in that it seems in many ways less like an alliance between two distinct political-theological trends than it does like a pact between ecclesiastical and secular lords. Struggle or even dialogue around key political-theological differences was kept to a minimum or even suppressed entirely as both sides acquiesced in a theoretically incoherent formula which distinguished between historical materialism as a "method of social analysis" and "dialectical materialism" as a global worldview. The political leadership of the left parties or the "political-military organizations" would hold sway in the secular realm; the Church would retain control over such critical ideological tasks as the intellectual, moral, and spiritual formation of the people.

It would be a serious mistake to underestimate the progressive potential of this alliance, which is still not entirely spent. Throughout the period between 1968 and 1989 the Catholic left helped to deliver the mass base which made the Latin American left such a powerful force for social progress. And the alliance with the left served the Church well. By showing that it comprehends the underlying cause of the social disintegration affecting the planet, the Church has maintained its credibility among the masses and positioned itself effectively to carry out the critically important task of conserving humanity's faith in and relationship to the cosmic order. Most important, however, has been the cultural milieu created by the collaboration between Catholics and Communists. This milieu contains many of the elements necessary to a new synthesis which transcends the limitations of both objective idealism and dialectical materialism, a synthesis which is already present in poetic form in the work of Ernesto Cardenal.

There were, however, serious limitations to the concordat negotiated between the secular and religious left. The most serious of these limitations derives from the fact that in the absence of principled struggle, none of the internal contradictions of either liberation theology or secular socialism were addressed. Because of this, the synergistic synthesis which existed *in potentia* in the Christian-Marxist dialogue was never really worked out. We will have an opportunity to examine the problems of socialism later on in this chapter. But we will do well to say a few words about the problems of liberation theology.

At base these are simply the same problems which have affected Christianity all along. Liberation theology fails to break definitively with both objective idealism (organization comes to matter from the outside) and the specifically Christian form of objective idealism, salvation through divine self-sacrifice, and its inner-worldly imitation. This is apparent from the way liberation theologians ground their fundamentally progressive claim regarding the unity of human history and salvation history.

> God makes history. Why? Because by becoming one with the lot that every person has in history (Gaudium et Spes 22) he converts history, seemingly profane history ... into the road by which the individual has access to transcendence and therefore salvation (Gaudium et Spes 22) ... The historical work of all people will lead, by the grace of God to eschatological metahistory (Segundo 1985:69).

Segundo grounds the meaningfulness of human history not in the creation, and thus in the immanent teleology of human nature, as did the Thomist tradition, but rather in the incarnation.

There is, furthermore, a tendency in liberation theology to mystify the salvific character of the historical process. If, on the one hand, human history itself is salvific, and if, on the other hand, faith in the crucified messiah reveals to us something unique about the salvific character of this history, then we are forced almost ineluctably to conclude that the salvific character of history derives from human suffering, and specifically from the suffering of the poor and the oppressed, who as it were continue the crucifixion. Thus liberation theology speaks of the revolutionary role not of the working class (which is revolutionary first and foremost in virtue of its creative power) or of the peasantry (which is revolutionary because of the window on the cosmic order provided by the experience of life in the village community) but rather of the "poor" who are revolutionary because they suffer. Furthermore, from this perspective, revolutionary virtue reaches its highest level of development not in the philosopher, pastor, or organizer, who develops human capacities, but rather in the revolutionary hero or martyr who risks, and perhaps sacrifices, his life for the revolution.

It should come as no surprise if the practical implementation of this theology has yielded highly ambiguous results. The Catholic left has been most effective in struggles against brutal regimes in which the demand for revolutionary self-sacrifice was at a premium. It has been much less effective in mobilizing the energy of the people in advanced capitalist societies where the main task is one of reawakening the creative energies of the working class and catalyzing discontent with the hegemonic consumerism.

As liberation theology gradually gained influence in the Catholic Church during the 1960s and 1970s, a right opposition emerged which argued that communism remained the Church's principal adversary, and that while building or rebuilding a base among the poor of the Third World generally, and Latin America in particular, was vitally important, this work must be carried out within the context of a geopolitical alliance with the bourgeois states of the West, and an intense ideological struggle against dialectical materialism, feminism, and other forms of secularism. This is the course preferred by the present pontiff and by the international theological movement organized around the journal *Communio*.

The theological key to *Communio*-theology can be found in *Love Alone*, a small book published more than twenty-five years ago by one of the trend's most creative theologians, Hans Urs von Balthasar (1968). Von Balthasar distinguishes between three approaches to theological reflection: the cosmological, the anthropological and the "aesthetic." The cosmological approach is the method of traditional Catholic theology, which used the concepts of Greek philosophy, Platonic or Aristotelian, as a criterion for the interpretation of the scriptures and the teachings of the church. Thus, in cosmological theology, the doctrine of God or the Trinity is explained in terms of philosophical categories of being, essence, person, etc. The anthropological approach is the method of most modern theology, which takes its categories from modern philosophy, or, by extension, from the social sciences, and interprets the tradition in terms of these categories. According to this perspective God is the perfectly good will of the liberals, the "ground of authentic being" of the existentialists, ——or the liberator of the oppressed.

The difficulty of both of these approaches, von Balthasar argues, is that they reduce God, and thus divine love, to a postulate of human reason, something understandable, and in a sense necessary in human terms ——something other than the fully free and unmerited love through which God reveals himself to us in the scriptures.

> Christianity is destroyed if it lets itself be reduced to a transcendental presupposition of man's self-understanding, whether in thought or in life, in knowledge or in action (1968:43) ... The moment I think that I have understood the love of another person for me ——for instance on the basis of laws of human nature, or because of something in me—— then this love is radically misused and inadequate, and there is no possibility of a response. True love is always incomprehensible, and only so is it gratuitous. (1968:44).

This is a critical point. What von Balthasar is suggesting is that any attempt to understand revelation in terms of rational, human criteria, be they Platonic, Aristotelian, Thomist, Kantian, existentialist, or Marxist, has the result of reducing the love which is revelation to merely a necessary, and in some sense merited reflection of our own human nature, or of the structure of being in general. To put this in another way, the *communio* created by divine love becomes simply a community of mutual interdependence, in which cooperation is rationally comprehensible on the basis of definite natural or social laws, and the ultimate purpose of which is the satisfaction of individual desires ——rather than a communion based on self-sacrificial love which is spontaneous, gratuitous, and incomprehensible in terms of anything which we know about human nature. Such a rational harmony does not really overcome the egoism of the individual, and thus is not genuinely or fully redeeming in character. It is this danger which makes von Balthasar and Ratzinger so cautious about any rationalistic hermeneutic. Dialectical and historical materialism is simply the most radical variant of the rationalism they seek to combat. Their structures would apply as fully to Rahner as they do to Segundo or the Boffs, and, perhaps, more fully than they do to Guttierez.

Communio-theologians, furthermore, understand love first and foremost as self-sacrifice, modelled on the substitutionary work of Jesus on the cross.

> The sign of Christ can only be deciphered if His human love and surrender 'even unto death' is read as the manifestation of absolute love ... His task, in love is to allow the sins of the world to enter into Him who is 'dispossessed' out of love of God ——to become the 'lamb of God who bears the guilt of the world (1 John 1:29) and my sins ... This is the dogma ——the dogma of vicarious suffering, of bearing the guilt of others' which in the last analysis determines whether a theology is anthropological or christocentric ... For it is precisely with this act that really unaccountable, inconceivable love begins and ends; a love moreover which *qua* love is self evidently divine (1968:81-82).

At the strategic level, *Communio*-theologians opted for a geopolitical alliance with the bourgeois states of Europe and North America, and played a critical role in providing organizational and ideological support to the anticommunist opposition in Central and Eastern Europe——particularly in Poland. This anticommunist campaign has been accompanied by a frontal assault on the womens' movement, which *Communio* seems to regard as fully as dangerous as communism.

This latter point merits further comment. We have already seen, in the first chapter, that the ancient cult of the *Magna Mater*, and other forms of goddess worship, were intimately bound up with the conviction that matter itself is self-organizing——and thus not in need of "formation," much less "redemption" from without. Feminist philosopher Mary Daly (1984) points out that Christianity has, historically, adopted two distinct strategies in relation to survivals of this cult. Protestant theologians have historically attempted to repress the cult altogether in the interests of safeguarding an extreme doctrine of divine transcendence. Catholic theologians, on the other hand attempted to co-opt the cult, in the form of Marian devotion, as a way of integrating peasant communities into the Church. More recently, the Church has used Marian devotion as a means of countering socialist immanentism. It is no coincidence that major Marian dogmas were proclaimed in the wake of the revolution of 1848 (the Immaculate Conception), and the Communist victories in Eastern Europe in the immediate postwar period (the Assumption), or that Marian apparitions (Fatima, Medjugorge) have figured prominently in anticommunist campaigns.

The Vatican recognizes in the women's movement its most potent adversary, for it is ultimately the women's movement, even more so than communism, which is in a position to expose the profound contradiction between the historic Catholic insights into the self-organizing dynamic immanent in matter, heir of the philosophical tradition and ultimately of the cult of the *Magna Mater* and patriarchal and idealist theologies which regard matter as the passive recipient of form, and humanity as merely a receptacle for infused divine grace, and more especially between the historic Catholic teachings regarding the underlying goodness of creation and of human nature, and the whole Pauline/Augustinian problematic centered on sin and redemption through divine self-sacrifice——a problematic which is emphasized by the *Communio* trend, but which is evident in liberation theology as well.

It should be apparent at this point why Social Catholicism generally, it all its forms, has, despite all its very many real strengths, ultimately failed as a strategy for social transformation. Historically Catholicism has conserved important elements of the archaic philosophical synthesis: recognition of the cosmos as an organized totality in which humanity plays a meaningful, even creative role. At the popular level these elements were reflected in the persistence of the ancient cult of the *Magna Mater*, albeit in sublimated, Marian form. It was precisely this conservative character which permitted the Church to take up a progressive role vis-a-vis the bourgeoisie. At the same time, the Christian theological problematic, with its insistence on divine transcendence and the inert passivity of matter, the sinfulness of humanity, and salvation through divine self-sacrifice, remains in fundamental contradiction with these progressive elements. And yet it is precisely the Christian theological problematic which provides the rationale for the authority of the celibate clerical corporation. Both progressive and "authentic conservative" elements perceived quite correctly that Catholic values were deeply at odds with the market system and that they had little or no common ground with the bourgeoisie, and thus opted for an alliance with the working class and the peasantry. Not surprisingly, however, implementation of this alliance policy unleashed the progressive, immanentist, *mater/ia*list elements within the tradition, and threatened the authority of the clergy——right and left. It is little wonder that there has been a backlash, and that even progressive elements in the clergy have been hesitant to resist this backlash. Nothing less is at stake than the existence of the clergy itself. This means, however, that much of the progressive potential in the Catholic tradition may long remain unrealized, and that important avenues of historical development may be closed off.

C. The Crisis of Objective Idealism

Ultimately objective idealist politics failed not because the values which it conserved and developed were not the authentic values conserved and developed by the cosmohistorical evolutionary process, but rather because it understood these values as something external to matter, which must be "formed" from the outside by the intervention of a superior authority: God, or His representatives in the traditional clergy, the leaders of some millenarian sect, or a modernizing nationalist or religious intelligentsia acting in alliance with either or both of these sectors. On the one hand, this idealist theory of organization drew attention away from the internal contradictions of capitalist society, contradictions which at once hold back the development of human social capacities and which might serve as a catalyst for development. On the other hand, by attempting to form human society from above, idealist movements have tended to adopt an authoritarian stance which itself becomes an obstacle to social progress.

This failing is most apparent in the case of Christian and Islamic movements, which remain committed to the notion of a transcendent creator God. Here there has been a marked tendency for regressive elements within the religious tradition to reassert themselves: clericalism and antifeminism in the case of Christianity, militarism, neoimperialism, and antifeminism in the case of Islam. But the difficulty is apparent even in movements which draw on American, Asian, or African traditions, where collectivist attempts to restrict the penetration of the market relations were not complemented by a successful strategy for tapping into the self-organizing potential of the masses.

The internal contradictions of objective idealism have been intensified by the action of market forces themselves, which gradually destroy the village community and its residues in the urban neighborhood and thus undermine the social basis for the religious traditions which objective idealist political movements attempt to rationalize and restore. In some cases this leads to outright secularization. In other cases it leads to an intensification of otherworldliness and of doctrines of salvation through self-sacrifice. As the social fabric disintegrates under the pressure of market forces, people come increasingly to regard material world as a place of chaos and disorder, subject to forces beyond their control, and to seek wholeness not in the realization, but rather in the negation, of their own individual telic projects. Thus the rise of fundamentalism, which stands in the same relationship to objective idealism/religious socialism as nihilistic postmodernism does to the dialectical materialist tradition.

Fundamentalist trends have emerged in several different salvation religions, including Hinduism and Judaism as well as Christianity and Islam. It is the two latter trends, however, which are most dangerous, both in the intensity of the repressive dynamic which they represent, and in the scope and depth of the mass base they have been able to build up. We consider each movement briefly in turn.

1. Christian Fundamentalism

We have already outlined in the preceding chapter the principal characteristics of Protestant Christianity. The Protestant doctrine of a sovereign god, entirely withdrawn from the world, which he uses as a passive instrument in the realization of an eternally inscrutable cosmic plan, represented from the very beginning a religious reflex of industrial capitalist society. At the same time, we have also seen that certain forms of Protestantism made an important contribution to the tasks of civilization building, mobilizing human productive energy in a way no previous social movement had been able to do, combating the tendency inherent in market economies to drain off the surplus extracted into low wage, low technology activities, and encouraging instead a program of systematic rational investment in infrastructure, education, research, and development. Unlike their liberal allies, Protestant reformers were never hesitant to call on the powers of the state when these seemed necessary to the continued health and progress of the body politic.

Christian fundamentalism, far from being simply a survival of traditional Reformed Protestantism, in fact represents a fundamentally new phenomenon, which traces its origins to late nineteenth century England, where the Plymouth Brethren developed a new form of "dispensational premillenialism," later popularized in the United States by John Nelson Darby and Cyrus Scofield, whose Scofield Reference Bible soon became "standard equipment" for the soldiers of the new fundamentalism.

Far from representing a revolt against science in general, this new doctrine was based on a rigid adherence to the positivistic theoretical problematic favored by the Anglo-American bourgeoisie. George Mardsen quotes Arthur Pierson, an important fundamentalist leader.

> I like Biblical theology that does not start with the superficial Aristotelian method of reason, that does not begin with an hypothesis, and then warp the facts and the philosophy to fit the crook of our dogma, but a Baconian system which first gathers the teachings of the word of God, and then seeks to deduce some general law upon which the facts can be arranged (Pierson in Marsden 1980: 55).

The new fundamentalism differed from historic Reformed Protestantism in two key respects. Where historic Reformed Protestantism tended towards a postmillennial eschatology, hoping that the church, through a strategy which integrated conversion and social reform, would be able to establish the thousand year reign of justice on earth before the Second Coming, the new fundamentalism teaches that the world can be set aright only by Christ himself, whose coming will precede any turn for the better in human society. Mardsen writes

> The area where dispensationalists were perhaps most out of step .. was in their view of contemporary history, which had little or no room for social or political progress. When they spoke on this question, dispensationalist premillennialists were characteristically pessimistic (Mardsen 1980: 66).

Second, the doctrine of the dispensations itself tended to radicalize even further the historic Protestant rejection of "salvation by works" in favor of a narrow focus on "faith" and submission to God's inscrutable cosmic plan. Though there are several different variants of the theory, most divide history into seven different periods, during each of which God relates to humanity in a specific way.

> The current age or dispensation is sharply separated from all teachings of Scripture having to do with he Jewish people, whether in the Old Testament or in the age of the kingdom to come. Even Christ's ministry was set in the era before the church age began. Thus his teachings in the Sermon on the Mount and the Lord's Prayer proclaim righteousness on legal grounds, (being still part of the Jewish "dispensation of law") rather than a doctrine of grace (which characterizes the church age or "dispensation of grace (Mardsen 1980: 52)."

Unlike historic Reformed Protestantism, which saw the church as a force for social reform, albeit reform flowing from conversion, the new fundamentalism rejected such efforts as hopeless, "judaizing," and a distraction from the sole task of Christians in the present period: the salvation of souls, the creation of a holy remnant which would "meet the Lord in the air" in the great moment of rapture which (in most versions of the theory) would precede the final tribulations.

It should be apparent by this point that Christian fundamentalism represented, from the very beginning, a reflex of the alienated world brought into being by the marketplace. Initially a minor current, the trend has gained strength as market relations have penetrated every sphere of social life, undermining the integrity of the social fabric and plunging our society into a downward spiral of crime and social disintegration. Unable to find any basis for order within the world of nature and society, or even to find there traces of God's saving intervention, Christian fundamentalism withdraws into a pure otherworldliness which finds hope only in submission to the inscrutable will of a sovereign God who, it seems, has seen fit to simply destroy, rather than to save and redeem, most of his creation.

It is one of the paradoxes of Christian fundamentalism that a movement so pessimistic about the possibilities of political life should have posted at least moderate successes in the political arena. Indeed, there was a trend during the 1980s, high point of fundamentalist political power, away from strict premillenialism and toward a new variant of postmillenialism, known as "reconstructionism" which seeks to amend the U.S. constitution to acknowledge the "divine monarchic sovereignty of Jesus Christ," and to impose biblical law on U.S. society. This latter current has failed to catch hold, however, in part because the internal theological dynamics of Christianity, especially Protestant Christianity, don't really support this kind of political theology, but more importantly because the pressure of market forces makes it impossible for fundamentalists to sustain even the modest gains they have made in recent years. The real danger of Christian fundamentalism, in this sense, is not that the movement will succeed in establishing hegemony by itself, but rather that, in alliance with the dominant neoliberalism, it will succeed in destroying one civilizational achievement after another as it pursues its reckless struggle against "works righteousness" and "secular humanism," while drawing off the support of constituencies troubled by the social disintegration engendered by the marketplace. In this sense it remains vitally important to expose the real basis of the social chaos affecting our society, and to compete vigorously with the fundamentalists for the support of "authentic conservatives" whose principal concern is in the integrity of our social institutions, and who are drawn to fundamentalism partly because their own understanding of social organization has taken on an alienated form, but partly also for the lack of alternatives.

2. Islamic Fundamentalism

Islamic fundamentalism traces its roots back to the founding of the Muslim Brotherhood by Hasan al-Banna in Egypt in 1928, and of the Jamaat-i-Islami in Pakistan by Mawlanda Mawdudi in 1941 (Esposito 1984: 131, 144). Like earlier Islamic revivalist movements, the Muslim Brotherhood rejects the synthesis of Islam with Hellenistic philosophy on the one hand, and popular devotionalism on the other hand, which emerged during the period of the great Islamic empires. They insist, rather, on a return to the Quran, the Sunnah, and the practice of the first four "rightly guided" caliphs (Pipes 1983: 124ff). Unlike Islamic socialists, they see no need to reinterpret Islam in the light of the industrial, democratic, and scientific revolutions, or in the light of the national liberation struggle. The *Shariah* is, as it was expressed in the practice of the 'ummah in the earliest days, a comprehensive guide to every aspect of life and must be implemented in its totality. Islam is

> a divine vision that proceeds from God in all its particularities and its essentials. It is received by man in its perfect condition ... He is to appropriate it and implement all its essentials in his life (Sayyid Qutb in Esposito 1984: 136).

Several principles are especially important to the fundamentalist vision. First, God is the sovereign ruler of the universe and thus the only authentic legislator. It is interesting to note in this regard that while Islamic socialists emphasized the unity of God, fundamentalists emphasize his sovereignty. While a variety of different political systems are possible, they exist only to organize the interpretation and implementation of *Shariah*, not the actual creation of new legislation. Second, fundamentalists are strongly committed to the patriarchal family structure characteristic of Islamic societies, including the specific emphasis on the segregation of women and the restriction of women to a primary role in the home and family. Third, Islam prescribes definite economic principles. While private property and differences in wealth based on hard work and greater talent are acceptable and even mandatory, all wealth ultimately belongs to God. Private property rights are thus limited by the common good, and all Muslims are required to pay a tax on capital as well as income to provide for the poor and support other public functions. Accumulation of wealth through usury is prohibited (Esposito 1984: 139-141).

Jamaat-i-Islami had a similar orientation. According to Mawdudi,

the belief in the unity and the sovereignty of Allah is the foundation of the social and moral system propounded by the Prophets ... The *Shariah* is a complete scheme of life and an all embracing social order (Mawdudi in Esposito 1984: 145-146).

The state is not an actual legislator, but rather God's vice-regent. *Shariah* embraces all areas of life. Mawdudi explicitly acknowledged that Islam was totalitarian. "No one can regard any field of his affairs as personal or private (Mawdudi in Esposito 1984: 147)." Only those who embraced Islam, furthermore, could interpret *Shariah*, and thus only Moslems could enjoy the full benefits of citizenship——though certain rights were guaranteed to dhimmi who paid the poll tax and refrained from proselytization (Esposito 1984: 148-149). In this sense Islam was as radically incompatible with nationalism as it was with liberalism or socialism. Membership in a certain ethnic group, birth within a certain territory, or naturalization are not sufficient for citizenship.

Islamic fundamentalism has found a broad constituency among the urban masses marginalized by the policies of both the conservative Arab regimes and the Arab socialist states. This base has grown rapidly as the crisis of socialism has deepened, and the masses which formerly looked to the Soviet Union for leadership search for some new liberator. The core of the movement, however, has come from young members of the intelligentsia, drawn to the cities from rural areas, but deprived of an outlet for their skills and confused by the disintegration of the social fabric under the pressure of penetrating market relations.

The typical social profile of members of militant Islamic groups could be summarized as being young (early twenties), of rural or small-town background, from middle and lower middle class, with high achievement motivation, upwardly mobile, with science or engineering education, and from a normally cohesive family (Ibrahim 1982: 11, in Esposito 1984: 203-204).

There is also some evidence of significant traditional petty bourgeois support for Islamic fundamentalism, especially among bazaar merchants, who played a key role in the Iranian revolution.

The Muslim Brotherhood has branches throughout most of the Arab world, and has served as an inspiration for organizations not formally linked to it, such as the Algerian *Mouvement de la Nahda Islamique* (MNI) (Roberts 1991: 137). Most of these organizations reflect the influence of Mawdudi as well as the Brotherhood's own ideologues.

Earlier fundamentalist organizations, such as the Muslim Brotherhood and *Jamaat-i-Islami*, argued that society had to be thoroughly Islamicized before any attempt was made to impose *Shariah*. In its early years the Muslim Brotherhood, for example, established hospitals, educational, social welfare, and cultural projects, and concentrated its energies on the recruitment and ideological formation of its membership (Esposito 1984: 132). After 1967, however, a new generation of fundamentalist organizations, influenced by the more radical ideas of Sayyid Qutb, emerged. The Egyptian *Takfir wal Hijra* (Excommunication and Emigration) and the Jamaat al-Jihad——responsible for the assassination of President Anwar al Sadat——are typical in this regard. These new organizations began increasingly to argue that in so far as the legitimacy of Muslim governments was based on *Shariah*, those which did not implement *Shariah* are illegitimate and a proper object of jihad. The same was true of the official *'ulamma* who downplay the importance of armed struggle and interpret *jihad* as merely the struggle to live a virtuous life. Authentic Muslims are under an obligation to overthrow such governments and the religious establishments they support, to make war on all who do not share a total commitment to the *Shariah*. Non-muslims, meanwhile, are no longer to be regarded as protected "people of the book" but rather as infidels whose lives and property are forfeit. The entire planet must be brought under Islamic rule (Esposito 1984: 202-203). The Movement of Islamic Resistance, (usually known by its Arabic acronym, HAMAS) operating within Israel and the territories, shares this general orientation (Legrain 1991: 71-80), as does to some extent, the Algerian *Front Islamic du Salut* (FIS) (Roberts 1991: 134-136).

Outside the Arab world fundamentalism has developed somewhat differently. In Pakistan, where the military government led by Zia ul-Haq carried out a systematic program of Islamicization, fundamentalist organizations such as *Jamaat-i-Islami* gave their critical support to the government, while arguing that a military government which placed itself above the *Shariah* could never fully realize their program (Esposito 1984: 166-178). Throughout the 1980s, Pakistan served as a staging area for Islamic fundamentalists fighting the progressive Afghan government and its Soviet supporters. After the return of the Islamic Socialist Pakistan Peoples Party to power under Benizar Bhutto, the fundamentalists mounted a vigorous opposition, which contributed to her temporary removal from power (Piscatori 1991: 21). She now governs with great difficulty, her every move facing scrutiny and resistance from the powerful fundamentalist opposition. The long term prospects for fundamentalist hegemony, however, remain unclear, as the Pakistani situation is further complicated by the existence of pro-Sufi, Wahabi, and Shi'ite groups along with the mainstream of Islamic fundamentalism represented by Jamaat-i-Islami (Ahmad 1991: 161-170).

In Iran fundamentalism has been shaped on the one hand by specifically Shi'ite motifs, and by the experience of over thirteen years of state power. Unlike Mawdudi and the Muslim Brotherhood, Iranian fundamentalists like Ayatollah Ruollah Khomeini argued that society need not be Islamicized before the *Shariah* is applied. Application of the *Shariah*, on the contrary, is to be the *means* of Islamicizing society (Esposito 1984: 189). Similarly, where other Islamic fundamentalists have focused primarily on the application of *Shariah*, while remaining open to a variety of political structures, the Iranians gradually evolved a system of "Rule by the Jurisconsult" in which a Supreme Guardian of the Revolution (a post Khomeini filled until his death), assisted by a Council of Guardians, played the leading role in the state apparatus (Esposito 1984: 196). Even more than Pakistan, Iran has witnessed the systematic application of Islamic law, including attempts to create an "Islamic economic system" distinct from both capitalism and socialism, and rigid enforcement of the segregation and veiling of women, etc. Like Pakistan, Iran has actively supported fundamentalist movements throughout the Islamic world.

The attraction of Islamic fundamentalism to the masses of the Third World is not difficult to understand. Since the crisis of the socialist states, there has been no other major global force willing to speak up for the oppressed masses of Asia and Africa. Unless the progressive forces are able to counter fundamentalist Islam with a new doctrine which taps into the creative potential latent in the villages and barrios of the planet, we can expect this trend to continue to grow, so that the planet becomes increasingly polarized between postmodern nihilists and religious fundamentalists, both armed to the teeth, neither terribly invested in either the ecosystem or the civilization of the planet we call home.

II. Dialectical Materialism

Where objective idealism sought an organizing principle outside of matter, attempting to secure an adequate ground for the cosmic order even if this meant sacrificing the possibility of ever realizing that principle within the confines of finite human social existence, dialectical materialism insisted instead on the self-organizing nature of matter and on the possibility of realizing latent human potential——even if this left the drive towards organization inadequately grounded at the ontological level, thereby endangering the ultimate meaningfulness of human history on a cosmic scale. It is above all the conviction that human society *can* be reorganized in such a way as to unleash the full potential latent in humanity, and that both the organizing principle of the new social order, and the power necessary to carry out the reorganization, are latent in human society itself, which has made dialectical materialism such an attractive theoretical alternative and the socialist movements which embraced it such a potent historical force. And it is the failure of dialectical materialism to establish an adequate ontological ground for organization which ultimately undermined the ability of the socialist movements to work out an adequate strategy for socialist construction. For the complex, self-organizing dynamism which drives human history simply cannot be derived from the fundamentally mechanical interactions of the physical world unless these interactions are themselves comprehended as the manifestation of a still more fundamental organizing dynamic.

In this section we will analyze the unfolding of the socialist project. We will begin by examining in some detail Marx's formulation of the socialist problematic. We will identify some critical questions, relating to the nature of the socialist transition and socialist construction, which Marx left unanswered, and examine several concrete attempts to answer these questions. We will explain why none of these attempts were adequate and why it was ultimately impossible to solve the problems of the socialist transition within the framework of the dialectical materialist problematic.

A. Marx's Synthesis

There is considerable debate within the dialectical materialist tradition regarding the precise character of Marx's sociological and philosophical position. It is not our purpose in this context to enter into an elaborate defense of one or another interpretation of Marx. Suffice it to say that by and large we uphold the continuity of Marx's thought, regarding the "early writings" as the outline of a research program which Marx then carried out systematically, but which he was ultimately unable to complete. It is precisely because he left so much unresolved that there has been continuing debate within the tradition regarding very fundamental questions. We will begin, rather, with a brief account of what we take to be the principal elements of Marx's system, pointing out vitally important unanswered questions, and then explore in some detail the very different ways in which the socialist movements have historically attempted to resolve these questions.

Where objective idealism regards complex organization as rooted in an ontological ground prior to matter, Marx, along with the whole dialectical materialist tradition, rejects the notion of an organizing principle prior to matter, and treats complex organization as something which emerges gradually out of simpler——mechanical, electromagnetic, biological——interactions within the material world. Unlike objective idealist theories, therefore, for which elaboration of the laws of the dialectic constitutes an ontology as well as a logic, for dialectical materialism any claims regarding such laws are always and only abstractions from more specific natural scientific and social scientific formulae.

While it is not really possible, within the context of a radical materialism, to regard matter in general as teleological, it is probably fair to say that for Marx the teleological dynamic which emerges for the first time with human labor power represents the crowning achievement of the process of evolution, and is a higher expression of less fully purposeful organizing dynamics which are already at work at lower levels. It is this view which later found expression in Engels' *Dialectics of Nature.*

It is not entirely clear from Marx's own writings just how, *in general*, he saw development from less complex to more complex levels of organization taking place. At least at the level of the social form of matter, however, the categories of production, contradiction, and the "negation of the negation," recovery of archaic holism, seem to be central. It is first and foremost humanity's capacity to produce——to create new forms of organization more complex than that given by our physiology——which separates it from lower animals.

> Men can be distinguished from animals by consciousness, by religion, or anything else you like. They themselves begin to distinguish themselves from animals as soon as they begin to *produce* their means of subsistence, a step which is conditioned by their physical organization. By producing their means of subsistence men are indirectly producing their actual material life (Marx 1846/1978: 150).

And it is this capacity which is the basis of the drive towards holism which finds expression in law, morals, and religion.

> The aggregate of these productive relationships constitutes ... the real basis on which a juridical and political superstructure arises and to which definite forms of social consciousness correspond (Marx 1859/1966: 217).

At the same time, as human productive capacities develop, the way in which production is regulated by the prevailing relations of production gradually becomes an obstacle to the full development of human social capacities.

> At a certain stage of their development the material productive forces come into contradiction with the existing productive relationships, or what is but a legal expression for these, the property relationships within which they had moved before (Marx 1859/1966: 218).

This has happened several times in the course of human history, and is happening again in capitalist society. On the one hand, Marx demonstrated, the market system holds back the development of the productive forces because, other things being equal, as the level of technological development, and thus the organic composition of capital rises, the rate of profit declines, setting in motion economic crises and eventually a general crisis of the capitalist system (Marx 1861). At the same time, the commodity production (the marketplace), because it makes the essentially social activity of human labor into merely a means of realizing essentially individual ends ——i.e. consumption—— constitutes an obstacle to the full realization of human sociality.

> In estranging from man (1) nature, and (2) himself ... estranged labor estranges the species from man. It turns for him the life of the species into a means of individual life ... labor, life activity, productive life itself, appears to man merely as a means of satisfying a need ——the need to maintain its physical existence ... (1978: 75-76).

Resolution of these contradictions is possible only when the marketplace has been transcended. Unlike earlier socialists who located the source of capitalist exploitation in the existence of private property, and thus advocated a crude communism in which private would be replaced by state or collective property forms, (and therefore merely negated, or negatively transcended), Marx is quite clear that the underlying structure of capitalist exploitation is located first and foremost in the phenomenon of commodity production, out of which alienation and exploitation necessarily emerge (Meikle 1985: 90-91). He thus makes a careful distinction between crude communism, which only generalizes the relationship of private property to the society as a whole (Marx 1844/1978:81-84), and

> Communism as the positive transcendence of private property, or human self-estrangement, and therefore as the real appropriation of the human

essence by and for man; communism therefore as the complete return of man to himself as a social (i.e. human) being, a return become conscious, and accomplished within the entire wealth of previous development. This communism as fully developed naturalism, equals humanism, and as fully developed humanism equals naturalism; it is the genuine resolution of the conflict between man and nature and between man and man ——the true resolution of the strife between existence and essence, between objectification and self-confirmation, between freedom and necessity, between the individual and the species. Communism is the solution to the riddle of history and knows itself to be this solution.

The entire movement of history is, therefore, both its actual act of genesis (the birth act of its empirical existence) and also for its thinking consciousness the comprehended and known process of its coming to be (Marx 1844/1978:84).

For Marx Communism is the realization of the "social individuality" of humanity: an end to economic exploitation and the creation of a social order in which "love can be exchanged for love and trust for trust" (Marx 1844/1978: 105).

In this sense, Marx does not so much seek to abolish religion as to transcend it in the Hegelian sense, making the solidarity which religion embodies, albeit in alienated form, into the very substance of human existence. This is the "negation of the negation," the recovery of humanity's archaic holism, albeit at a higher level of differentiation and integration, which has given Marxism such a potent appeal, and which has constituted the historical basis for dialogue between Marxism and the salvation religions.

Marx nonetheless left certain very important questions unresolved. Chief among these was the process by which socialist consciousness develops within the working class. How do workers, who are submerged in the particularities of their day to day struggle to survive, serving the needs of others in order to realize their own consumption interests, develop to the point that they are able to adopt the standpoint of totality, working, organizing, thinking, loving, and struggling to reorganize society in order to promote the self-organizing activity of the cosmos?

Marx recognized that workers develop socialist consciousness very gradually, through their struggle with the bourgeoisie.

At first ... they direct their attacks not against the bourgeois condition of production, but against the instruments of production themselves; they destroy imported wares ... they smash to pieces machinery, they set factories ablaze ....

But with the development of industry the proletariat not only increases in number; it becomes concentrated in greater masses, its strength grows, and it feels that strength more ... Thereupon the workers begin to form combinations (Trades Unions) against the bourgeois; they club together to keep up the rate of wages; they found permanent associations in order to make provision beforehand for these occasional revolts...

The real fruit of their battles lies, not in the immediate result, but in the ever-expanding union of the workers... (1848/1978: 480-1)

Eventually the workers realize that "every class struggle is a political struggle" and form a political party through which to prosecute the interests of their class.

The final stage in the development of socialist consciousness, however, takes place not through the spontaneous action of the proletariat, but through the intervention of the intelligentsia, which Marx sees not as a stratum of the proletariat, but rather as a section of the bourgeoisie.

Finally ... a portion of the bourgeoisie goes over to the proletariat, and in particular, a portion of the bourgeois ideologists, who have raised themselves to the level of comprehending theoretically the historical process as a whole (Marx 1848/1978: 481-2).

It is this last stage which is critical for the formation of communist, as opposed to merely trade union or political class consciousness.

The Communists, therefore, are, on the one hand, practically, the most advanced and resolute section of the working class parties of every country, that section which pushes forward all others; on the other hand theoretically, they have over the great mass of the proletariat the advantage of clearly understanding the line of march, the conditions, and the ultimate general results of the proletarian movement (Marx 1848/1978: 484).

A close reading of Marx's analysis of the development of socialist consciousness points out an important key to understanding the current crisis of the international communist movement and of socialism generally. Marx recognizes implicitly that the industrial working class *never* develops spontaneously past the point of trade unionism, or at best of political class consciousness, to a fully communist level. While he did not spell out the reasons for this, they are already implicit in his analysis of alienation.

While industrialization creates a high level of material interdependence, which forms the basis for the emergence of socialist consciousness, the mediation of this interdependence through the marketplace obscures our underlying solidarity and makes every relationship appear simply as a means for individual advantage. As workers begin to organize, they do so not based on a fully developed dialectical grasp of human interdependence, but rather for the simple reason that by organizing they can improve their lives. The alienation generated by generalized commodity production prevents them from achieving knowledge of, and thus commitment to "the historical process as a whole."

This is the central problem which has occupied the workers movement throughout its history. Broadly speaking, three solutions have been advanced. Social Democracy has argued that the quantitative growth of the working class would ultimately issue in a qualitative leap both in the level of organization and the level of consciousness of the workers themselves, making possible a transition to socialism and the full development of human capacities. Sometimes, as in the case of Engels, this development has been conceived in materialist terms. At other times, as in the case of Bernstein, and revisionist social democracy, it has been conceived in terms of intellectual and moral progress directed by the social democratic party. Leninism, on the other hand, opts to build a small leadership core, which grasps not only the "line of march, conditions, and ultimate general result" of the historical process, but also the specific contradictions of the present conjuncture, and which use these contradictions as a lever for seizure of state power, and the gradual construction of socialism and communism. There is, finally, a cluster of tendencies, including populism, "theomachy," and socialist humanism which comprehends socialism as the "negation of the negation," the recovery at a higher level of humanity's archaic holism, and which regards socialist construction as first and foremost as a problem of organizing the working classes and developing their intellectual, moral, and spiritual, as well as their productive and political capacities. We need now to examine each of these approaches in turn.

B. Development of the Socialist Tradition

1. Social Democracy

a) Materialist Social Democracy

Marx himself was well aware of the limits to the spontaneous development of socialist consciousness within the working class. Most of the leadership of the workers movement, on the other hand, particularly in the most advanced countries, neglected or rejected Marx's insights in this regard. Marx's understanding of human labor as an expression of the larger capacity of matter to develop towards ever higher levels of organization, a capacity which was, however, constrained by the disintegrating effects of the market system, was displaced by a technological and economic determinism which regarded the development of socialism as a more or less inevitable result of the development of industrial technology and of the contradiction between industrial technology and capitalist property relations. Indeed, it would not be too much to say that it was increasingly the dynamic of industrialization itself, even more so than the contradiction between industry and the marketplace, which was regarded as the principal basis for the emergence of socialist consciousness. This is the perspective which, with some modifications, dominated European social democracy.

This tendency first became apparent in the writings of Frederick Engels. At the philosophical level Engels undertook the first effort to systematize the laws of the dialectic and to apply them comprehensively to the process of natural and social evolution. He identified three principal laws:

a) the tendency of quantitative differences to gradually accumulate and to transform themselves into qualitative differences,

b) the principle of the unity and contradiction of opposites, and

c) the principle of the negation of the negation.

Engels illustrated at some length how these principles were operative at every level of organization, from the simplest natural processes to the most complex forms of social organization. Ultimately, however, it was the concept of the transition from quantitative to qualitative difference which governed Engels' thinking on political questions, and which shaped the politics of the social democratic parties during the latter half of the nineteenth century (Engels 1880/1940).

The material basis for socialism, according to Engels, is the industrial system created by capitalism, which gave to the productive forces a uniquely social character.

The bourgeoisie could not transform these puny means of production into mighty productive forces without transforming them at the same time into social means of production worked by a collectivity of men (Engels:

1880/1978: 702).

The advanced social division of labor means that realization of the value of products depends on exchange. At the same time, however, the appropriation of the surplus product has remained individual, i.e. capitalist appropriation.

> The separation was made between the means of production concentrated in the hands of the capitalists, on the one hand, and the producers, possessing nothing on the other (Engels 1880/1978: 704).

The relative or absolute impoverishment of the working class leads ultimately to a permanent tendency towards underconsumption, fomenting first economic and then political crises, exacerbating the contradiction between the bourgeoisie and the proletariat, and creating the material conditions for social revolution.

According to this "theory of the productive forces" the socialization of the means of production is the precondition for the continued development of the productive forces.

This theory had definite strategic implications.

> German Social Democracy occupies a special position and therefore, at least in the immediate future, has a special task. The two million voters whom it sends to the ballot box, together with the young men and women who stand behind them as non-voters form the most numerous, most compact mass, the decisive "shockforce" of the international proletarian army. This mass already supplies over a fourth of the votes cast ... Its growth proceeds as spontaneously, as steadily, as irresistibly, and at the same time as tranquilly as a natural process ... If it continues in this fashion, by the end of this century we shall conquer the greater part of the middle strata of society, petty bourgeois and small peasants and grow into the decisive power in the land, before which all other powers will have to bow, whether they like it or not (Engels 1895/1978: 571).

Within this context the task of the party was primarily to organize and educate the working class. As Karl Kautsky put it, "Our task is not to organize the revolution, but to organize ourselves for the revolution; not to make the revolution but to take advantage of it (quoted in Laclau and Mouffe 1985)." Social democratic parties worked to develop a socialist political culture within the working class, while eschewing any attempt to form alliances with other social forces which might hasten their conquest of power. Indeed, all social forces outside the working class were regarded as backward and reactionary.

This strategic conception is a simple application of Engels first law of the dialectic. Quantitative differences (e.g. differences between workers and capitalists over the division of the value added) gradually built up to the point that they became qualitative (i.e. a difference over the proper organization of human society) making possible a qualitative transition from one mode of production to another, rather than merely quantitative changes within the capitalist system (reforms).

Soon, however, it became apparent that the expectations of the Socialist International were not to be realized. While proletarianization proceeded apace, the working class was undergoing rapid internal economic differentiation and political fragmentation. Socialist propaganda could contain this to some degree, but it was very far from raising the working class from trade union to communist class consciousness. On the contrary, the development of the British and North American trade union movements was very much in the direction of reformism and class collaboration.

b) Idealist Social Democracy

One solution to the growing internal contradictions of the working class was to seek a principle of unity outside the economic arena, in the realm of politics and ideology. Thus the revisionist paradigm developed by Eduard Bernstein. Bernstein argued that

> Sciences, arts, a whole series of social relations are today much less dependent on economics than formerly ... the point of economic development attained today leaves the ideological, and especially the ethical factors greater space for independent activity than was formerly the case (Bernstein 1909/1961: 15-16).

If the political and cultural spheres are relatively autonomous, then it is possible for the social democratic party to become

> an organ of the class struggle which holds the entire class together in spite of its fragmentation through different employment ... that is the Social Democracy as a political party. In it, the special interest of the economic group is submerged in favor of the general interest of those who depend on income for their labor, of all the underprivileged (quoted in Laclau and Mouffe 1985: 32).

The party accomplishes on the political and ideological level what the economic struggle itself failed to achieve: the unification of the working class around socialist aims. At first sight Bernstein's conclusions appear to be strikingly similar to those at which Lenin arrived based on his analysis of Russian conditions. There is, however an important difference. For Lenin, we will see, the leadership of the party is based on its superior grasp of the "line of march, conditions, and ultimate general results of the proletarian struggle," and specifically of the real, material contradictions which drive development forward. Bernstein, on the other hand, rejects outright the notion of a *materialist* teleology, and argues instead that the option for socialism is an *ethical* decision which, therefore, is never *necessary* in the strict sense of the word. Bernstein understands this ethical decision in largely Neo-Kantian terms; thus his tendency to identify socialist policies with those which serve "all of the underprivileged" ——i.e. which meet the Kantian ethical criterion of universalizability. The only evolutionary process which Bernstein recognizes is one of gradual emancipation from natural necessity, and thus of the development of the autonomous moral subject.

Strategically, Bernstein's formulation points in the direction of the reformist politics which have characterized the social democratic parties throughout most of this century. The party, as the expression of an ethical ideal, acts on the terrain of the state to promote social policies which benefit the "underprivileged in general:" development of a social safety net, promotion of trade union demands for higher wages, partial or complete nationalization of key industries. There is little or no strategy here to raise the masses to a fully communist level of development.

c) Social Democratic Practice

There was no one point at which idealist, Bernsteinian social democracy clearly gained the upper hand over the materialism of Engels and Kautsky, though clearly after the Russian Revolution most materialistically minded socialists were drawn to the Leninist parties. Still, even to this day the social democratic parties, especially in Latin Europe, harbor a materialist left.

The development of social democracy has not, however, been driven by internal ideological struggles so much as by the shifting dynamics of alliance between the industrial proletariat, the middle strata, and the progressive bourgeoisie. Just as Engels predicted, social democratic parties gradually gained strength throughout the period leading up to the Second World War, and together with social liberal and Christian Democratic parties, played the predominant role in shaping postwar European society. Social Democracy was also influential in Australia and New Zealand.

What, concretely, has social democracy achieved?

At the organizational level, strong social democratic parties have tended to reinforce, and be reinforced by, strong trade union organizations. Wolfgang Merkel notes that in fact social democratic parties have related to trade unions in a variety of ways. Sometimes, as in the United Kingdom, the trade unions have effectively controlled the party. Elsewhere, in Finland, Switzerland, and the Netherlands, the party has dominated the unions. In Germany and Scandinavia, there has been a tendency for the parties and the unions to cooperate closely, without either dominating the other. Only in the countries of Latin Europe, which have historically had strong communist parties, have social democratic parties had weak relations with the trade unions (Merkel 1992: 142).

Social democratic parties have, however, done very little to develop this organizational base politically and ideologically. At most, they have managed to conserve the socialist subculture which developed among workers during the early years of this century——a task which has become increasingly difficult as the more labor intensive industries are redeployed to Asia, Africa, and Latin America, and social democrats look increasingly towards the intelligentsia and service workers for support. Nowhere have social democrats cultivated a mass movement to transcend market relations.

Social democratic parties have deployed most of their organizational resources in the struggle to win elections, and once in power, to win social reforms. Electorally, social democracy has been consistently successful throughout this century, and was especially so throughout the postwar period (1945-1973). On the policy front, the achievements of social democratic parties have been quite significant, but fall far short of really socialist transformation. After a wave of nationalizations in the postwar period, social democratic parties began increasingly to concentrate on welfare state measures (Patterson and Campbell 1974: 64-68). Unlike the U.S., however, where welfare measures tended to benefit only the poorest segment of the population, and to take the form of income (individual consumption) supports, social democratic governments in Europe (usually with the active and enthusiastic cooperation of social liberals and Christian Democrats) have directed funds first and foremost to the creation of a social infrastructure which provides all citizens with child support payments, access to free or low cost health care, education through trade school or university, decent, affordable housing, and an excellent public transportation system. The European welfare state has, in other words, attempted and at least partly succeeded in replacing the old networks of social support which disintegrated under the pressure of the marketplace, while actively investing in the development of human social capacities.

At the same time, the existence of a large complex of nationalized firms has assisted in the creation of a coherent capitalist planning process. State firms (e.g. airlines, automobile manufacturers) provide a stable or even guaranteed market for private high technology firms, while stricter state control of credit ensures these firms with a steady stream of capital.

What is missing here, of course, is any strategy for realizing the historic socialist project of transcending the marketplace in order to promote the all sided development of human social capacities, of increasing humanity's capacity to participate consciously in the self-organizing activity of the cosmos. The interests which social democracy has helped to realize remain those of workers in so far as they are consumers——i.e. the empirical, alienated, interests of workers whose consciousness has been deformed by their insertion into commodity relations. European consumerism is certainly more "social" than its North American counterpart, and often more refined. But it is consumerism nonetheless. Social democracy has no strategy for realizing the interests of workers as producers, as participants in the self-organizing activity of the cosmos, nor does it have any strategy for developing workers theoretically to the point that they can even become aware of these interests. It restricts, but does not even begin to transcend, the operation of market relations.

The future of social democracy remains unclear. Beginning in the late 1970s a series of neoliberal parties, supported by growing antitax sentiment within the working class, swept the countries of Northern Europe in particular. This phenomenon was most marked, of course in the United Kingdom, but it has affected even the very progressive countries of Scandinavia. At the same time, national chauvinist sentiment has grown rapidly, and not only on the right. Neofascist parties have gained considerable strength in France, Germany, and even Italy, while the French Communist party has taken up a position which can only be regarded as chauvinistic and protectionist, actually encouraging working class resistance to immigration, international integration, etc.

Behind these developments lies much the same social process which undermined the hegemony of the Democratic Party in the United States. Penetration of market relations into every sphere of life makes all activity simply a means of realizing individual consumer interests. Gradually political socialism and even trade union solidarities begin to erode. Workers are less and less willing to pay taxes in return for collective benefits, much less to promote the all sided development of human social capacities. To the extent that they support the expansion of the welfare state, it is a welfare state which reserves its benefits for their own kind: a Europe for the Europeans, rather than a Europe which represents the first step in the development of a unified planetary state.

Clearly it would be a mistake to write off social democracy as a force for social progress. The social democratic parties of Europe remain an important element in the global progressive bloc. At the same time, it would also be a serious mistake to look to social democracy to help chart the next steps in the human civilizational project. Social democracy represents the deformation of the socialist project under the influence of the marketplace, and is thus largely indistinguishable from the sort of bourgeois reformism which attempts to reassert the progressive dynamic of the industrial, democratic, and scientific revolutions within the framework of a society still regulated, in the final instance, by market relations.

2. The Communist Movement

Lenin was already aware of the dangers of the social democratic strategy in the early part of this century. He recognized that the capitalist development did not automatically lead to the emergence of socialist consciousness within the working class.

The history of all countries shows that the working class, exclusively by its own effort, is able to develop only trade union consciousness, i.e. it may itself realize the necessity for combining in unions to fight against the employers and to strive to compel the government to pass necessary labor legislation, etc.

The theory of Socialism, however, grew out of the philosophic, historical, and economic theories that were elaborated by the educated representatives of the propertied classes, the intellectuals ... (Lenin 1902/1929: 32-33)

... there can be no talk of an independent ideology being developed by the masses of the workers in the process of their movement (Lenin 1902/1929: 40-41)."

Socialist "consciousness could only be brought to" the working class "from without."

Hence the task of Social Democracy is to combat spontaneity, to divert the labor movement, with its spontaneous trade unionist striving, from under the wing of the bourgeoisie, and to bring it under the wing of revolutionary Social Democracy (Lenin 1902/1929: 40-41).

It was thus necessary to build an organization of professional revolutionaries (1902/1929: 114, 159ff) which could organize and direct the working class movement, and develop within it an understanding of the ultimate aims of the revolution. Let us examine the Leninist problematic in more detail.

At the foundation of Leninism is the so called "theory of reflection." According to the reflection theory, all matter is an interconnected system and each element in the system bears the imprint, as it were, of its interactions with other matter. These imprints or reflections gradually accumulate, leading first to quantitative and then to qualitative change. Lenin, however, puts more emphasis than Engels on the interconnectedness of matter, and on the "principle of contradiction" or the "principle of the unity and struggle of opposites." For Lenin, it is the actual interaction between the elements within material systems that change and development takes place. Qualitative change takes place when contradictions accumulate to the point that the system must develop fundamentally new structures in order to maintain its integrity and continue to grow. In order to promote development, it is necessary to grasp the principal contradiction characteristic of any particular system, and use that contradiction as a catalyst for growth and development.

Politically, this meant that the revolutionary vanguard, which already understood the "line of march, conditions, and ultimate general result" of the historical process, should neither wait passively for the quantitative growth of the working classes to reach a critical mass and lead to qualitative change, nor should they engage in moralizing polemics about the virtues of socialism. Rather, they needed to analyze the concrete contradictions characteristic of the present period, and use those contradictions to build power.

Lenin took as his point of departure what most Russian social democrats saw as an overwhelming obstacle: the fact that in Russia, as in most countries outside of Western Europe and North America, the development of capitalism and the democratic revolution were far from complete, and were in fact blocked by the collaboration of foreign capital and entrenched feudal elites. In so far as Russia was still at least partly feudal in character, the emerging Russian revolutionary movement had a dual character. On the one hand, the Russian people as a whole sought to abolish the remnants of feudalism, and to emancipate themselves from the Tsarist autocracy. At the same time, the emerging proletariat was already beginning to struggle for socialism.

According to the classical Marxism of the Second International, it was impossible for the Party to undertake socialist tasks until capitalism and democracy were fully developed and the masses of the Russian people were fully proletarianized. Furthermore, the "democratic revolution" was largely a matter for the bourgeois and petty bourgeois parties. Socialists should devote themselves to patient organizing and propaganda among the most advanced segments of the working class, working quietly until conditions were ripe for more profound transformation.

Lenin saw the matter differently. The bourgeoisie, Lenin reasoned, however much it might include democratic demands within its program, will never be consistently democratic. "The very position the bourgeoisie holds as a class in capitalist society inevitably leads to its inconsistency in a democratic revolution (1905/1971:78)." There are, in effect, many demands which, while very far from being fully socialist, are beyond the capacity of the bourgeoisie to fulfill. This makes it possible to win the support of the masses without developing them politically or ideologically to the point that they support, or even understand, the long-range aims of the Communist Party

This presented a unique opportunity for the party.

> From the premise that a democratic revolution is far from being a socialist revolution, that the poor and needy are by no means the only ones to be "interested" in it, that it is deeply rooted in the inescapable needs and requirements of the whole of bourgeois society ——from these premises we draw the conclusion that the advanced class must formulate its democratic demands all the more sharply (Lenin 1905/1971:69)

The Communist Party, Lenin reasoned, could come to power quickly in a semi-feudal country like Russia, by demonstrating that it was the most consistent advocate for the democratic demands of the broad masses of the people. In the Russia of 1917 this meant prosecuting with greater vigor than any of the other parties the peasants' demand for land, the workers' demand for bread, and the demand of the whole people for Russian withdrawal from the First World War.

After the success of the Russian revolution, many communists in the advanced capitalist countries assumed that since their countries had already completed their democratic revolutions, they could organize the working class around directly socialist demands. Lenin rejected this notion, pointing out that the "broad masses" of the working class were "still, for the most part, apathetic, inert, dormant, and convention ridden" (Lenin 1920/1971:573) and would be won over not by ideological struggle, but rather on the basis of concrete demands which, while not fully socialist, could not be met by the bourgeoisie.

This is, in effect the formula which has been followed by all Communist parties since Lenin's time, the innovations of Dimitrov, Mao, Gramsci,

Mariategui, Fidel and Che notwithstanding. On the one hand, the party develops a rich internal culture, and forges a highly developed sense of revolutionary class consciousness on the part of its members. On the other hand, it analyzes the concrete contradictions which make it impossible for the ruling class to continue to rule, at least as it has been doing, and then draw on these contradictions as a way to leverage power. In some cases this means forming a broad popular front, drawing on the spontaneous collectivism of the peasantry, using it as a bridge to the formation of socialist consciousness, or positioning themselves as the most effective defenders of the (trade union) interests of the working class and/or building popular fronts around the struggle against fascism, for broad structural reforms, investment in human capital, etc. In other cases it meant rejecting all alliances in order to seize on an economic, political, or cultural crisis as an opportunity to sharpen the struggle for socialism. This was the case during Stalin's campaign for forced collectivization, and Mao's cultural revolution.

Lenin's analysis has the merit of recognizing that communist ideology does not develop spontaneously within the working class in the context of capitalist society, but must be carefully cultivated. Lenin did not, however, develop a strategy for raising the masses to a fully communist level of development, but rather a strategy which would permit Communists to secure the support of the masses and "seize state power" without requiring the masses to develop into real communists. Leninism has proven itself to be an effective strategy for power in countries where the democratic struggle has been blocked by imperialism or set back by the emergence of an indigenous fascist movement. This is the real secret of the long wave of revolutionary upheavals which began in 1917 and which ended only in 1989. Leninism permits a revolutionary elite to seize power in a predominantly agrarian country, break the back of reactionary landed elites, and undertake a systematic rationalization of the social order.

Not surprisingly, however, Lenin was less successful in developing a strategy for socialist construction——for developing human society to the point that the marketplace and commodity production could be transcended once and for all. Indeed, as it approached this question, Leninism revealed itself to be a tremendously unstable political formation, and soon gave birth the plethora of trends and tendencies which has characterized communist movement throughout most of this century. We need now to examine each of the principal variants of Leninism in some detail.

a) **Struggles in the Soviet Union and in the Third International**

Not all of Lenin's comrades in the Bolshevik party shared, or even fully understood his interpretation of the dialectic as a theory of development through interaction and contradiction. Some of these——Bogdanov and Theomachists, Lukacs and the socialist humanists, and Gramsci and the cultural hegemonists——in fact advanced creative alternatives to Leninism, which we will have occasion to examine later on. Others, however, simply adhered dogmatically to one or another of the specific tactical formulae which Lenin had developed in response to specific situations: the worker peasant alliance, uninterrupted revolution, etc., and made these into the basis for a global interpretation of Leninist politics. This was true of both the right opposition within the Soviet Union, and of the "left" or Trotskyist opposition. As a result, these trends were quickly eliminated by Stalin who, his many failings notwithstanding, clearly understood better than his rivals the Leninist art of building power through contradiction and struggle.

i) **The Right Opposition**

We have already noted that the Russian revolution tapped into and unleashed the revolutionary, even millenarian communitarian aspirations of the Russian peasantry. The peasants had little commitment to the larger ambitions of the socialist project: i.e. to centralize the resources necessary to develop human social capacities. Indeed, the peasants did not even understand clearly the need to supply an army in order to defend the gains of the revolution. The party, for its part, never having invested any effort in learning how to catalyze the development of this level of commitment on a mass scale, responded to its need for grain by conducting what amounted to random and arbitrary confiscations of grain from those determined to be "rich" peasants. Similarly, the Bolsheviks needed for the peasantry to supply recruits for the army (Wolf 1969: 92-93).

The party, in other words, soon ran up against the contradictions of Lenin's strategy. Having won power by promising "land, bread, and peace," Lenin was unable to defend that power without fighting a war and depriving the peasants of the very things he had just given them. The result was a series of peasant revolts and mutinies. Indeed, some peasants during this period believed that

> a new party had come to power in Moscow. They were, they proclaimed, for the Bolsheviks who had given them the land, but they were against the Communists who were now trying to rob them (Footman 1962: 270).

The end of the civil war permitted the party to relent somewhat in its pressure on the peasantry. But the same contradictions appeared at a higher level in the debates around socialist construction which raged throughout the course of the next decade.

Lenin's New Economic Policy (NEP) permitted the *mir* to flourish, along with small private enterprise. The state retained control of the "commanding heights" of the economy. While for Lenin NEP represented a strategic retreat, it became for a section of the Russian Communist Party, and of the international communist movement generally, the cornerstone of a complete strategy for the construction of socialism. This faction, led by Nicolai Bukharin, eventually came to be known as the "right opposition."

Philosophically, the right opposition tended towards a "mechanistic" interpretation of the dialectic. Rather than grasping reality as first and foremost a system of relationships, the structure of which defined the various elements within the system, members of the right opposition tended to regard reality as a system of interacting elements the relations between which tend toward equilibrium. In this regard, though they drew some of their ideas from the far more revolutionary work of Alexander Bogdanov, their position was not unlike that of the general system theory which was beginning to emerge in Austria and the U.S. during this period. While sometimes contradictions became so great that re-establishment of equilibrium require a complete reorganization of the system, it was harmony and not contradiction which dominated their perspective. And certainly a society such as Russia, which had just undergone a fundamental reorganization, ought to be able to achieve equilibrium without further structural modifications, at least for a relatively long period of time (Susiluoto 1982).

This philosophical perspective had definite political implications. Radical land reform would lead to rising incomes among the peasantry, and thus the kind of internal demand for manufactured goods which capitalist development had obstructed. State owned industry would produce primarily for this market, with agricultural implements increasing agrarian production and cheap consumer goods maintaining the political good will of the peasantry. The communitarian traditions of the peasantry, together with a cultural revolution led by the party, would lay the groundwork for the gradual voluntary collectivization of the peasantry, and the industrialization of the work force. Peasant communitarianism would, in effect serve as a bridge to the development of socialist consciousness. Here we see the old populist strategy recapitulated at a higher level of rationalization.

In the area of revolutionary strategy the right opposition devoted special attention to countering the fascist threat which was already growing in the

1920s. The members of this faction argued that it was necessary to counter the fascists' hold on the hearts and minds of the people by demonstrating that Communism was in continuity with——and was in fact the highest expression of——the human civilizational project and thus with their most deeply held national, popular, and religious traditions.

This tendency dominated from 1921 until 1928. By the end of this period, however, it was running up against a serious obstacle. The very measures which had been introduced in order to increase rural incomes (radical land reform, low taxes, favorable prices) had undermined the centralization of surplus necessary for industrialization. By the autumn of 1927 there were widespread shortages of industrial goods. This shortage removed the principal incentive for the peasantry to produce and sell surplus grain.

Before the revolution landlords had produced 12% of the total grain, 47% of which they marketed outside of the villages, contributing 21.6% of the grain available for export or for consumption in the cities. The *kulaks* or rich peasants produced 38% of the total grain, of which they marketed 34%, contributing 50% of the grain available for export or urban consumption. The poor or middle peasants produced 50% of the total grain, but marketed only 14.7% of this outside the villages, contributing only 28.4 percent of grain available for export or urban consumption. Total production was 5000 m. poods, of which 1300 m. poods were marketed outside the villages.

By 1926-1927 total grain production had fallen to 4747 m. poods, of which only 630 m. poods were marketed outside of the village. State and collective farms (replacing the landlords) produced only 1.7% of the total grain, marketing 47.2% of this outside the villages, but contributing 6% of the grain available for export or urban consumption. The *kulaks* produced 13% of the total grain, of which they marketed 20% outside the villages (significantly less than before the revolution) and contributed only 20% of the grain available for export or urban consumption. The poor and middle peasants produced 85.3 percent of the total grain, but of this marketed only 11.2%, contributing 74% of the total available for export or urban consumption. It was difficult to escape the conclusion that the *kulaks* were hoarding grain badly needed to finance industrialization, and that the transfer of land from the landowners to the peasantry had undermined the centralization of surplus, and set in motion what amounted to a capital formation crisis (Osband 1982: 19, Dobbs 1948: 216-217).

Bukharin and his allies argued that investment should be temporarily reduced, and that more industrial goods should be directed towards the countryside, increasing the incentive for the peasants to produce and market surplus grain (Osband 1982: 19). They were, however, increasingly in the minority.

ii) Trotskyism

There had been elements within the party which had opposed the NEP from the very beginning. The most important of these tendencies was the emerging Trotskyist opposition. Trotsky had only partially assimilated Lenin's theoretical, organizational, and strategic innovations. He shared with most social democrats the conviction that development of the productive forces was the precondition for the successful construction of a socialist society. And he believed that the establishment of state and collective property forms was essential to this process. Indeed, during their early struggles, it was Trotsky, and not Stalin, who expressed opposition to continuing Lenin's New Economic Policy, which permitted significant levels of private accumulation on the part of the peasantry and the petty bourgeoisie. They differed, however, regarding the possibility of accomplishing this task in a Soviet Union which was rapidly becoming isolated in a hostile capitalist world, and regarding the long term effects of the process of "primitive socialist accumulation" on "socialism in one country." Trotsky argued that the kind of rapid technological development required for socialist construction was possible only with the assistance of the advanced industrial countries, something that was unlikely short of an international socialist revolution. Primitive accumulation within the context of a backward society like the Soviet Union, would lead inevitably to bureaucratic degeneration.

> The basis of bureaucratic rule is the poverty of society in objects of consumption, with the resulting struggle of each against all ... When the lines are very long, it is necessary to appoint a policeman to keep order. Such is the starting point of the power of the Soviet bureaucracy (Trotsky 1937/1972).

The construction of socialism in a backward country such as the Soviet Union could succeed only if there was technical, economic, and political support from the working classes of the advanced capitalist countries——i.e. only if there was an international proletarian revolution.

> Backward Russian capitalism was the first to pay for the bankruptcy of world capitalism. The law of uneven development is supplemented through the whole course of history by the law of combined development. The collapse of the bourgeoisie in Russia led to the proletarian

dictatorship——that is to a backward county's leaping ahead of the advanced countries. However, the establishment of socialist forms of property in the country came up against the inadequate level of technique and culture. Itself born of the contradictions between high world productive forces and capitalist forms of property, the October revolution produced in its turn a contradiction between low national productive forces and socialist forms of property.

... isolation and the impossibility of using the resources of world economy even upon capitalistic bases ... entailed, along with enormous expenditures upon military defense, an extremely disadvantageous allocation of productive forces, and a slow raising of the standard of living of the masses ...

... The longer the Soviet Union remains in a capitalist environment, the deeper runs the degeneration of the social fabric. A prolonged isolation would inevitably end not in national communism, but in a restoration of capitalism (Trotsky 1937/1972: 300-301).

Unlike Stalin, who stressed the importance of active ideological education, and development of a "socialist realist" culture which promoted socialist values, Trotsky argued that "Culture feeds upon the juices of industry, and a material excess is necessary in order that culture should grow, refine, and complicate itself (1937/1972: 179)." Trotsky regarded attempts to forge a "proletarian culture" in an economically backward country such as the Soviet Union as artificial and doomed to failure.

While certain elements of Trotsky's analysis have considerable merit, their implications are overwhelmingly pessimistic and did not really offer a credible road forward for the Soviet Union. The conditions for international proletarian revolution of the kind envisioned by Trotsky did not and never have existed, and for an impoverished and isolated socialist country to make the promotion of world revolution its principal strategic goal would have been irresponsible adventurism. It is hardly surprising that Trotsky's ideas never attracted significant popular support.

iii) Stalinism

Stalin, like Lenin, believed that contradiction and struggle drive social development. Concrete social contradictions open up new opportunities to build power, or the carry out revolutionary social transformations. Up until roughly 1928, this meant a policy of class alliances. Thus his opposition to Trotsky's demands for collectivization and world revolution during the early 1920s, and his support for Lenin's NEP and for the united front strategy.

Beginning in the middle and late 1920s, however, new contradictions began to surface, and new opportunities presented themselves. On the one hand, Stalin interpreted the rise of fascism as a sign that the world capitalist system was approaching its final crisis. If this was true then only socialism could resolve the contradictions, and it was important to sharply differentiate the Communist Parties from social democratic reformism, which did not offer an authentic alternative to capitalism, and which had little interest in taking advantage of the crisis to build working class power. Thus the "left turn" and the "third period line" promulgated in 1928. Stalin directed Communists internationally to "direct the main blow against social democracy," which, he argued, was simply the "left wing of fascism."

On the home front similar opportunities presented themselves. The working class, and the industrial sector generally, needed a much larger and more reliable supply of grain than the peasants were willing——or indeed had any good reason——to provide. One way to increase this supply was to collectivize agriculture. The shortage of grain thus created for Stalin an opportunity to mobilize the workers around an aim in which they had previously had no interest: the further extension of state and collective property forms, and thus of the political power of the Communist Party. Thus the campaign for forced collectivization.

> The party's immediate response to the falling off of grain collections was "emergency measures" which in fact harked back to the forcible requisitioning of the 1918-1921 period. In the early months of 1928 large quantities of grain were confiscated, illegal searches were conducted of peasant households, trials of speculators and hoarders were convened ... This campaign was ostensibly aimed at the *kulaks* and rich peasants. In fact, the Party leadership knew very well that the real target was the middle peasants. Mikoyan, who was in charge of grain procurement through the Commissariat of Trade, admitted that "the main masses of grain surpluses were in the hands of the middle peasants," and that wheat taken from them was confiscated by means of measures which were "harmful, illegal and inadmissible (Costello 1982: 6)."

After a few months, grain procurement fell off again, as the peasants' supplies were depleted. Peasants resisted by reducing the area under cultivation and by strikes and sabotage (Costello 1982: 6). In 1929 harsh new measures were imposed.

A whole new series of state administrative agencies was established to deal with the rural areas and a new law was passed which made the failure to meet grain quotas punishable by imprisonment, confiscation of property, and even imprisonment (Costello 1982: 9).

Finally, during the final months of 1929, a massive campaign of collectivization was instituted. "By March 1930 59.9% of peasant households were collectivized (Costello 1982: 9)," though this figured dropped off considerably over the course of the next year as pressure was relaxed somewhat.

The results of the process were ambiguous. From the standpoint of the rural masses collectivization was nothing short of a disaster.

The First Five Year Plan forecast an increase in gross agricultural output from 1927-1928 to 1932-1933 of 50%. Instead gross agricultural output declined throughout the First Five Year Plan. Taking 1928 output as 100%, in 1929 it fell to 98%, in 1930 to 94.4%, in 1931 to 92%, in 1932 to 86%, and in 1933 to 81.5%. Gross output for 1937 (a very good year) was only 8% above the 1928 figure, but the output for 1938 and 1939 was still below that of 1928 (Costello 1982: 10).

Industry fared somewhat better. While growth rates fell short of the unrealistic goals set by the State Planning Commission (21.7%, 32%, 45%, and 36% for the years 1929 - 1932 respectively), the results were, nonetheless, quite remarkable: 20%, 21.8%, 22%, and 18%. Still, most of this growth was extensive rather than intensive. Labor productivity actually fell during this period.

Too wide of an investment front swallowed up more resources than could be effectively used ... The distribution network was so bad that the Chairman of the State Planning Commission admitted that 30% of the consignments of supplies went bad before they even reached the consumer (Costello 1982: 10).

The Soviet Union, Stalin argued, was transformed as a result of his strategy, into a socialist society governed by the "basic law of socialism"

the securing of the maximum satisfaction of the constantly rising material and cultural requirements of the whole of society through the continuous expansion and perfection of socialist production on the basis of higher techniques (1952/1972: 41).

The policy of forced collectivization and rapid industrialization was thus at least a mixed success from the standpoint of the CPSU; the third period line was not. After the Nazi victory in Germany, and not without significant pressure from Bulgarian Communist leader Georgi Dimitrov, Stalin acquiesced in a broader interpretation of the popular front. According to this new interpretation, Communists would cooperate with Social Democrats, Liberals, and Catholics in the struggle against fascism, while attempting to win over the people on the basis of their superior organization and effectiveness. The heroic leadership of the Communist Parties in the struggle against fascism won them state power in most of Eastern Europe, and status as the principal opposition in Italy and France. From Stalin's point of view, though the popular front strategy was simply a different way of doing what he had earlier attempted to accomplish through the medium of the third period line: use the contradictions created by the struggle against fascism to build the power of the communist parties. Soviet policy in Eastern Europe after the war demonstrated that the popular front was simply a tactical move, intended to secure state power for the party——not a strategy for securing the long term support of the peasantry, petty bourgeoisie, and the progressive antifascist sectors of capital. This was especially true during the period of "Zhdanovism," during which the popular front was increasingly identified with those forces which explicitly supported the "socialist camp" in the emerging cold war.

What Stalin's strategy of primitive socialist accumulation permitted was the rapid development of large capital goods industries based on the technology of the first and second industrial revolutions. This in turn permitted the development of a military-industrial complex which was able to defend socialism against the developing fascist threat and to make a major——perhaps the decisive contribution——to the struggle against fascism. The centralized redistributional systems which emerged made it possible to centralize the resources necessary to build up among the finest health and educational systems in the world, and to put the Soviet Union in the forefront of many fields of scientific research and cultural production.

What the system could not do was to tap into the creativity and the self-interest of the broad masses of the Soviet people, or to catalyze the kind of open-ended theoretical work among the intelligentsia which was necessary to make the Soviet Union a real leader in the Third Industrial Revolution or in the development of protosynergistic forms of production and social organization. What the Soviet State did, in effect, was to extract surplus from a dependent worker and peasant population through a system of tributes legitimated ideologically and guaranteed by force. In this respect it was much like the rationalized tributary states which emerged from earlier social revolutions. Mobilized by fear, or by a willingness to sacrifice themselves for Mother Russia and the Revolution (but not by an actual understanding of and interest in the human civilizational project) the Soviet people could only carry out (or resist carrying out) the directives of their leaders. They could not participate actively in the reorganization and thus in the actual development of the system.

Two points are in order here. First, industrial forces of production are not in fact sufficient for the achievement of communism. The industrial worker is never a full, conscious participant in the self-organizing activity of the cosmos. Something more like what has come to be called postindustrialism, in which all of the less creative tasks have been automated is required. Development of postindustrial technologies, however, requires a more complex method of organizing human interdependence than that developed by the Soviet state. For industrialization it is enough to centralize surplus and allocate it to strategically important development projects. For "postindustrialization" it is necessary to mobilize the virtually infinitely diverse talents and abilities of hundreds of millions of highly skilled workers. This in turn requires a relational, rather than a linear understanding of power. The coercive power of the Soviet state was sufficient to suppress counterrevolution and defeat fascism, and to mobilize the labor of millions of unskilled laborers. Mobilization of a quarter-billion skilled workers, on the other hand requires forms of leadership which appeal to, and at the same time raise the level of, the interests of those workers. This requires not so much "ideological education" as a high level of theoretical development which permits the entire working class to grasp——and appreciate the tremendous beauty of——humanity's long march towards communism.

At the same time, the system of generalized commodity production remained in place. Soviets worked for a wage, or for a share of the profits of the collective farm, and used that income to purchase commodities in the marketplace. This meant that the alienating dynamic of the marketplace operated in the Soviet Union much as it did in the capitalist countries. Gradually all activity became simply a means of realizing individual consumer interests.

This placed definite limits to the further development of the Soviet Union, at least so long as its development was guided by Stalinist strategy. Writing in 1952, Stalin identified several tasks facing Soviet society in the coming years.

... a relatively higher rate of expansion of means of production ...

... to raise collective farm property to the level of public property ... and

.. to replace commodity circulation by a system of products-exchange, under which the central government, or some other social-economic center, might control the whole product of social production in the interests of society ...

... to ensure such a cultural advancement of society as will secure for all members of society the all-round development of their physical and mental abilities, so that the members of society may be in a position to receive an education sufficient to enable them to be active agents of social development ... for this, it is necessary ... to shorten the working day at least to six, and subsequently to five hours ... (1952/1972: 68-70)

Having built an industrial infrastructure and won the war against fascism the CPSU was ready to use its capacity to centralize and allocate resources to lead humanity into a brilliant new era of scientific, technological, and cultural development. And in many ways, the postwar Soviet Union succeeded in this regard. Just before its collapse, the old Soviet Union had the largest number of scientists and engineers per million population (5,387) and the highest research and development expenditures per GNP (6.2 per cent) of any major nation in the world (1988 figures, CRS Review 1992). The disintegration of the Soviet Union has put Russia even more clearly in the lead in the number of scientists and engineers per million population (9,398). Russian universities continue to be highly rated. According to the Gourman Report, one of the principal rankings of educational institutions, the former Soviet Union had 24 of the 74 universities outside the U.S. with a rating of 4.0 out of 5.0, or higher, with Moscow State University receiving a rating of 4.93, equal to the University of Michigan, and behind only Harvard, the University of California, Berkeley, and the University of Paris——and ahead of Oxford, Cambridge, . and Heidelberg, Yale, Stanford, Chicago, and Princeton (Gourman 1987: 159-160, 207).

This powerful scientific and cultural infrastructure translated into a important contributions in vitally important fields of science and technology. In the field of mathematics, Russia leads in such specialties as mathematical logic, topology, differential equations, and functional analysis. Russia is also a leader in several areas of physics and astrophysics, especially particle physics, low temperature physics, aerodynamics, hydrodynamics, mass dynamics, and the molecular structure of solids (Medish 1984: 176:ff). Perhaps most important, however, are the Soviet contributions in the field of cosmonautics, where they were almost unquestionably the world's leaders, particularly in the areas of space propulsion and the human exploration of space. The Defense Department has recently purchased the Soviet Topaz nuclear reactor, designed for use as a space propulsion system, and the Soviet space station *Mir* represent's humanity's first attempt to establish a permanent human presence outside the earth's gravity well.

It is often assumed that political controls prevented social scientists in the Soviet bloc from making important contributions. This is not really true. Eastern Europe has long been leader in the field of linguistics and the Soviet Union has played an important role in the development of systems theory (Susiluoto 1982), building on the historic contribution of Alexander Bogdanov (1928/1980). Eastern Europe has even nourished some critical tendencies in dialectical and historical materialism, such as the teleological humanism associated with Lukacs (1971) and the Praxis group in Yugoslavia.

Soviet bloc achievements in the arts and in international athletic competition are well known. Russia in particular has nourished a vibrant literary tradition, including what is probably the most artistically developed tradition of science fiction writing on the planet. Soviet science fiction served as an important vehicle for exploring the problems of socialism, while contributing to the elaboration of an enormously hopeful vision of humanity's future.

Oliver Wendell Homes once wrote that "When I pay taxes I buy civilization." By centralizing the entire social surplus product in the hands of the state, the citizens of the Soviet bloc countries, even in the face of enormous military pressure from the United States——and in the face of the real limitations of the state structure as a catalyst for promoting human development——purchased a truly enormous quantity of civilization.

But Stalin and his successors faced a problem. The Soviet people did not share the party's vision for the future. On the contrary, having built an industrial infrastructure and satisfied their basic consumption needs, they were interested in expanding individual consumption——in achieving the same kind of consumerist lifestyle which was becoming possible during this period in the advanced capitalist countries. The next "contradiction" which emerged in the Soviet system was the contradiction between the country's socialist economic system and the consumerist aspirations of the Soviet people, generated spontaneously by residual commodity relations in Soviet society. This contradiction did not permit of a socialist resolution.

iv) *Perestroika* **and the Crisis of Soviet Socialism**

It is not possible in this context to trace the complex process leading to the decline and fall of the Soviet Union. Suffice it to say that in spite of significant popular discontent, the system remained not only stable but even dynamic up until the middle 1970s. Increasingly, however, there was a tendency for production to stagnate, and for the people to grow ever more unhappy as their consumerist aspirations went unsatisfied. There had always been elements within the party——heirs of the tradition of the right opposition——which believed that this stagnation could be overcome only through decentralization and by giving greater scope to market forces. Kruschev had attempted reforms along this line during the late 1950s, only to be beaten back by the party and state bureaucracies. Gradually these forces gained strength, and in 1985 they came to power under the leadership of Mihail Gorbachev.

Initially, Soviet advocates of *perestroika* argued simply that socialist property relations, and in particular the centralized planning apparatus were too advanced for the still rather low level of development of the productive forces. Gradually this initial thesis was displaced by the more radical claim that the command economy itself had become an obstacle to development of the productive forces. While extraordinarily effective in mobilizing the resources for rapid industrialization, an economic system structured around state ownership of the means of production and allocation of resources by centralized planning agencies could not organize the complex activities which characterize a complex industrial society. "Socialist" norms defending the rights of workers, furthermore, became an obstacle to labor discipline. Soviet workers, in effect became lazy and inefficient, at the same time they were demanding improved individual consumption standards. Gorbachev's own analysis is typical in this regard.

> At some stage,——this became particularly clear in the latter half of the seventies——something happened that was at first sight inexplicable. The country began to lose momentum. Economic failures became more frequent. Difficulties bean to accumulate and deteriorate, and unresolved problems to multiply. Elements of what we call stagnation and other phenomena alien to socialism began to appear in the life of society. A kind of "braking mechanism" affecting social and economic development formed. And all this happened at a time when scientific and technological revolution opened up new prospects for economic and social progress.

> Analyzing the situation, we first discovered a slowing of economic growth. In the last fifteen years the national income growth rates had declined by more than a half and by the beginning of the eighties had fallen to a level close to economic stagnation ...

> The gross output drive, particularly in heavy industry, turned out to be a

top-priority task, just an end in itself. The same happened in capital construction.

It became typical of many of our economic executives to think not of how to build up the national assets, but of how to put more material, labor and working time into an item to sell it at a higher price ...

The country became "accustomed to giving priority to quantitative growth in production" over qualitative growth, extensive accumulation priority over intensive (1987: 18-20).[1]

To remedy this situation Gorbachev proposed a return to market norms, in the context of mixed economy.

Our aim is to create a mixed, multiform economy in which all forms of property ownership develop freely ... in order to give the biggest number of working people the opportunity to become owners (Gorbachev 1991).

Agriculture and much of industry would be reprivatized, and price controls and other mechanisms of centralized planning gradually abandoned. The Soviet Union would be preserved as a unified market in which a mixed economy could develop, while the Soviet state continued the generous support for education, research and development which have been one of its greatest strengths.

Gorbachev's professed fidelity to the socialist tradition notwithstanding, *perestroika* clearly represented a rupture in the process of socialist construction. More faithful to Lenin than to Marx, Gorbachev read, and attempted to resolve, the internal contradictions of the Soviet system. But he never really took seriously Marx's critique of commodity production. In the process he ended up retheorizing the socialist experience as a strategy for industrializing a backward country and then returning it to the capitalist road.

But the worst was yet to come. Gorbachev calculated, quite correctly, that if he was to defeat entrenched bureaucratic elites in the party and state bureaucracies he would need to the support of the people. This meant an

---

[1]. Gorbachev has an important insight here. Value, as it has historically been understood by the dialectical materialist tradition, is simply the quantitative expression of the complexity of a commodity --i.e. the quantity of average socially necessary labor time it embodies. While superior to price, it still tells us relatively little about the relative contribution of various activities to the qualitative complexity of the ecosystem, the development of human social capacities, and the self-organizing activity of the cosmos as a whole.

end to repressive strategies for political control, and ultimately the establishment of a multiparty system. Gorbachev did not, however, count on the impatience of the masses. When he proved himself unable to deliver on the consumer demands of the Soviet people he was swept away in a tide of popular discontent which took the Soviet state and the socialist system with it.

For the theoreticians of *perestroika* and the right opposition generally the communist movement represents simply an extension of the rationalizing, humanist project of the Enlightenment, or even of human civilization generally. From Lenin's insight that only communists could realize the democratic demands of the masses, the right opposition drew the (rather different) conclusion that communism was, first and foremost a means of realizing those demands. While this analysis correctly points out the *continuity* between the socialist project and the larger project of human civilization generally, and lays the groundwork for strategic initiatives towards sectors which understand that larger project in terms of popular religious or democratic ideologies, it has also leads to a tendency to liquidate the long range aims of the communist movement——transcendence of commodity production and creation of a society in which humanity realizes fully its participation in the self-organizing activity of the cosmos. It is not surprising, given this theoretical orientation, that attempts to reform and restructure socialism have led ultimately to the rejection of the socialist project itself.

b) Maoism

Maoism represents first and foremost an attempt to come to terms with the limits of Soviet Marxism, and in particular to find a way to restrict the corrosive impact of residual commodity relations on the process of socialist construction. In order to understand the nature of the Maoist strategy we need to examine in some detail the history of the Chinese Communist movement.

We have already discussed the Taiping and related rebellions of the nineteenth century, which must be regarded as the early stages of the Chinese Revolution. Indeed, as Wolf (1969: 120) points out, the Communist Party of China was first able to establish itself precisely in this regions in which the Taiping had met with the greatest success. Defeat of the uprisings of the nineteenth century entailed the creation of vast regional armies, and the emergence of regional power brokers, or warlords, who soon outdistanced the imperial government in power (Wolf 1969:124, 127). By 1911 the Empire had fallen and a weak republic largely subordinate to warlord pressures and to foreign capital had taken its place. A class of *compradores* or large merchants emerged in the major coastal ports, serving, in effect, as agents of foreign capital (Wolf 1969: 125-126).

The result of these developments was a general crisis in the countryside. There were a number of factors involved in this crisis. To be sure, population pressure on a limited land base played a significant role. Rationalization of production and investment of the surplus in increasing agricultural productivity was obstructed by the parasitic regime of foreign capitalists, *compradores*, and warlords. Indeed, the collapse of the imperial bureaucracy meant an end to its role in building and maintaining water management and flood control projects, and to the system of public granaries which it had maintained. The surplus generated by the peasants went instead to support the military exploits of the warlords, to the luxury consumption of the *compradores*, or else was transferred to foreign capitalists (Wolf 1969: 127-128). Traditional village handicrafts were more or less finally wiped out in this period (Wolf 1969: 131), and the eastern coastal region began to experience some industrialization.

> By 1919 the number of industrial workers had reached 1,500,000. Three quarters were engaged in transportation or in light industry, especially textile production. Three fifths worked for Chinese-owned enterprises, two fifths in enterprises owned by foreigners (Wolf 1969: 136).

This modest industrialization, however, was insufficient to make up for the destruction of traditional Chinese agriculture and handicrafts. China was clearly on the road to underdevelopment.

Both the countryside and the city witnessed significant ferment during this period. Peasant organizations were generally more successful in the remote mountainous regions, or in coastal areas distant from major ports. Around Canton, on the other hand, where the land was productive, peasants were reluctant to endanger their chances for private accumulation. And penetration of market relations into the villages had transformed many of the secret societies which, in the mountains became cells of the revolutionary party, into a kind of *mafia* (Wolf 1969: 138-140). But workers movement was remarkably successful in the major port cities during this period.

> In 1918 the first industrial union made its appearance ... and only a year later workers were already on strike in support of nationalist students. By 1925 one million workers went out on strike in support of political causes. In 1927 union membership numbered three million, and an attempt at urban insurrection relying heavily on worker support came close to success

in seizing power in May of that year (Wolf 1969: 137).

Early Communist strategy in China centered on a very conservative application of the popular front strategy.

In 1923 formal liaison was established between the KMT (Kuomintang) and the Communist Party of the Soviet Union. Under this agreement the Soviet Union sent advisers to shape the KMT into a disciplined party organization with an organized mass following. At the same time, the nascent Chinese Communist party was pressured to yield its autonomy and to merge its forces with the KMT. The aim was to create an organization capable of mounting an effective anti-imperialist struggle and to introduce liberal reforms within China, to create a national democratic state, but to eschew revolution (Wolf 1969: 141).

The principal contributions of this period were the Whampoa Military Academy, which transformed what would otherwise have been a disorganized peasant uprising into a disciplined revolutionary army, and the Peasants Training Institute, which trained leaders for the peasant organizations.

This results of this strategy were disastrous. No sooner had the peasant armies and worker strikes organized by the Communists brought the KMT to power, than the KMT turned on the party, expelling the Communists and murdering over 5000 Communist leaders (Wolf 1969: 144-145).

The defeat of the party in 1927 led to formulation of a new strategy under the leadership of Mao Zedong. Mao shared with Lenin and Stalin an interpretation of the dialectic which gave priority to contradiction and struggle. Indeed, with Mao, the other "laws of the dialectic" tended to disappear altogether, or to fade entirely into the background, and the Leninist formulation "unity and struggle of opposites" gave way to what Mao called the "principle of contradiction."

The law of contradiction in things, that is the law of the unity of opposites, is the fundamental law of nature and of society and therefore also the fundamental law of thought. It stands opposed to the metaphysical world outlook ... According to dialectical materialism contradiction is present in all processes of objectively existing things and of subjective thought and permeates all these processes from beginning to end; this is the universality and absoluteness of contradiction. Each contradiction and each of its aspects have their respective characteristics; this is the particularity and relativity of contradiction. In given conditions, opposites possess identity and consequently can coexist in a single entity and can transform themselves into each other; this again is the particularity and relativity of contradiction. But the struggle of opposites is ceaseless, it goes on both when the opposites are coexisting and when they are transforming themselves into each other, and becomes especially conspicuous when they are transforming themselves into one another ... (1937/1971: 128).

"Contradiction" for Mao thus embodies both interconnectedness and conflict, both unity and struggle. Contradiction is the dynamic relationality which at once brings the universe generally, and particular systems within it, into being, and then inexorably transforms them into something radically new and different. It is contradiction itself——active vigorous struggle——which creates unity, and thus organization, and makes social progress possible.

This interpretation of the dialectic had definite political implications. Mao upheld a broadly Leninist understanding of the role of party and of the popular front, but argued that successful application of the popular front strategy could only be based on a strong, independent Communist party, and on an understanding of the critical importance of struggle within the popular front. On the one hand the party built a strong base among the peasantry, pursuing a relatively liberal policy towards rich peasants and even towards landlords (1969: 147-149). On the other hand, they stressed the importance not only of rationalizing production, but of mass participation in the revolutionary struggle through the Red Army, and of ideological transformation.

At the core of this strategy was Mao's recognition that it was necessary to "build the party on the proletarian ideological plane." Not a fundamental break with Leninism, this Maoist formulation nonetheless put far greater stress on cultivating among party members not only strict organizational discipline, but also, and more importantly, a profound commitment to Communism. Being a Communist meant undergoing a profound process of self transformation. This is reflected in the "five requirements" for party membership specified in the party's 1973 Constitution.

1. Conscientiously study Marxism-Leninism-Mao Zedong Thought and criticize revisionism;

2. Work for the interests of the vast majority of the people of China and of the world;

3. Be able at uniting with the great majority, including those who have wrongly opposed them, but are sincerely correcting their mistakes; however special vigilance must be maintained against careerists,

conspirators, and double-dealers so as to prevent such bad elements from usurping the leadership of the Party and the state at any level and guarantee that the leadership of the Party and State always remains in the hands of Marxist revolutionaries;

4. Consult the masses when matters arise;

5. Be bold in making criticism and self criticism. (Communist Party of China 1973/1976: 169).

At the core of this process of self transformation was a struggle against "liberalism."

Liberalism stems from petty bourgeois selfishness; it places personal interests first and the interests of the revolution second, and this gives rise to ideological, political, and organizational liberalism ... A Communist must have largeness of mind and he should be staunch and active, looking upon the interests of the revolution as his very life, subordinating his personal interests to those of the revolution; always and everywhere he should adhere to principle and wage a tireless struggle against all incorrect ideas and actions, so as to consolidate the collective life of the party and strengthen the ties between the party and the masses; he should be more concerned about the party than about any individual, and more concerned about others than about himself. Only thus can he be considered a communist (1937/1971:137).

This process of ideological struggle was, furthermore, extended beyond the boundaries of the party to the popular front itself. Unlike Stalin, Mao cautioned against "directing the main blow" against center forces, such as the social democrats, at least in the countries of the Third World where such forces were relatively weak and could actually be won over to socialism. On the other hand, he rejected sharply rightist interpretations of the popular front which tended to dissolve the long range aims of the socialist project into vague humanistic ideals. Mao formulated the problem in this manner, when discussing the party's strategy in the struggle against Japan.

The basic condition for victory in the War of Resistance is the extension and consolidation of the anti-Japanese united front. The tactics required for this purpose are to develop the progressive forces, win over the middle forces, and combat the die-hard forces; these are three inseparable links, and the means to be used to unite all the anti-Japanese forces is struggle. In the period of the anti-Japanese united front, struggle is the means to unity and unity is the aim of struggle. If unity is sought through struggle it will live; if unity is sought through yielding it will perish ... Some think that struggle will split the united front or that struggle can be employed without restraint ... this must be corrected (Mao Zedong 1940/1971: 184).

In this dynamic understanding of the popular front, the process of ideological struggle which shapes the party itself is extended, albeit in less intense form, to the complex alliance of forces struggling against fascism and for national liberation. The party begins from the assessment that there is real continuity between the democratic aspirations of the peasantry, petty bourgeoisie, and even the national bourgeoisie, and the long range aims of the socialist project. It aims not simply to neutralize, but actually to win these sectors over to socialism ———or at least raise them to a higher level of development. At the same time, it recognizes that this is possible only through struggle over the real conditions for realization of these democratic aspirations, and ultimately around the necessity of communism. As a result of this policy, when the party developed to the point that it could exercise power throughout China in 1949 it at once enjoyed a broader social base than had the Bolsheviks in Russia, and it had initiated a much deeper process of social transformation. After a brief period during which the party followed an essentially Soviet model of development, Mao broke sharply with what he called the "revisionist theory of the productive forces" and rejected the strategy of primitive socialist accumulation. The Chinese model integrated a modified form of Bukharin's strategy of rural demand led industrialization (Bettelheim 1976, 1978, Amin 1982), with a commitment to voluntary collectivization (which, however, was completed after only a few years).

These policies by themselves were, however, very far from guaranteeing the socialist character of the resulting postrevolutionary society. "The struggle of opposites is ceaseless ..." The Chinese party, unlike the CPSU, remembered that persistence of "Small production engenders capitalism and the bourgeoisie continuously, daily, hourly, spontaneously, and on a mass scale (Yao Wen—yuan 1975)." This, precisely, is what the CCP argued had happened in the Soviet Union. The persistence of commodity relations had created a new state bourgeoisie, which exploited the working class and peasantry, and which sought to extend its hegemony to the countries of Asia, Africa and Latin America where, under the pretense of supporting national liberation struggles, the Soviet Union was gradually building a world empire. Under Kruschev this state bourgeoisie finally seized power, and began to gradually dismantle the remaining elements of working class rule.

It was vitally important, the party reasoned, to prevent a comparable process from taking place in China and in the other Third World countries where communist-led national liberation movements were coming to power. For the Communist Party of China, rural demand led industrialization was not simply a means of developing the productive forces necessary for socialism (as it had been for the right opposition), but a way of winning the peasantry, petty bourgeoisie, and national bourgeoisie over to the socialist project. This meant that struggle formed an essential moment of the process. Specifically, the party initiated a struggle against "bourgeois right" and "bourgeois ideology." It was impossible, in a predominantly agrarian country like China, to suppress commodity relations completely. The party took steps, however, to radically restrict wage differentials, so that persons at the highest grade of the country's eight grade wage scale earned only 1.5 to 2.0 times as much as those at the lowest grade (as against wage differential ratios of 4.0 in the advanced capitalist countries, 5.0 in the Soviet bloc, and 10 in most of the Third World) (Amin 1982: 57). The party also insured that equal exchange prevailed between the city and the countryside. The party could not rely on its strategy of economic development to automatically develop socialist consciousness (Amin 1982: 53). Members of the intelligentsia were required to spend time working side by side with the peasantry, and workers were involved in the management of their enterprises. At the same time, the party initiated an effort to actively develop socialist consciousness among the masses. The masses were encouraged to study the sayings of Chairman Mao. Collectives at every level practiced criticism and self-criticism.

This strategy has yielded mixed results. On the one hand, China has been able to achieve food self-sufficiency and to lay the groundwork for industrialization without subjecting the peasantry to a period of intense exploitation. Communist leadership has rendered China self-sufficient in the production of food stuffs, and laid the groundwork for industrialization. In 1984-1986 China imported only 2.4% of its food as against 14.4% for countries at a comparable level of development (United Nations Development Program 1990:150). Annual growth rates in agriculture, light industry, and heavy industry respectively averaged 14.1%, 29% and 48.8% between 1949-1952; 4.5%, 12.9%, and 25.4% between 1953 and 1957, 2.4%, 33.7%, and 78.8% between 1957-1958; -2.4%, -21.6%, and -46.6% between 1960-1961 (the years of the disastrous Great Leap Forward); 8.3%, 47.7%, and 10% between 1964-1965; and -2.5%,-5%. and -5.1% between 1967-1968 (the years of the Cultural Revolution). Labor productivity increased at an average rate of 4.6% between 1950 and 1978 (Amin 1982: 54-65). China's per capita growth rates compare favorably not only with those of other Third World countries, but with those of the U.S.S.R. and the U.S. For the period 1960-1974 the World

Bank gives the following percentages for annual per capita growth:

China	5.2
India	1.1
Pakistan	1.4
Bangladesh	-0.5
USSR	3.8
United States	2.9

(Amin 1982: 75)

Unlike the Soviet Union, the Chinese recognized that the people themselves are the "primary productive force." This led to high levels of investment in education. A recent report published by the National Center for Education Statistics demonstrate that Chinese students have achieved an extraordinary proficiency in mathematics.

Mathematical Attainment: Age 13

	China	Canada	U.K.	France	Switz.	Taiwan	Korea	USA
Numbers/Operators	84.9	65.6	58.5	65.0	73.6	74.7	77.4	61.0
Measure	71.3	49.9	51.2	52.7	62.0	63.7	59.5	39.5
Geometry	80.2	68.1	70.3	73.1	76.6	76.6	77.4	54.3
Data Analysis	75.4	76.4	79.5	79.3	81.8	81.2	81.2	72.2
Algebra	82.4	52.7	54.0	57.0	62.7	69.2	70.8	49.2
Conceptual Undstg.	81.6	65.1	62.0	67.4	71.7	74.7	78.3	57.4
Procedural Knowledge	83.0	61.9	59.0	65.7	69.0	74.7	73.4	56.0
Problem Solving	75.6	58.9	60.8	59.3	71.9	68.6	68.5	52.3

Source: Lapointe et al 1992

Millions of Chinese have participated significantly in shaping policies at the commune and enterprise level. Careful attention to meeting the economic needs of the peasantry has guaranteed a strong base of support within this sector, and provided a context in which struggle can take place. This has meant that, on the whole, the Chinese Communist Party has had to rely far less on repression *towards the masses* than was the case in the

Soviet Union, where the policy of primitive socialist accumulation almost necessarily involved development of a massive repressive state apparatus.

And clearly the ideological struggles of the Cultural Revolution bore fruit in a powerful ethic of devotion to the common good. China was able to secure among the masses a higher level of commitment to the socialist project than has been the case in most other socialist countries. This commitment has survived many shifts in economic policy, and not a few serious errors on the part of the party and the state.

And yet Maoism has run into hard times. After Mao died, his faction was unable to hold onto power. The new leadership relaxed ideological constraints and began to permit the renewed operation of the marketplace. This is how Chinese reformers see the crisis.

> Because our socialism has emerged from the womb of a semi-colonial, semi feudal society, with the productive forces lagging far behind those of the developed capitalist countries ... we are destined to go through a very long primarily stage. During this stage we shall accomplish industrialization and the commercialization, socialization and modernization of production, which many other countries have achieved under capitalist conditions ...

> Beginning in the late 1950s, under the influence of mistaken Left thinking, we were too impatient for quick results and sought absolute perfection, believing that we could dramatically expand the productive forces by relying on our subjective will and on mass movements, and that the higher the level of social ownership the better. Also, for a long time, we relegated the task of expanding the productive forces to a position of secondary importance ... A structure of ownership evolved in which undue emphasis was placed on a single form of ownership, along with a corresponding political structure based on over-centralization of power. All this seriously hampered the development of the productive forces of the socialist commodity economy (Zhao 1987: 26-27)

The current line of the Communist Party of China differs from that of the CPSU Gorbachev in two respects. First, there is a tendency for Soviet advocates of *perestroika* to argue that the strategy of primitive accumulation, including state ownership of the means of production, played a progressive role during the earlier stages of industrialization, but has now become obsolete (Abalkin 1986: 32). The CCP, on the other hand, in keeping with its historic rejection of primitive socialist accumulation, tends to argue that conditions in China, at least, have never justified such "high" levels of property ownership——but leave open the possibility that, as the productive forces develop, state ownership might someday become appropriate. In this sense, the Chinese do not go as far as the Soviet advocates of *perestroika* in retheorizing the socialist experience as an inadvertent strategy for *capitalist* development.

Second, the current leadership, unlike the deposed Zhao Ziyang who we quoted above, rejects calls for democratization and cultural pluralism. They seem to believe that the political power of the CCP, and the ideological purity of its leadership, are the best guarantee for the future of socialism. In this sense, unlike the *perestroika* group in the Soviet block the leadership of the CCP remains within the framework of a classically Leninist "uninterrupted revolution by stages"——the only difference being the prolongation of the first, New Democratic stage of the revolution.

But we should not conclude from this that the Chinese leadership is somehow more faithful to socialism than were the advocates of *perestroika*. On the contrary, the Chinese party seems to be willing to permit the most reactionary and backward aspects of a market economy——i.e. a capital market, which will inevitably begin to channel capital into low wage, low technology activities on which it can find a high rate of return——while systematically blocking the formation of a progressive industrial bourgeoisie which might challenge the party for power. This is a dangerous tactic. In the long run an industrial bourgeoisie dependent on the state for capital allocations, but possessing the freedom necessary to innovate seems like a more reliable ally for the working classes than a stratum of *rentier* elements and a mass of petty bourgeois entrepreneurs concentrated in low wage, low technology service and manufacturing activities. Similarly, where Gorbachev, if not his successors, seemed intent on integrating the Soviet Union into the world market as an exporter of high technology goods and services, the Chinese leadership seems intent on exploiting its unmatched comparative advantage in cheap, unskilled labor, and seems willing to use the most draconian methods available in so doing.

From this point of view, even if the Chinese breach with socialism is less complete or at least less explicit, both Soviet and Chinese Leninism turn out in fact to have been strategies for *capitalist* industrialization. And in so far as the Soviet strategy allows for entry into the world market at a higher technological level, it must be regarded as being, on the whole, more progressive.

Where did Mao go wrong? I would like to suggest that, as in the case of the Soviet Union, it is ultimately an incorrect understanding of the dialectic which is behind the strategic errors of the Chinese Communist Party. Mao correctly perceived the persistence of social contradictions within Chinese society as a result of residual commodity relations due to small commodity production and the market in labor power. And he correctly understood that if China was to radically transcend the market system, it would be necessary to unleash antimarket constituencies. Perhaps because Chinese society was less commodified than Stalinist Russia he was able to find such constituencies to support his cultural revolution, where the left within the CPSU, such as it was, could not.

But contradiction and struggle do not by themselves yield a higher synthesis. And this, precisely is what was needed: a new nonbureaucratic, nonmarket method of centralizing and allocating resources. Instead, Mao counted on the "principal aspect of the contradiction," the socialist forces of production, to "win out" over the "secondary aspect of the contradiction," residual commodity relations, bourgeois right, bourgeois individualism, etc. The result was merely the *restriction* of these forces rather than their transcendence. And we have already seen, in our analysis of Soviet society, that even rapidly developing socialist productive forces cannot, by themselves, out weigh the powerful constituency for consumerism created by those rudiments of the marketplace which even Mao could not abolish——the market in labor power. As Marx himself noted, so long as human beings work for a wage, in order to consume, they will never be whole. The new leadership has now unleashed these long restricted forces, with the result that China is back on the capitalist road. And the result may ultimately be far more horrific than anything wrought by Stalin.

3. The Third Force: Dialectic as Holism, Organization, and Teleology

If social democracy understood the dialectic as simply the gradual, quantitative accumulation of proletarian forces, which subtly and imperceptibly as it were yields a qualitative change in social structure, and Leninism understood the dialectic as the "unity and struggle of opposites," or as the "principal of contradiction," comprehension of which enables the Communist Party to seize state power and carry out a socialist revolution from above, then the third principal trend within the socialist movement tended to understand the dialectic as the "negation of the negation." What is being negated, of course, is the negativity of capitalism itself——the ecological and social disintegration, the loss of holism, self-organizing dynamism, and telic centeredness engendered by the market system. In this sense, the "third force" within the socialist movement represents a resurgence of many of the themes of objective idealism and religious socialism, with the single, but very important difference that for "third force" socialism, holism, self-organization, and teleology are first and foremost a product of *human* activity, the culmination, perhaps, of a long cosmic evolutionary process, but not in any sense the expression of an organizing principle prior to matter.

We have grouped three trends together under this heading:

a) the populism and guild socialism,

b) the "theomachist" or "god-builder movement" which emerged on the left wing of the Bolshevik Party during the first part of this century, and

c) the various forms of socialist humanism associated with Lukacs, the Frankfort School, the Praxis Group, and the Latin Communist tradition of Gramsci and Mariategui.

While these trends are in many ways quite different from each other, it will become apparent in the course of our discussion that they share a common logic, and point in a common strategic direction. We will begin by analyzing each of these trends in some detail. Then we will examine the unique revolutionary experience of Nicaragua, which brings together elements from all three trends, and which, in many ways, represents the most advanced expression to date of the socialist experiment.

a) The History of the Third Force

i) Populism and Guild Socialism

Both the social democratic and the Leninist theories of the socialist transition presuppose the existence of a fully developed industrial capitalist society. From the earliest days of the international workers' movement, however, there has been intense debate regarding the possibility of a "non-capitalist" road of development, a passage from precapitalist feudal or "asiatic" forms of social organization directly to socialism, without a protracted period of capitalist development. Marx himself showed significant interest in the possibility that traditional communal institutions such as the Russian *mir*, or village community, might serve as an alternative basis for the emergence of socialist consciousness (Marx 1881/1978: 675). Since then, we have witnessed the emergence of a whole series of movements which have advanced socialist aims, but which have located the social basis for the emergence of socialism in various national, popular, communal, and religious traditions. The Russian populist movement represents, perhaps, the most theoretically self-conscious manifestation of this phenomenon, but the Italian peasant leagues of the late nineteenth centuries, *Zapatismo, Sandinismo*, and certain elements in the Salvadoran uprising of 1932 all reflect this general orientation. Closely associated with this tendency is the largely urban movement known as guild socialism, which attempted to found a postmarket social order on the institutions of the craft guild, the journeymens association, or the urban mutual benefit society. Finally, in the past 25 years, we have witnessed the emergence of a neopopulist trend associated with "dependency theory" and the *campesinista* trend in the Third World socialist movements.

While populist movements have generally been less theoretically driven than Leninism or even social democracy, and while they have rarely felt obliged to adhere to some fixed Marxist orthodoxy, such populist theory as has been developed *presupposes*, even if it does not always accept, important elements of Marxist analysis. Of particular importance in this regard is Marxist political economy and the Marxist theory of class struggle, to which populists have made important contributions. Because of this, it is proper to consider populism under the heading of movements influenced by dialectical materialism, even though many populists never read, or even explicitly opposed, Marx.

In this section we will examine briefly the theoretical foundations and historical records of these trends.

aa) Rural Movements

aaa) Russian Populism

The development of socialism in Russia followed a trajectory determined, in large part, by the relative isolation of the region from the pressures of capitalist development. Where the peasant community had largely disintegrated, or at least lost its economic functions, throughout most of Europe by the end of the feudal period, the Russian *mir*, as we have noted, remained a vigorous institution at least up through the 1920s. There are a number of reasons for this persistence. First of all, the existence of vast tracts of arable, but as yet uncultivated land gave the Russian peasant considerably greater leverage *vis a vis* the military aristocracy than his European counterpart. Second, the emerging Russian state always regarded the peasant community as a bulwark against both the penetration of Western individualism, and against the independence of the nobility (Wolf 1968: 58). Wolf describes the workings of the Russian *mir*.

It usually was formed by former serfs and their descendants settled in a single village ... each household had a right to an allotment. Before the emancipation, each household within the commune was entitled to an allotment of commune land; in addition, each household held its house and kitchen garden in hereditary tenure ... each household farmed its allotment on its own. Rights to pasture, meadows, and forest, however, were held jointly by the commune. Finally, in Great Russia and Siberia the commune had the power to reallot land at intervals among its constituent households ...

... most repartitioned on the basis of the males in the family (59 percent), with a minority repartitioning on the basis of working adults (8 percent), or the number of souls in the household (19 percent), while 2 percent repartitioned only partially ... The peasant could not sell, mortgage, or inherit land without consent of the entire commune. Nor could the peasant refuse to accept a new allotment, less productive than the one before. The commune also limited the right of the peasant to grow what crops he wanted by enforcing a rigid cropping system ...

It was governed by a council of all heads of households, called the *shkod* from *shodit'*, to come together. At the head of the council stood the village elder or *starosta* whose function it was to formulate the consensus of the village assembly and to represent it in dealings with outsiders ...

... Jointly responsible for taxes since 1722 ... it voted to admit new members, and issued permits for those who desired to leave ... it could remove an ineffective head of household and appoint another to be head in his place ... it exercised fierce social controls over the conduct of its members, ranging from corporal punishment to public shaming. (Wolf 1968: 58-62).

The collective institutions of the *mir* provided the institutional base for deeply rooted communitarian religious traditions. Particularly important in this regard were the *raskolniki* or the "Old Believers" who had broken with the Orthodox Church over a variety of ritual questions, but who became a center of gravity for millenarian tendencies in the peasantry generally.

The Old Believers were strongly antistate, identifying the tsar with the Antichrist. They came to believe in a Kingdom of Earth in the mythical White Waters, governed by a white tsar who would one day come forward to rule over Russia ... they gave ready asylum to escaped serfs (Wolf 1969: 70).

According to Trotsky the revolution owed a great deal to

the work of the sectarian ideas which had taken hold of millions of peasants. " I know many peasants," writes a well-informed author, "who accepted ... the October Revolution as the direct realization of their religious hopes (Trotsky 1932: III: 30)."

This does not mean, to be sure, that the Slavic domain escaped incorporation into the world market. On the contrary, the seventeenth, eighteenth, and nineteenth centuries witnessed the gradual transformation of what had been a relatively autonomous, and relatively primitive, tributary social formation into a wheat and rye exporting periphery of the developing West (Anderson 1974: 257-259, Konrad and Szeleny 1979: 95). As in the colonized regions, however, incorporation into the world market did not, at least initially mean the abandonment of tributary forms of exploitation and their displacement by capitalism, but rather, the intensification of the forcible extraction of surplus through rents, taxes and forced labor, in order to increase the production commodities for sale by the nobility (Andersen 1974: 259). This meant that Russia and the Slavic east preserved the institution of serfdom far longer than did the west. Indeed, the enserfment of the Slavic peasantry was only being completed at roughly the same time the peasantry of Europe was undergoing at least formal emancipation. "After passage of laws ever more restrictive of the peasants' right to free movement, the peasant was finally bound in full serfdom to a given estate in the legal code of 1649 and flight was made a criminal offense in 1658 (Wolf 1968: 52)."

The state which this enserfed peasantry supported was furthermore, very different from the emerging Western state. The Russian and other slavic monarchies circumscribed the liberties of their nobles as carefully as that of their peasants. Patrimonial claims on the land were broken: all land belonged ultimately to the state, with taxing rights assigned to particular nobles for a limited period in return for civil or military service (Konrad and Szelenyi 1979: 90). Gradually the nobility was transformed from a class of semiautonomous warlords into a military and civil service bureaucracy (Konrad and Szelenyi 1979: 88-90).

It was above all this state, and not the marketplace, which became the principal engine of industrialization in the Slavic east. Surplus extracted from the peasantry was used by the state to finance investment, primarily in capital goods and armaments. This system "placed the industrial bourgeoisie, whose chief customer was the state, in a position of dependence. What the state could not buy went sooner for export than to the domestic market (Konrad and Szelenyi 1979: 97)." Konrad and Szelenyi note that

It was a constant goal of government economic policy, through the whole era of early rational redistribution and indeed on into the age of socialist rational redistribution, to maintain state control over the labor market. Control over the peasantry was achieved through the village commune ... Service nobility and hereditary aristocracy alike were obligated to perform state service (Konrad and Szelenyi 1979: 98).

State led industrialization required the development of an enormous bureaucracy, and thus the creation of a secular intelligentsia employed almost entirely by the state. While much of this intelligentsia was co-opted into the nobility (the upper 7 of the 14 civil service ranks conferring noble status), the intelligentsia as a whole did not share the nobility's interest in preserving serfdom. On the contrary, both opposition and "police" intellectuals agitated for the abolition of the system and for development of more rational models of cultivation.

In this sense the development of capitalism in Russia followed the statist road. As in France, however, the state was unable to modernize fast enough, and to limit its exploitation of the peasantry sufficiently, to contain the pressure for revolution. On the one hand the peasantry had an interest in throwing off its centuries long domination by the tributary state and restoring fully the rights of the peasant communities. This meant not only an end to serfdom, but also an end to rents, taxes and forced labor. The intelligentsia, on the other hand was interested primarily in the rationalization of a social structure and a development strategy in which it already held a relatively favorable position. This meant, first of all, eliminating the "irrational" claims on both economic surplus and economic authority advanced by the autocracy and the hereditary aristocracy, and replacing them with a fully meritocratic system, and second, developing more efficient methods of extracting and investing the surplus produced by the Russian peasantry, in order to promote the all sided development of Russian civilization. Neither the bourgeoisie, which was very small and almost entirely dependent on the state, nor the industrial working class, had significant social weight.

There were, broadly speaking two ways to carry out this process of rationalization. One possibility, of course was an opening to the West and the comprehensive commodification of Russian society. This would have meant exploiting the antifeudal, rather than the communitarian aspect of the peasant movement. Had Russia followed this pattern the result would have looked something like the society which emerged from the French revolution. The second was rationalization of the existing centralized redistributional structure. The choice of this second option was determined largely by the balance of forces in Russian society. No significant sector of Russian society was favorably situated to profit from the development of a free market. The Russian bourgeoisie, even more than its French, Prussian, or Japanese counterparts, was dependent on state purchase of its products, and was hardly interested in competing with cheaper, higher quality European consumer goods.

Russian populism emerged as an attempt on the part of the Russian intelligentsia to mobilize the peasant communities for this process of rationalization. For the *narodniki*, or populists, participation in the communitarian structure of the village provided the Russian peasant with a basis in experience for the development of socialist consciousness. The peasant already had *both* a history of participation in the collective management of economic resources, and a basis in experience for understanding the underlying unity of human society, and indeed of the cosmos as a whole. From this standpoint capitalist development, which would entail not only proletarianization of the peasantry, but also the destruction of the *mir*, and the development of private property relations, seemed to be a step backwards (Radkey 1958:passim, Venturi 1966: passim).

Capitalist development, the *narodniki* argued, was, in any case impossible. On the one hand, they reasoned, the development of capitalism presupposed the ruination of the peasantry and the proletarianization of the work force. On the other hand, such an impoverishment of the rural population in a rural country like Russia would drive effective demand so low that there would, in effect, be no home market for the goods the capitalists produced. Capitalism was possible, the populists reasoned, only on the basis of a foreign market for manufactured goods --in effect only for a dominant imperialist country-- something which Russia, having entered on capitalist development too late, could never aspire to be. (Radkey 1958: passim, Venturi 1966).

The real basis for socialism lay in the long standing collectivist traditions of the peasantry, and in particular, in the redistributional land tenure characteristic of the *mir*, which constituted a distinct "popular mode of production" and which would form the basis for a "non-capitalist road" to socialism in Russia. Transfer of the land to the peasantry would put an end to the exploitation of the countryside, leading to rising rural demand, which would in turn provide a market for manufactured goods --both "soft" consumer goods and agricultural implements which would, in turn, increase agrarian productivity and open up the road to development, which would proceed within a collectivist framework guaranteed by the institutions of the peasantry.

The populists did not, to be sure, expect socialist consciousness to develop spontaneously out of the communitarian traditions of the *mir*. On the contrary, it was a populist, Sergei Nechaev, probable author of *The Catechism of a Revolutionary*, who first formulated the doctrine of the professional revolutionary and of the revolutionary party. According to Nechaev the revolutionary is

> a man set apart. He has no personal interests, no emotions, no attachments; he has no personal property, not even a name. Everything in him is absorbed by the one exclusive interest, one thought, one single passion——the revolution ... All gentle and enfeebling sentiments of kinship, love, gratitude, and even honor must be suppressed in him by the single cold passion for revolution. ... Revolutionary fervor has become an every day habit with him, but it must always be combined with cold calculation. At all times and everywhere he must do what the interest of the revolution demands, irrespective of his own personal inclinations (in Wolf 1969: 81).

These professional revolutionaries were to be organized into a tightly knit cadre, led by a Committee which would

> combine the scattered and fruitless revolts and so transform the separate explosions into one great popular revolution.

Nechaev's ideas first found expression in the populist faction known as *Narodnya Volya*, or "the people's will" which engaged in terrorist actions intended to catalyze peasant revolt. This trend dissipated after the capture and trial of 152 Nechaevists in 1871. Other populists, especially those in the *Chornyi Peredel* of "black partition" faction, devoted increasing attention to the task of actually organizing the peasantry, and training them in both legal and illegal means of struggle, eventually building the Socialist Revolutionary Party, which was the largest party in Russia at the time of the Revolution.

The populist strategy did, nonetheless, have serious weaknesses. On the one hand, populism represents an incomplete rationalization of traditional communitarian structures. This obstructs the development of the kind of organization which is necessary in order to build and exercise power effectively in a rapidly industrializing society, and in order to develop human social capacities towards synergism. Concretely this means that even where populist movements have succeeded in the principal aim——protection of the land rights of the peasant communities——they tend, like most of the religious socialist tendencies we analyzed above, to founder on the question of industrialization. On the other hand, populist movements rarely get this far. The operation of the marketplace tends to undermine the unity of the peasant community, leading to economic differentiation and eventual disintegration as impoverished workers flee the countryside, seeking work in the cities, and undermining the mass base for the populist strategy.

This precisely is what happened in Russia. We have already noted above that emancipation came on terms extremely unfavorable to the peasantry, with most of the best lands being taken over by the nobility, and with the peasants forced to pay off massive debts. In order to pay off these debts, or simply to support themselves on their miserable allotments, growing numbers of peasants were forced to work in the factories. From the beginning Russian workers labored in large factories.

Russia as early as 1895 had surpassed Germany. In that year the wage earners in Russian factories with more than 500 employees constituted 42 per cent of all workers, whereas in Germany these large establishments accounted for only 15 percent of the working population ... By 1912 the workers in Russian factories with more than 500 employees were 53 percent of the whole ... Even more striking is the comparison with the United States. Of all employees in establishments with more than fifty hands the workers in enterprises of five hundred hands or more were 47 per cent in the United States in 1929. They were 61 per cent in Russian in 1912. (Gordon 1941: 354).

Even so, populism remained a potent force, which profoundly affected the development of Russian socialism generally. The workers were, furthermore, organized collectively from the very beginning into *artel'*. Members of an *artel* "contracted with each other, and collectively with an employer, to work at a fixed rate in cash and perquisites to share the proceeds equally (Dunn and Dunn 1967: 10)." Even urban socialism in Russia developed along populist lines.

And the revolution itself, by restricting the operation of market forces, led to a reassertion of traditional communitarian forms of social organization.

... the revolution of 1917 was a resurgence of old customary land-tenure (Owen 1963: 245)

The land settlement of the previous decade was wiped out in many parts of Russia by the revival of the *mir*. The total extent of land seized by the communes in 1917-1918 was put at about 70 million desiatins from peasants and about 42 million desiatins from large owners. About 4.7

million peasant holdings, i.e. about 30.5 per cent of all peasant holdings, were pooled and divided up. The effect of the agrarian revolution, therefore was in the first place to wipe out all peasant property, but also and no less to do away with the larger peasant property ...(Mitrany 1961: note 7, 231-232).

Many peasants formed radically egalitarian communes in which

members worked together without pay, ate at a common table, and lived in a dormitory. They had no use for money; everything but clothing, and sometimes even that, was collective property (Wesson 1963: 8).

This kind of radically egalitarian communism represented a resource for revolutionaries organizing against capitalism. But as we have seen it presented serious obstacles to a socialist government trying to centralize the surplus necessary for the further development of the country.

bbb) Latin American Populist Movements:

Russia was not the only place where movements of a populist character developed. A similar tradition emerged in Latin America during the first decades of the twentieth century, as peasant communities resisted the penetration of capitalist relations into the countryside, and Latin American intellectuals searched for a way to tap into the potential latent in the peasant movements and to chart a distinctly Latin American road to development.

Clearly the most important of these movements was the Mexican Revolution, which is of special interest because it was at least partly successful, and thus gives us an opportunity to analyze what happens when a populist movement comes to power, or at least gains significant influence within the revolutionary coalition. We have analyzed in some detail in the last chapter the impact of the penetration of capitalist relations of production on the Mexican economy. The *Reforma* of 1857 had effectively dissolved the old peasant communities, prohibiting any "civil or ecclesiastical corporation" from buying or owning land, other than buildings used exclusively for the purpose for which the corporation existed. The result was the rapid and almost complete transformation of the Mexican peasantry into a class of *peones* living on *haciendas*. At the same time, the last half of the nineteenth century witnessed a rapid infusion of foreign capital, and the development of a significant, but largely foreign owned, industrial sector (Wolf 1969: 12ff).

Pressure for revolution came from three main directions. The peasants, first of all, were demanding land, and more specifically the restoration of the rights of the peasant communities. The *Plan de Ayala*, official program of the southern peasant movement, led by Emiliano Zapata, represents perhaps the purest formulation of the demands of a peasantry under attack by the marketplace and the capitalization of landed property.

> be it known: that the lands, woods, and waters which have been usurped by *haciendados, Cientificos,* or *caciques* through tyranny and venal justice, will be restored immediately to the *pueblos* or citizens who have the corresponding titles to such properties, of which they were despoiled through the bad faith of our oppressors. They shall maintain such possession at all costs through force of arms (in Wolf 1969: 31).

The Zapatistas fought under the banner of Our Lady of Guadalupe, the ancient goddess *Tonantzin,* who had been a symbol of peasant revolt since the time of the Aztec empire.

Second, there was a small but growing urban proletariat, which at this time was still largely under the influence of anarcho-syndicalist ideas, and had little understanding the necessity of an worker-peasant alliance.

Finally, there was an emerging intelligentsia and national bourgeoisie, the development of which was held back, on the one hand by the exploitation of the peasantry, which restricted the formation of an internal market for manufactured goods, and by the penetration of foreign capital, which controlled the "commanding heights" of Mexican industry (Wolf 1969: 38ff).

These three sectors were linked together in an overarching nationalist ideology, the most complex manifestation of which was the "aesthetic monism" of Jose Maria Vasconcellos. Often regarded as a Bergsonian, Vasconcellos in fact traced the roots of his system to Pythagoras and Plotinus——Pythagoras because he grasped the fundamentally aesthetic character of the cosmic order, Plotinus because he was the first (in Vasconcellos' mind at least) to interpret the Pythagorean tradition in a rigorously monistic manner (Romannell 1969: 110).

According to Vasconcellos, the universe is one great emanation, or "fulguration" of cosmic energy. This cosmic energy manifests itself in three cycles: physical, biological, and psychic (115), each more complex than that which precedes it, each increasingly more beautiful. Human beings can grasp the universe fully only through aesthetic intuition. In this regard Vasconcellos sides with the later Schelling against Hegel and his leftist interpreters (119). Our aesthetic intuition of the development of the cosmos towards increasingly complex forms of organization provides the basis for a theory of value to guide our actions. Our task as human beings is first and foremost to create beauty, or rather to participate in the drive of the cosmos itself which is always and everywhere giving birth to new and more beautiful forms of organization (121-129).

Vasconcellos believed that Latin America generally, and Mexico in particular had a unique role to play in this process. The process of *mestizaje* had joined together in the peoples of Latin America the heritage of all the peoples of the planet. Latin Americans constituted a true *raza cosmica*, called to lead humanity into a new and brilliant future.

The development of the Mexican Revolution indicates particularly well the limits of the populist model. The *Zapatista* armies had little interest in seizing control of state power, or for that matter in anything which took place in the cities. Their almost exclusive focus was on restoring the historic rights of the village community. This orientation had the effect of ceding hegemony over the revolutionary process to the national bourgeoisie. The regime which emerged from the revolution did, after much hesitation, and only under the pressure of an ever vigilant peasant movement, carry out radical land reforms, establishing a system of *ejidos*. Under this system land was controlled by the villages and redistributed periodically to individual families for cultivation. It also nationalized the oil fields, and in general restricted the role of foreign capital in the Mexican economy (Wolf 1969: 44-45). This set in motion a process of rapid economic growth and development. Paradoxically, however, it also laid the groundwork for the eventual sabotage of the revolution. As the economy developed, an authentic Mexican bourgeoisie emerged, which resented the restrictions imposed upon them by the Mexican government. After the Second World War the pace of reform slowed. Renewed ferment during the 1960s prevented the collapse of the populist system, but by the 1980s the structure which had made possible the development of a Mexican national economy was being dismantled. The *ejidos* were under pressure to dissolve, and the North American Free Trade Agreement (NAFTA) threatened to complete the integration of Mexico into the world market.

A similar process unfolded in the cultural arena. Vasconcellos' aesthetic monism became a real national ideology in Mexico. Vasconcellos himself served as Secretary of Public Education in the early 1920s and developed a Socialist Education system which at once tapped into and rationalized Mexican popular traditions, and laid the groundwork for an authentically socialist culture. This system survived, with some modifications, up through the middle 1980s. The result was an intellectually active and theoretically developed population, the intelligence of which no doubt contributed significantly to the pace of economic growth in Mexico. As the economy developed, however, and was reintegrated into the world market, the schools and universities felt growing pressure to produce fewer philosophers and more skilled workers. Currently the Mexican educational system is undergoing a "reform" based on U.S. models, that hardly bodes well for the future.

Alongside the Mexican revolution there were other populist trends in Latin America: Sandino's movement in Nicaragua, with its philosophy borrowed from the "Magnetic-Spiritual School of the Universal Commune" and its political program centered on the establishment of worker and peasant communes (Hodges 1986: 23-107) and Feliciano Ama's *Confradia del Espirito Santo* which fought alongside the *Partido Communista Salvadoreno* during the great uprising of 1932 (Berryman 1984), among them. Most of these movements were successfully repressed, and so we have no way to evaluate just what kind of society they would actually have constructed.

bb) Urban Populist and Guild Socialism

Populism is not an exclusively rural phenomenon. We have already noted the role of the workers' *artel* in the development of socialism among the Russian industrial proletariat. William Sewell (1980) has argued at great length that French socialism emerged out of the struggle of the *compagnonages*, associations of journeyman artisans, against the penetration of market relations into French society, in the years following the revolution of 1789 ——long before "socialized" industrial forces of production had become important in the French economy, which retained an agricultural and artisanal character.

The most important theoretician of guild socialism was unquestionably Emile Durkheim. Durkheim began by noting, like Hegel, that the development of capitalism and the growing division of labor had vastly increased the level of objective material interdependence or what he called "organic solidarity." But he also noted the existence of serious contradictions in the economic system: what he called the "anomic" and "coercive" forms of the division of labor. The anomic form derives from insufficient economic regulation, and results in a loss of meaning. The forced form results from the existence of inequalities between contracting parties, and leads to exploitation (Durkheim 1893/1964). These contradictions were further reflected in rising suicide rates in the most advanced industrial countries (Durkheim 1897/1951), which he attributed to a deepening moral crisis. People seemed to find less and less meaning in their lives, and to feel less and less connected to each other and to society as a whole. While capitalism had increased the level of material interdependence, it was also undermining our ability to understand the social significance of our work and to feel like members of a cohesive social group.

In response to this situation Durkheim advanced a far reaching proposal for the development of occupational groups or "corporations." This proposal, often misunderstood as a kind of protofascist corporativism, amounts to a comprehensive attempt to adapt the tradition of the medieval guilds and journeymen's associations to the new and much more complex conditions of the late 19th and early 20th centuries. These corporations were not only to regulate wages, hours, and working conditions; they were eventually to collectively control the means of production, and connect them to the "directing and conscious centers of society." They were, further, like the guilds, to serve as the center of a rich social life, and the locus of new moral forces which would combat the egoism and anomie which was gradually eating away at the social order. (Bellah 1973: xxxi).

Durkheim argued, furthermore, that the social basis for the implementation of his proposal lay in the popular religious traditions of the masses. His study of Australian religion had convinced him that religious symbols were "collective representations" of the structure of human society. Indeed, God *is* the community, in transcendent form, binding the individuals together into a social being which is greater than themselves, to which their ties are stronger than any tie of self-interest, and which has the moral authority, but also the compelling beauty, to command self-sacrifice. In ritual gatherings he found a "collective effervescence" which catalyzes the formation of a sense of unity and oneness which transcends the existing empirical forms of social order and which opens up the possibility for the emergence of radically new social forms which later on become embodied in new economic and political institutions.

> In such moments of collective ferment are born the great ideals upon which civilizations rest. These periods of creation or renewal occur when men for various reasons are led into a closer relationship with each other, when reunions and assemblies are most frequent,relationships better maintained and the exchange of ideas most active. Such was the great crisis of Christendom . . . in the twelfth and thirteenth centuries. Such were the Reformation and the Renaissance, the revolutionary epoch and the Socialist upheavals of the nineteenth century. At such moments this higher form of life is lived with such intensity and exclusiveness that it monopolizes all minds to the more or less complete exclusion of egoism and the commonplace. At such times the ideal tends to become one with the real, and men have the impression that the time is close when the ideal will in fact be realized and the Kingdom of God established on earth (Durkheim in Bellah 1973: l)

Durkheim hoped fervently for the renewal of such collective effervescence in his own time, and had little doubt concerning its probable source.

> Who does not feel ... that in the depths of society an intense life is developing ... We aspire to a higher justice which no existing formulas express ... One may even go further and say with some precision in what region of society these new forces are forming: it is in the popular classes (Durkheim in Bellah 1973: xlvii).

It is interesting to note that the development of urban working class socialism incorporated significant "guild" elements even in countries such as the United States where there was no formal guild structure. This suggests that Durkheim's vision, far from being utopian, in fact reflected a profound grasp of the real material conditions. The case of Italian immigrant socialism is instructive in this regard. The most important of the institutions which the immigrants brought with them from the towns and villages of Sicily and the *Mezzogiorno* was unquestionably the mutual benefit society. At the most basic level, the mutual benefit society or *societa* provided what amounted to health insurance, life insurance, and a decent burial for its members and their families. These benefits were financed on the basis of membership subscriptions. In addition to these assistential functions, the *societa* was the religious center of the community, which preserved the distinctly anticlerical popular Christianity of the immigrant communities. Joachim Martorano tells of how his grandfather, who never set foot in a church, and who claimed that the church bells were crying "bring money, bring money," used to take him from time to time to the lodge of the *Societa della Santissima Crocifissa di Cimina*.

> There was a large cross, a table with a Bible, candles, a cup, and a dish, for bread and wine. I remember they would say the Our Father, the Hail Mary, maybe the rosary, and then there would be drinking and eating and cigar smoking. It was a fraternal gathering (oral testimony).

We can see here a radical laicization of the ritual of the mass, and a transposition of the symbolic solidarity of the Eucharist into a real solidarity among friends.

On the one hand this made the *societa* a center of what Durkheim called "collective effervescence." Ritual, Durkheim believed, served to recreate such collective effervescence, and to preserve it during the long periods of passivity between great upheavals. At the same time, even where the *societa* encouraged activism, it was of a markedly nonrational and even insurrectional character. The emergence of a revolutionary socialist movement, as opposed to a movement of resistance to capitalism, presupposed a real transformation in the religious life of the immigrant communities --what Gramsci called a moral and intellectual reform (Gramsci 1949c: passim). It was necessary to focus the immigrant communities on the necessity for organization and carefully planned political interventions.

Proletarianization and the formation of a trade union movement among the immigrants did not, by itself, accomplish this. Consider, for example, the testimony of Maria Valiani, who was involved in the unionization struggle at Kuppenheimer in 1919. Ms Valiani is in her seventies now, is deeply religious, and especially devoted to the Virgin Mary, with whom she claims to share meals on a regular basis.

> The owners of the companies they don't care. They want to do all the touring and all the spending, and all the enjoyment. And that's no good. Jesus said. . . He told the rich man "You shall never enter the Kingdom of Heaven. . . because if you're rich that means you didn't pay your subjects enough to live on." So those are the laws of God. And that's the way it is. It says it in the Bible. Didn't He say to the rich man "if you're rich that means you didn't pay you're subjects enough? And that's why they're suffering. So you're gonna go down in the pit, and they're gonna come to heaven with me." So you gotta learn (oral testimony: Valiani).

Its radical anticapitalism notwithstanding, Ms Valiani's outlook is hardly an adequate basis for socialist political organization.

During the early years of the twentieth century some of the mutual benefit societies, like the *Circolo Socialista "Camillo Prampolini"*, in Latrobe, Pennsylvania, transformed themselves gradually into trade union, educational, and ultimately into political organizations. It was only in the years immediately preceding the First World War, however, that the socialist circles were gradually won over to Marxism, and one by one, united into a single federation affiliated, albeit tenuously, with the Socialist Party of America. Thanks largely to the efforts of men like Giuseppe Bertelli, this work of unification was completed, with the foundation of an Italian language socialist newspaper, *La Parola del Popolo*, in 1908 and the foundation of the FSI in 1910 (Velona 1958: 23). In 1922, the only year for which figures are available, after the left as a whole had begun to decline dramatically, and after some sections of the FSI had been absorbed into the new Communist Party, the FSI had 54 sections nationally, including 8 in Chicago, with members in 17 out of 50 Wards, 4 in New York City, 13 in Pennsylvania, and 7 in New Jersey. Sections often numbered between 100-150 members. (*La Parola* 21 October 1922).

The political strategy of the FSI centered first and foremost on the rationalization of the popular religious traditions of the immigrant community and on the development of rational forms of collective organization. In the 24th/Oakely neighborhood, in Chicago, for example, the party seems to have grown out of the *Societa Lovagnini*, one of the several mutual benefit societies in the community. The local section might gather there, over wine, or else in the barber shop, traditionally a center of political debate in the Italian immigrant communities, which the socialists in this case managed to make their own. The local section was also involved in the foundation of the West Side Cooperative which drew into their periphery many who were suspicious of socialist ideas, and offered to them a concrete example of the superiority of socialist forms of organization (oral testimony: Valiani, Tarabori).

As they transformed the *societa* into an effective organ of struggle, the organizers of the FSI transformed the religious ideology of the immigrants into a bridge to socialism. The fact of the matter is that nearly all of the immigrant socialists, whether atheists or believing anticlericals, saw socialism as in a very real sense in continuity with the Christian tradition ... as in fact the realization of the ideal of a society based on fraternal love, which the Church itself had long ago betrayed. Issue after issue of *La Parola* contains at least one article on the religious question, and at Christmas and Easter the paper published "socialist interpretations of the birth, life and execution of Jesus ... and of the terrible suffering endured by his proletarian father and his peasant mother. These stories fostered a fervent cult of the *Gesu Socialista*, and a profound revolutionary spirituality which in many ways marked a real departure from the festive "aestheticism" of the traditional *societa*, in favor of an ascetic identification with the Jesus who was born the son of a carpenter, worked side by side with his father, and struggled and eventually died for the redemption of labor.

Writers in *La Parola* reminded their readers continuously that the teachings of Jesus were first and foremost a "system of moral principles without any link to an external cult (Crivello 1922)." Jesus preached "universal brotherhood" and a struggle against egoism --essentially the same values for which the socialists themselves were struggling (F.M. 1922). Domenico Saudino writes:

> Now there had appeared a new star which will guide humanity along the road to emancipation. This star is socialism. Laboriously, but incessantly, humanity proceeds towards the abolition of class privilege and the establishment of collectivism (Saudino 1921).

At the end of this long road lies the fulfillment of promise made long ago by Jesus himself. Socialism is the "kingdom of heaven, the kingdom of justice invoked by Jesus in his immortal prayer. The resurrection of the working people will be the resurrection of Christ" (F.M. 1922).

Ultimately the political strategy of the FSI was vulnerable to the same forces of social disintegration which destroyed rural populist movements. The emergence of a mass consumer society, had a profound effect on the fabric of North American society, and on the Italian American community in particular. Higher wages rendered non-commodity production unnecessary. As incomes rose, Italian Americans began to purchase goods and services which had once been produced in the household, or provided cooperatively through the neighborhood community. Leisure time centered less and less around the *societa*, the barber shop, and the front porch or the local tavern, and more and more at home, around the radio, and later the television set. This in turn undermined the experience of community which served as the basis in experience for the emergence of a socialist politics, and ultimately undercut resistance to the increasingly hegemonic bourgeois culture.

Only a vision of the future in which the principles of organization, development, and value are firmly grounded in the structure of being itself can withstand the corrosive impact of market forces.

cc) Neopopulism

As the socialist movement has matured, it has come to understand better and better Marx's insight that while the industrial, democratic, and scientific revolutions give humanity an enormous new dynamism, the penetration of market relations into every sphere of life simultaneously undermines the social fabric, and ultimately human social capacities in general, in a way which is fatal for the emergence of socialist consciousness. Because of this, there has been growing emphasis, as the crisis of capitalism has deepened, on the search for a "noncapitalist road" to development. This is apparent to some extent in the religious socialisms which we examined earlier in this chapter. Asian, African, and Arab socialisms, however, understand their goal to be merely the restoration of traditional communitarian institutions as a basis for development. They have not yet grasped fully need to radically transcend the marketplace. It is only when mediated through historical materialist analysis that both the full potential and the limitations of the noncapitalist road become apparent. Since the middle of the 1960s a number of Third World Leninist parties, especially in Latin America, have undertaken what amounts to a comprehensive re-appropriation and transformation of populist theory which has fundamentally altered the character of communist strategy.

In the economic arena the "dependency theory" which developed among students of underdevelopment during the 1960s and 1970s has taken up and elaborated a number of traditional *narodnik* themes. The dependency theorists (cf. e.g. Frank 1973, Wallerstein 1981, Amin 1975, 1980), like the populists, have argued that capitalist development of the productive forces is impossible in the Third World because of the low level of effective demand created and sustained by imperialist exploitation. This means that, on the one hand, the traditional "national liberation front" strategy associated with the parties of the Third International, which looked to an alliance with the anti-imperialist national bourgeoisie to break the bonds of foreign exploitation, and unleash development of the productive forces, thereby creating the material basis for the emergence of a mass socialist movement, has to be discarded, because the pattern of dependent development blocks the formation of such a national bourgeoisie, and even of a large industrial working class.   On the other hand, the intense exploitation of the countrysides, together with the growth of marginalized urban sectors, has created a new basis for revolutionary socialism in the Third World.

Peasant production, many dependency theorists argue (Amin 1975, Stavenhagen 1977, Dias Polanco 1977, Esteva 1978), is characterized by relations of production which incline the peasantry more or less spontaneously towards socialist or at least collectivist ideas. Peasant production is rarely oriented towards accumulation, and rarely employs wage labor. Indeed, much peasant production is not commodity production at all but production for subsistence. The peasantry, further, has a particular affinity for forms of ownership which integrate "social property, collective organization, and individual or family use of parcels," (Esteva 1978:708) or what we earlier called "communitarian land tenure." The struggle of the peasantry for land is a struggle to defend its communitarian social forms from the ravages of capitalist development. This struggle can form the basis for a popular movement directed towards the restriction and elimination of capitalist relations of production generally, and thus the basis for the emergence of mass socialist movements in the countrysides of the Third World.

In recent years dependency theorists have been forced to modify earlier claims that imperialism blocs capitalist development entirely. The experience of newly industrialized countries like Brazil and Korea leads them to speak instead of "dependent development" and a "peripheral capitalism" (Amin 1980) which is characterized by a high level of industrialization, with exports to the advanced capitalist countries making up for the lack of internal demand. There has also been a growing recognition that industrialization, while not undermining the peasant community entirely, leads to significant disruption, as family members migrate to the cities to work in factories while retaining more or less intact ties to their villages of origin. This has meant a growing strategic emphasis on urban organizing, and on trade union and electoral, as against armed struggles.

Broadly speaking this kind of neopopulist analysis has been associated with two distinct strategic lines. Initially, during the late 1950s and early 1960s, revolutionary strategies influenced by dependency theory assumed that the underdevelopment of the Third World would create a more or less permanent revolutionary situation, and that only a small "spark" would be necessarily to light the revolutionary fires. Thus the formation of revolutionary *focos*, composed almost exclusively of urban intellectuals, who hoped by audacious political-military tactics to ignite a peasant revolution. The success of the Cuban revolution, which (probably incorrectly) understood itself as an implementation of the *foco* strategy, lent credibility to this line, and it was only after dozens of defeats throughout Latin America that the perspective was finally abandoned.

What emerged in its place was the strategy of prolonged popular war. Drawing on the theory and practice of Lin Biao and General Giap, the Latin American advocates of *guerra prolongada* pointed out that the peasantry, unlike the urban proletariat, and was relatively insulated from the alienating pressures of the marketplace, and thus constituted a more likely constituency for the revolutionary organizations. Gradual, patient organizing and education, which attempted to transform the spontaneous communitarianism of the peasantry into a commitment to scientific socialism, was integrated with a cautious political military tactics centered on the passive accumulation of forces, the development of mass peasant organizations, and the use of military force in a primarily defensive fashion. The approach brought much better results, and laid the groundwork for the emergence of the new generation of political military organizations and mass popular organizations of the 1970s——organizations such as the *Frente Sandinista de Liberacion Nacional* and the Salvadoran *Fuerzas Populares de Liberacion*. It was only the crisis of the late 1970s, which created a revolutionary situation in many countries of the Third World, which brought the gradual progress of this strategy to an abrupt halt, as more opportunistic forces took the lead, in some cases mounting successful bids for power (as in Nicaragua), in other cases simply escalating the intensity and scope of guerilla warfare and undermining efforts at mass education and organization.

ii) The Theomachists

The Theomachist or God-builder movement had a very different origin. The group took shape in Russia as the revolution of 1905 began to ebb and Lenin, realizing that the possibility of insurrection had passed, urged the party to participate in the Duma or legislature convened by the Tsar. A small group of Bolsheviks, led by Bogdanov, resisted this move, arguing that the party should concentrate instead on organizing and educating workers.

Bogdanov and the left Bolsheviks ... firmly believed in the continuing revolutionary spirit of the Russian proletariat. They wanted to combat the pall of reaction that had descended over Russia in 1906 by immediately initiating an armed struggle against the autocracy and summoning the proletariat to join them in socialist and not simply democratic revolution (Rowley 1987: 130-131).

This group came to be known as the boycotters, or "*otzovists*" because they advocated boycotting the Duma.

The struggle with Lenin soon spread to other questions. The *otzovists* regarded human knowledge as a product of the active, organizing capacity of the human mind.

[Scientific knowledge] first is a simple description of the facts, an empirical elaboration of the question ... then an "explanation" that proceeds from this description——in other words, a unifying and simplifying grouping of the facts upon the basis of an established scientific view. If such a process of working out of experience succeeds in embracing all that is repetitive and typical in the phenomena, without contradictions or important exceptions, then a whole scientific understanding has been attained (Bogdanov in Rowley 1987: 49).

Lenin argued against Bogdanov, in what is without question one of his theoretically vacuous and politically most mean spirited polemics, that our knowledge derived from the reflection of external objects on our sense organs (Lenin 1908/1970), and attempted to portray Bogdanov as a subjectivist. He apparently feared that Bogdanov's theory of knowledge, which stressed the active role of the human mind in organizing information, provided a theoretical basis for his political strategy, which stressed the role of proletarian consciousness in organizing the revolution, and that it could eventually provide the basis for a reconciliation of communism with religion.

In fact Bogdanov sharply rejected the view that the organizational patterns which science discovered within nature and history were simply a projection of the structure of the human mind (Bogdanov 1928/1980: 27) or of language (Bogdanov 1928/1980: 30). On the contrary the human mind and "common language " are "compelled by the unity of organizational methods to express this unity (Bogdanov 1928/1980: 30)". Despite his skepticism regarding the possibility of any knowledge of "absolute reality" which transcends experience, he was, in fact, closer to epistemological realism than most of his critics have acknowledged. To use Hegelian language, knowledge was, for Bogdanov, the "truth" of reality, which, already implicitly organized when it remains at the level of mere object, is elevated through the medium of human experience, analysis, synthesis, etc. to a higher level, at which its organization becomes fully explicit.

Bogdanov argues that analysis of natural and human history reveals a process of the gradual development of more complex organizational forms.

Complete disorganization is, in reality, the same as naked non-being (1928/1980:5).

It would be strange to consider as "unorganized" the harmonious, titanically stable solar systems and their planets which were formed over myriads of ages. In contemporary theory, the structure of each atom, in its type, with its amazing stability based on the immeasurably fast, cyclically closed movements of its elements, is the same as that of the social system (5)

Today science destroys previously impassable boundaries between living and dead nature, filling the gulf between them. In the world of crystals were discovered some of the typical properties of organized bodies, which had been considered before as exclusive characterizations of life. For example, a saturated solution of the crystal changes its form by an "exchange of matter" ... And the life like crystals of Leman, obtained under known temperatures from ethylene ether, are capable not only of multiplication by division and "copulation" ... but also of feeding and growth by assimilation of matter ... (4-5)

The entire history of the evolution of anatomy and physiology is full of discoveries of mechanisms in the living organism, from the very simple to the most complex, which were previously invented independently by people

(7).

Thus the experience and ideas of contemporary science lead us to the only integral, the only monistic, understanding of the universe. It appears before us as an infinitely unfolding fabric of all types of forms and levels of organization, from the unknown elements of ether to human collectives and star systems. All these forms, in the interlacement and mutual struggle, in their constant changes, create the universal organizational process, infinitely split in its parts, but continuous and unbroken in its whole (6).

The highest stage on the ladder is, for us, the human *collective*; in our time it is already a many millioned system composed of individuals. In labor and in knowledge human kind creates its own "reality"——its own objective reality with its own strict orderliness and its systematic organization. The practical activity of the great social organism is none other than *world-building* (in Rowley 1987: 305).

All human activity for Bogdanov is organizational in character. Work is the "organization of the external forces of nature." Politics is the "organization of human forces." Knowledge is the "organization of experience." "Mankind has no other activity except organizational activity, there being no other problems except organizational problems (Bogdanov 1928/1980: 3)."

Bogdanov makes an interesting distinction between what he calls "centralist" and "skeletal" forms or "egression" and "degression (1928/1980:167-202)." By egression he means the tendency of organizations to develop a kind of organizing center——what in human organizations we refer to as leadership. This organizing center is characterized by a greater organizing capacity than the rest of the structure. This greater capacity generally derives from its "plasticity."

[Plasticity] denotes a mobile, flexible character of couplings of the complex, and ease in regrouping of its elements ... The more plastic is the complex, the greater the number of combinations which can be formed under any conditions which change it, the richer is the material for selection and the faster and more full is its adaptation to these conditions ... If life conquers dead nature, if the fragile human brain has mastery over fire and steel, it is precisely because of its plasticity. Plasticity of the living protoplasm is the basis of the entire biological and social evolution.

Tektological progress, based on plasticity, leads to complex organizational forms ... In its turn complexity is favorable to the development of plasticity, since it enlarges the richness of possible combinations.

Therefore, in general, the higher is the level of organization, the more complex and plastic it is (1928/1980: 185).

Degression, on the other hand, refers to the tendency of organizations to develop a kind of supportive or skeletal structure, which protects the organization against disorganizing tendencies in the environment.

> ... degression is an organizational form of a tremendous *positive* significance: only degression makes a higher development of plastic forms possible, fixing, securing their activities, and protecting tender combinations from their rough environment (1928/1980: 188).

The difficulty is that

> If there are no special conditions particularly favorable to the skeletal part ... the process of growth and complication will be stronger and faster in the plastic part which is more organized and better able to assimilate; the skeletal part, which is less able to assimilate, must lag behind (193).

Unlike science which is actively engaged in organizing new experience, the nature of ideologies is generally degressive (195)."

Bogdanov was clearly aware, in a way that the Leninist tradition was not, of the distinction between the self organizing capacity of the cosmos on the one hand, and the structures which regulate that activity on the other hand. This in turn made it possible for him to understand the difference between the development of human social capacities——actually raising the masses to a communist level of development, and merely securing their support or acquiescence through economic, political, or cultural sanctions.

This analysis led the group around Bogdanov to develop a unique and particularly advanced strategic line. They reasoned that if the working class was to become the ruling class of society, it would have to develop a high degree of organizational ability. Thus, rather than simply trying to win the political support of workers by advancing broad trade union and democratic demands, they argued, it was necessary to actually raise workers to a fully communist level of development.

> The struggle for socialism is not at all embodied in a single war against capitalism and in the simple collection of forces for that task. This struggle is a positive and creative work——the creation of the new elements of socialism in the proletariat: in its internal relations, in its unifying living conditions. It is the working out of a socialist, proletarian culture (Bogdanov in Rowley 1987: 284)

The already existing culture of the proletariat, which was essentially a religious culture, provided a starting point for this strategy. The "God-builders" Lunacharsky and Gorky argued that

> God is the complex of ideas worked out by tribes, nations, mankind, which awaken social feelings and give them organized form with the aim of linking the individual personality to society (Gorky in Rowley 1987: 157).

> The religious sense is the joyful and profound feeling of the knowledge of the harmoniousness of the bonds that unite humankind with the universe (Gorky in Rowley 1987:169)

Realization of this harmony means full development of the human capacity to participate in——indeed to consciously direct——the self-organizing activity of the cosmos. This in turn, however, presupposes the resolution of the contradiction between the developing force of production and the capitalist relations of production which obstruct that development.

> Scientific socialism resolves these contradictions, proposing the idea of the victory of life, the defeat of and the subjugation of chaos to reason by means of knowledge and labor, science and technology (Lunacharsky in Rowley 1987:170).

The task of the communist movement was to raise the spontaneous religious socialism of the masses to a scientific level, and by developing human organization capacities to bring the God humanity has hitherto only imagined into being. The first step in this strategy was the development of a popular culture which could serve as a kind of bridge between religious and scientific socialism. Gorky's novels, particularly *The Mother* and *The Confession* did this very directly. For Gorky, nature represents the feminine, and history the masculine dimension of God. Of the earth he writes

> You lie on her breast and your body grows and you drink the warm, perfumed milk of your dear mother, and you see yourself completely and forever the child of the earth.

Of the proletariat he writes

> You are my God, the creator of all gods, which you weave out of the beauty of your soul and the labor and agony of your seeking. (in Rowley 1987: 166-167)

Bogdanov took a somewhat different tack, writing science fiction novels which depicted the development of a socialist society on the older and more developed planet Mars (Bogdanov 1908).

Developing workers to a fully communist level, however, involved more than simply developing a commitment to communist ideals. It was necessary for workers to develop a fully scientific understanding of the unity of the cosmos and its development towards ever higher degrees of organization. In order to do this, Bogdanov, Lunacharsky, and their collaborators established a party school to train industrial workers to serve as full time party cadre.

> The second extremely important task of our faction during the current interrevolutionary period [is] the task of *broadening and deepening socialist propaganda* ... This necessitates, first, the creation of *propagandistic literature*, both legal and illegal, much more definitive and encyclopedic in content than that which has existed unto now ... It is necessary to work out a new type of party *school* which, in completing the party education of the worker, fulling the gaps in his knowledge ... and harmoniously systematizing that knowledge, would create a reliable and conscious leader who would be prepared for all forms of proletarian struggle (Bogdanov in Rowley 1987: 254).

Bogdanov

> characterized the propaganda and agitation of the 90s of the last century as elementary socialist and that of the first half-decade of the present century as primarily democratic-revolutionary. Purely socialist propaganda must now be implemented in this third period. Class deepening of Social Democratic propaganda is now the first task of the school. The party school must prepare not only political revolutionaries but conscious socialists (in Rowley 1987: 255).

After the revolution Lunacharsky, who had settled his differences with Lenin, served as Commissar for Enlightenment (Minister of Education and Culture) in the Soviet Government. Bogdanov established the *Proletcult* organization, which continued the struggle to raise the scientific and cultural level of the working class, and became the first President of the Soviet Academy of the Social Sciences, where he trained many of the Soviet Union's early economic planners. Like Gramsci, both were loosely aligned with the right opposition during the struggles of the 1920s.

In analyzing the struggles surrounding the God-builders, one cannot help but conclude that they were, in retrospect, several decades ahead of their time. There is little doubt that the revolutionary impulse of 1917 was predominantly democratic and agrarian in character and that the fastest road to power was, as Lenin understood, formation of a popular front based on the party's superior ability to deliver on these democratic and agrarian demands. At the same time, Bogdanov and the God-builders understood what Lenin and most Marxists of both social democratic and communist convictions have tended to forget. Communism is possible only when the human productive forces have developed to the point that human beings can consciously organize the entire production process without recourse to market mechanisms. Each individual in the society must have a fully developed scientific understanding of the optimum use of the resources he has at his disposal, and an active and powerful desire to use those resources in that manner. This requires an extraordinary level of scientific education. The education of the people thus always constitutes the principal *communist* task of the communist movement, whatever other tasks a particular political conjuncture may impose.

iii) Socialist Humanism[2]

From the very beginning there was a powerful impetus within the workers movement to interpret socialism as an attempt to recover, albeit at a higher level of complexity and differentiation, the lost harmony of the communitarian social order. This is true to some extent of all of the various tendencies within the socialist movement, materialist as well as idealist. But there has also been a tendency, as the workers movement became mired in trade union struggles or in the struggle for state power, to lose sight of this ultimate aim. Populism and theomachy both represent, at least in part a reassertion of this tendency. But nowhere was this impetus stronger than among the humanistic intelligentsia, who found themselves increasingly marginalized and their knowledge increasingly devalued as bourgeois society plunged into a downward spiral of alienation and social disintegration. Consider the following passage from Lukacs' *Theory of the Novel.*

Happy are those ages when the starry sky is the map of all possible paths——ages whose paths are illuminated by the light of the stars. Everything in such ages is new yet familiar, fully of adventure and yet

---

[2]. We include in this trend not only Lukacs and his followers, the Frankfurt School, the Praxis Group, and the Polish group which publishes <u>Dialogue and Humanism</u>, but also the whole tradition of Latin Communism associated with Gramsci and Mariategui.

their own. The world is wide and yet it is like a home, for the fire that burns in the soul is of the same essential nature as that which fuels the stars; the world and the self, the light and the fire are sharply distinct, yet they never become permanent strangers to one another, for fire is the soul of all light and always clothes itself in light. Thus each action of the soul becomes meaningful and rounded in this duality: complete in meaning——in sense——and complete for the senses; rounded because the soul rests within itself even while it acts; rounded because its action separates itself from it and, having become itself, finds a center of its own and draws a closed circumference around itself (Lukacs 1916/1971: 29).

But it is not only in capitalist societies that sentiments of this sort emerge. The contradictions of "actually existing socialism" have also sparked discontent among the humanistic intelligentsia. Unlike the technocrats and empirical revisionists who rallied uncritically to the banner of *perestroika*

> These critics take the elite to task for making too many concessions to the technocrats, for abandoning the long range goals of socialism, for substituting economic growth for the humanization of social relations, for giving undue importance to overtaking the developed capitalist countries at the expense of the specific goals of socialism, which inevitably brings with it the introduction into the socialist countries the values and habits of the consumer society. These critics of the ruling elite demand a greater equality of living standards instead of increasing differentiation of incomes; promotion of collective rather than individual consumption; development of collective forms of community life rather than a privatized life-style; training for political activism in genuine movement organizations rather than sanction for competitive, achievement oriented behavior; and continuation of the cultural revolution in the place of an apolitical cultural policy tailored to consumer wishes——a radical separation of culture and market. These critics opposed the technocracy, its interests, and its kind of rationality on every important social issue. They offer as an alternative a policy of putting *telos* before *techne*——a program which could only be carried out by the ruling elite, but not by the ruling elite as it is ... (Konrad and Szelenyi 1979: 241)

Socialist humanism has made three very important contributions to the socialist movements. First, it has refocused attention on humanity's underlying drive towards holism, and attempted to ground this drive, albeit with only limited success. Second, it has refocused attention on the corrosive impact of commodity relations (i.e. the marketplace) on the social fabric and on the development of human social capacities, and drawn attention to the need to transcend not only the market in capital, but also and perhaps even more importantly, the market in labor power. Third, this trend has given us the rudiments of an effective strategy for building socialism centered on the actual cultivation of socialist consciousness among the working classes——a strategy which integrates the strengths and avoids at least some of the weaknesses of both populism and theomachy.

At the very core of the humanistic interpretation of the socialist vision lies reassertion of a fundamentally teleological vision of human nature and of human history——if not yet of the cosmos as a whole. Erich Fromm, one of the most important members of this trend, argues for a return to the ethics of virtue advocated by Aristotle and Spinoza. We have already seen that for Aristotle matter was simply the potential for organization, a potential drawn forth by the prime mover, the beauty of which excited matter and acted on it as a final cause. The soul is the form of the body, encoding certain definite potentials. Virtue is simply the realization of this potential and a just society one which promotes the development of human virtue. For Spinoza, similarly, there is but one Substance——Nature or God. Individuals participate in this Substance to the extent of their power: to the extent that they are capable of affecting and being affected. Virtue, realized as power, is thus simply the ability to connect with the universe.

What dialectical and historical materialism has added to this view is an understanding of humanity's capacity for productivity, our capacity, through labor, to add something fundamentally new to the universe, so that our *telos* is not something fixed for all eternity by the prime mover, but rather something we help to design as well as to realize.

Productiveness, for Fromm is realized first and foremost in thinking and loving. Productive thinking, or

> Reason ... reaches to the essence of things and processes ... Its function is to know, to understand, to grasp to relate oneself to things by comprehending them. It penetrates the surface of things in order to discover their essence, their hidden relationships and deeper meanings ... Being concerned with the essence of things ... means being concerned with ... the generic and universal ... getting a whole consistent picture and seeing what the structure of the whole requires for the parts (Fromm 1947: 108-111).

In the process, it discovers latent, unrealized potential, and lays the groundwork for creative labor which transforms the world, making it richer and more complex, more beautiful and more powerful than it was before.

> The normal human being is capable of relating himself to the world simultaneously by perceiving it as it is and by conceiving it enlivened and

enriched by his own powers (1947: 97).

The second contribution of humanistic socialism is a reassertion of the historic socialist critique of the marketplace. There has been a tendency within the Soviet tradition to assume that it is above all the market in capital goods which much be transcended, so that surplus can be centralized and allocated rationally in such a manner as to best promote the development of human social capacities. Moves to restrict the market in consumer goods and labor power are deferred to "later stages" in the process of socialist construction. And it is true that the socialist states have accomplished a great deal on this front. Maoism goes somewhat further, cautioning against the dangers of "residual commodity relations." But Maoist practice was in fact directed as much against small producers (peasants, petty bourgeoisie) as it was against the market in labor power, and often resulted simply in further bureaucratic centralization.

The socialist humanist critique of the marketplace, on the other hand, has focused specifically on the tendency of market relations to generate both ideological and moral disorders. At the ideological level, insertion into market relations leaders to what Georgi Lukacs calls "reification," the transformation the complex system of relations which constitutes the universe, into a system of only external relationships between things.

> The chief changes undergone by the subject and object of the economic process are as follows: 1) in the first place, the mathematical analysis of work-processes denotes a break with the organic, irrational, and qualitatively determined unity of the product ...
>
> ... the fragmentation of the object of production necessarily entails the fragmentation of its subject ...
>
> ... Consumer articles no longer appear as the products of an organic process within a community (as for example in a village community) ... Only when the whole life of society is thus fragmented into the isolated acts of commodity exchange can the "free" worker come into being; at the same time his fate becomes the typical fate of the whole society ... (Lukacs 1971: 88-91).

At the moral level this alienation is reflected in a tendency to regard all activity into simply a means of realizing individual consumer interests. Erich Fromm points out that the dominant character type in market societies is what he calls the "marketing orientation." Losing sight of their own latent potential to be creative and powerful, knowing and loving, and thus to *add* something new to the complex organized totality we call cosmos, human beings instead become focused on selling themselves in order to secure the resources they need in order to consume. At the same time, the actual identity of the individual——the complex of *unique* qualities which make a person both different from everyone else and for this very reason essential to the social whole, is suppressed.

> Like the handbag, one has to be in fashion on the personality market, and in order to be in fashion one has to know what kind of personality is most in demand. (Fromm 1947: 78)

Radical individualism in the sense of utter disregard for the our vocation to contribute to cosmos and community, and radical conformism, in the sense of slavish subordination to market norms, thus triumph over authentic individuality in service to the common good.

Taken together this intellectual and moral alienation leads to a profound spiritual alienation——to what Fromm and his colleagues call the "authoritarian personality." Stripped of any ability to understand the world around them as a complex organized totality, human beings feel themselves at the mercy of forces beyond their control. Stripped of both a distinctive identity and a desire to make a contribution to the common good, human beings begin to feel radically depraved——to regard themselves as nothing. It is only by submission to the imperatives of the marketplace——by self-negation rather than self-affirmation——that meaning and hope can be secured. Thus the enduring constituency within capitalist societies for authoritarian social movements. And thus the deeply authoritarian nature of nominally liberal market societies.

The final contribution of the humanistic trend is in the arena of political strategy. The young Lukacs, together with Gramsci and other leaders of the humanistic trend, shared the proletarian maximalism of many of the young intellectuals attracted to Lenin's revolutionary movement (Heller 1983: 177-178). Gradually, however, partly because they developed a deeper sense of the inner continuity between socialism and the whole human civilizational project, and partly because the sobering experience of the struggle against fascism at once cured them of their romantic illusions regarding the working class and taught the value of class alliances, the socialist humanists became powerful advocates for a sophisticated popular front strategy, and in particular for the Christian Marxist dialogue, and the dialogue between religious idealism and dialectical materialism generally. In order to understand the full strategic significance of this dialogue it is necessary to examine the ideas of Antonio Gramsci, whose formulation of the strategy has had profound practical impact not only in Europe, but more especially in Latin America and the rest of the Third World.

Antonio Gramsci was one of the founders of the Italian Communist Party. Gramsci regarded socialism as the product of a rationalization of the popular Christianity of the European masses. Like most of the Marxist tradition, Gramsci recognized that Primitive Christianity was first and foremost an expression of the suffering of the subaltern classes of the Roman Empire, and of their aspiration for a classless and communal social order (Portelli 1974:57-9). Unlike Engels and Kautsky, however, who believed that the Christian project was doomed to fail because the low level of development of the productive forces made communism impossible, Gramsci believed that the failure of Christianity was due first and foremost to the success of the ruling classes of the feudal period in drawing on the Christian tradition as a source of legitimation, and in ultimately conforming Christian doctrine to the needs of the feudal social formation.

This was a gradual process and Gramsci argues that the popular character of the Christian Church endured well past the apostolic period. After the Edict of Milan, which granted toleration to Christianity (311 A.D.), the Church entered into an objective alliance with the state apparatus of the Empire, without, however, compromising the revolutionary character of Christianity as an ideology (1974:61) --something indicated by the well known patristic condemnations of private property (Saudino n.d.). It was only with the rise of feudalism, as the clergy became more and more integrated into the landed elite, that Christian theology became increasingly a means of legitimation for the feudal classes (1974:69ff). Thus the origin of the reactionary "Catholic-feudal bloc" which confronted the rising bourgeoisie, peasantry, and working class of the early modern period.

Further, this Catholic feudal—bloc did not go uncontested. On the contrary, the popular classes of town and country gave birth to a long series of movements which sought to restore the original popular character of the Christian tradition: the Valdese, the Joachites, the Franciscans, and the mendicant orders generally all sought in their own ways both to break the alliance between altar and sword which kept the peoples of Europe in thrall, and to purify Christian theology of feudal accretions. These movements were defeated, Gramsci implies, not because the material conditions for communism were not yet present, but because they all failed to develop an effective strategy for cultural hegemony --to unite all of the popular classes of the city and the countryside, bourgeois, peasant and proletarian, into a single revolutionary Christian bloc to break the feudal stranglehold and open up the road to democracy and ultimately to a classless and communal social order. Some, like the Valdese and the Franciscan Spirituals erred to the "left," becoming isolated revolutionary sects. Others, such as the main body of the Franciscan order and the mendicant tradition generally, erred to the "right," becoming reabsorbed into the Church and the general cultural milieu of the times.

It was only with the Protestant Reformation that the Catholic feudal bloc was finally ruptured. Unlike their predecessors, the Reformers were able to forge a new counter hegemonic bloc which united bourgeois, peasant, and proletarian. The Reformers, to be sure, did not succeed in restoring the original revolutionary communitarian character of the Christian tradition. On the contrary, the new Protestant Christianity joined the popular religion of the peasant masses to essentially bourgeois tasks: the creation of unified national states and the development of industrial forces of production. The Lutheran tradition, with its insistence on independent national churches, bound the German masses to the task of building a German nation, which they came to see as the embodiment of their fundamentally Christian ideals. Calvinism, similarly, bound the English masses to the tremendous project of the industrial revolution, by endowing work and economic growth with a new religious meaning (Portelli 1974: 106ff). The Protestant Reformation liberated Christianity from its feudal deformations only to recast it in a way that conformed to the needs of the rising capitalist relations of production.

Gramsci extended this analysis to the French Revolution, which he regarded as integral to the protracted struggle of the Reformation. The Jacobins succeeded, he reasoned, because they were able to present the Revolution as the defense, in fact the ultimate realization, of the ideals of liberty, equality, and fraternity, ideals with deep roots in the Christian tradition.

> Otherwise adherence to the new ideals and revolutionary politics of the Jacobins against the clergy by a population which was almost certainly still profoundly Catholic and religious would be inexplicable (Gramsci 1950: 48)

The Revolution was, therefore, far from being fundamentally anti-religious, or opposed to the beliefs and values of the Christian tradition, essentially an attempt to return to its primitive ideals. Gramsci, in fact, calls it the Liberal Democratic Reformation (1949b: 268).

> The Church as a community of the faithful conserves and develops political and moral principles which are in direct opposition to the Church as a clerical organization ... the principles of the French revolution are those of the community of the faithful as against those of the clergy, a feudal order allied with the king and the nobility (1949b: 294).

The entire modern revolutionary process is, therefore, nothing but the realization of the popular ideals of the Christian masses, through a protracted process of moral and intellectual reform which liberates those ideals from the ruling class political projects to which they have become attached. Marxism itself, or the philosophy of praxis as Gramsci called it, is simply the most advanced, and in fact the definitive, product of the prolonged Reformation. Indeed, Gramsci has even adapted the old theory of the "three sources and three component parts" (Lenin 1913/1971: 20-23) of Marxism to account for the roots of Marxism in the Reformation tradition. German idealism, he argues, was the final theoretical product of the Lutheran Reformation, summing up its aspirations for personal and political autonomy and the strong sense of German national identity which originally moved the Lutheran struggle. English political economy, similarly, is the heir of Calvinism, which first gave a religious meaning to economic rationality. French socialism is the heir to the French revolution. The "philosophy of praxis" contains within itself the contributions of each of these earlier stages of the prolonged process of the reformation, but goes further in that it promises to make the historic ideal of the Christian tradition, the aspiration for a classless and communal society, into a concrete social reality.

In analyzing the strategic options facing the working class, Gramsci distinguished sharply between "partial" and "totalitarian" approaches. A party pursuing purely partial aims simply attempts to realize specific consumer interests on behalf of the class or class fraction it represents. A totalitarian party, on the other hand, aims at the global organization——or reorganization——of human society. This task can be carried out in any one of three ways. If a ruling class rules primarily by coercive means, it may be said to exercise dictatorship. Dictatorship is an unstable form of class rule, and simple dictatorship, while it may be necessary under certain circumstances, is hardly a goal for a genuine revolutionary class such as the proletariat. It is also possible to rule through what Gramsci calls transformism——i.e. by co-opting potential adversaries, transforming their global aims into partial ones. A demand for socialist revolution for example, is transformed into a demand for jobs, for higher wages, for new educational opportunities, etc. A demand for restoration of the rights of the peasant communities is transformed into a demand for moderate land reform and credits for rural cooperatives. While more stable than pure dictatorship, transformism has the effect of draining energy away from the organic project of the ruling (or revolutionary) class.

A stable ruling class, on the other hand, is able to exercise what Gramsci called "hegemony" over the other classes of society. Essentially this means convincing all of the principal social classes that the interest of the ruling class is identical with that of the society as a whole. It is here, of course, that the difference between proletarian leadership and that of other ruling classes asserts itself. The interest of the proletariat is, of course, nothing other than the development of human productive capacity in general, and is thus identical with the interests of humanity as a whole. Where the hegemony of the warlords and the bourgeoisie involves mystification, the hegemony of the proletariat is founded on a process of rationalization——of helping both the proletariat and the other social classes to grasp their authentic vocation as human beings.

The struggle for hegemony is carried out by professional intellectuals. Gramsci distinguished between "organic intellectuals" who arise directly from the life conditions of a fundamental social class, and articulate its aspirations, and "traditional intellectuals" attached to institutions which had grown up under past social formations and which have become linked to a new ruling class as subordinate elements in their systems of class rule (Gramsci 1949c: passim). The organic intellectuals of the proletariat are the theoreticians and organizers of the Communist Party. The traditional intellectuals of greatest interest to Gramsci are the clergy, who have historic relationships with both the landed nobility and the village communities.

In order to establish cultural hegemony, the organic intellectuals of the proletariat must develop, disseminate, and begin to implement an authentically proletarian worldview. This means developing among the workers an understanding of the unique vocation of humanity in the cosmos——our role as the architects and engineers of complex organization. At the same time, the other popular classes, especially the peasantry, the petty bourgeoisie, and progressive elements in the industrial bourgeoisie, because the conditions of their existence differs from that of the proletariat, will not find this vision unambiguously compelling. Thus the vitally important role of traditional intellectuals——e.g. the clergy——who have been drawn to the side of the proletariat. These traditional intellectuals must reinterpret the religious traditions of the peasantry in such a way as to transform those traditions into a bridge to the development of socialist consciousness, or at least the basis for a stable strategic alliance with the working class. Thus the vital importance of the Christian Marxist dialogue, and the dialogue between dialectical materialism and objective idealism generally in the strategy of socialist humanism.

Its remarkable complexity and sophistication notwithstanding, there is a profound internal contradiction in the ideological line of socialist humanism. On the one hand this trend seems to have comprehended with a clarity we have not seen since Hegel the categories of totality, self-organizing complexity, and teleological purposefulness. It is precisely their grasp of these categories which makes them able to appreciate the revolutionary potential of the popular religious traditions. Indeed, as they write about religion, they often seem to have crossed the line between mere dialogue with religious idealism to the articulation of a new materialist spirituality. This tendency is especially pronounced in Fromm, but it is apparent even in thinkers like Lukacs who had less direct engagement with the Christian Marxist dialogue.

At the same time, even the most advanced socialist humanists retain the materialistic conviction that fully teleological organization is introduced into the universe only with the advent of human labor and the social form of matter. As Fromm puts it, Man (sic) must recognize that

> ... there is no power transcending him ... there is no meaning to life except the meaning man gives his life by the unfolding of his powers, by living productively (1947: 53).

As a result, as with the dialectical materialist tradition generally, human purposefulness remains without any real ontological ground. Teleological organization is a product of human intervention, something imposed by humanity on the natural world, and by more advanced social formations on less advanced, not the expression of a deeply rooted tendency of being itself.

This has important consequences. Most importantly, it has left socialist humanism open to the postmodernist assault. Indeed, Lukacs in particular is often targeted as an apologist for totalitarianism, and his philosophy used document the link between totalizing theory and totalitarian politics. At the same time, in the absence of a firm ontological ground for its teleological ethics, socialist humanism has been unable to defend itself against appropriations of Gramscian strategic thinkers by "radical democrats" such as Laclau and Mouffe, who reject the socialist vision entirely.

b) The Nicaraguan Revolution

Third force socialism has been an undercurrent in most of the great revolutionary upheavals of the industrial period. Rarely, however, has it become the dominant trend in the revolutionary coalition, and rarely have all three types of third force socialism flowed together into a single, coherent, revolutionary movement. This is, however, precisely what happened in Latin America in the period between 1968 and 1989.

Neopopulist dependency theorists, and *foco* and prolonged popular war strategists refocused attention on the revolutionary potential of the peasantry. Artists, song-writers, poets, and novelists labored to produce a new revolutionary culture. And a vigorous Christian Marxist dialogue gave birth to a strategic alliance between a new generation of communists struggling to build cultural hegemony and progressive elements in the religious institutions anxious to rebuild their base among the working class and the peasantry.

Among the various Latin American cases, that of Nicaragua is probably the most interesting, since it permits us to examine the work of a revolutionary organization both on the road to power and in the process of socialist construction. Roger Lancaster's recent study of the Nicaraguan revolution (1988) is especially useful in this regard.

Close inspection of events in Nicaragua allows us to formulate a picture of revolution and the proletariat distinct from Marx's predictions, with their reliance on a modern proletariat divorced from religion, and counter to all progressivist philosophy, which politicizes from the critical position of pure reason. What the fully modernized proletariat represents is a weak and ultimately alienated normative impulse, frequently divorced from any sort of class agenda or any sense of itself as a class. Under the reign of fully developed capitalism, the normlessness of the marketplace no longer incites envy, a desire to level one's superiors, a need to reclaim the tradition. Unable to mount a normative challenge to the flux and uncertainty of capitalist production, the modernized proletariat can only visualize itself at best as a partly hostile, partly integrated component in the production process. Trade unionism and social democratic reformism, then, are its highest forms of class consciousness, and it is frequently unable to attain even that level of self-organization.

Revolutionary class consciousness seems virtually impossible for a fully modernized proletariat reproduced in the womb of advanced capitalism. But it is possible in earlier stages of capitalist development, and in regions where advanced capitalism is structurally prevented from fully emerging. That is, revolutionary class consciousness is possible where modernity collides with tradition, and this collision is implicit in the nature of ongoing relations between the West and the Third World. This real revolutionary consciousness draws on rural or peasant traditions in such a manner that it invents a revolutionary proletariat out of vaguely articulated "popular classes," rationalizes its class outlook and agenda, and projects the image of a stable conservative normative society as its version of the good life (Lancaster 1988: 212-213).

Mass socialist movements in Russia or China developed popular ideologies which were only implicitly religious. "They declared their atheism to the world while at the same time reconstituting the idea of God in historical materialism as *History*. Hand behind back, then, these revolutions invariably posit a new religion on the ruins of the old ... Liberation theology in Nicaragua" merely "makes this relationship explicit (1988: 203-204)."

Indeed, Lancaster argues that socialism in fact *requires* some kind of religious legitimation.

> The socialist economy——with its aversion to pure markets, and its concomitant need for ideological incentives——inevitably promotes a consolidated state religion of a sort quite unnecessary, if not counterproductive, to capitalism (Lancaster 1988: 187).

What makes the Nicaraguan case most interesting is, on the one hand an attempt to draw consciously on the popular communal institutions, and especially on the popular religious traditions, as a bridge to the formation of socialist consciousness. At the same time, the FSLN was even more insistent than the Chinese Communist Party that these essentially precapitalist, communitarian solidarities be transformed into a catalyst for action on behalf of development and social progress——that they become an organizing, and not simply a unifying factor.

In order to understand how this revolutionary dynamic developed, and why it played itself out in the way that it did, it is necessary to characterize at least briefly the nature of Nicaraguan capitalism.

> Nicaragua, like other Central American countries, was incorporated into the international division of labor as a primary export economy. But these exports were of little relevance until the coffee period began during the last third of the nineteenth century. During the colonial period the production of indigo was the principal source of external income; this was combined with extensive livestock breeding and subsistence farming in an extremely backward productive structure ...

> Coffee production began later than in other countries of the isthmus and reached its peak between 1920 and 1940. The formation of an agrarian bourgeoisie is tightly linked to this coffee expansion, although coffee did not introduce great changes in the economic structure: the latifundist patterns of extensive stockraising easily adapted to the new export crop. The search for lands appropriate to coffee affected not the landowning oligarchy but the tenant farmers, indigenous communities, ... and others

who were dispossessed and pushed into marginal zones (Vilas 1986: 49).

Nicaragua, like much of Latin America, was the scene of intense struggles for national liberation during the early part of this century. As in Mexico, the national liberation struggle was led by populist forces which looked not simply to the recovery of national sovereignty, but also to the reorganization of Nicaraguan society. Cesar Augusto Sandino was an eclectic thinker who drew his inspiration primarily from the anarcho-syndicalist tradition. He looked to a new Nicaragua founded on self-governing peasant and artisan cooperatives and held together by a scientistic mysticism he imbibed while studying with the "magnetic-spiritual school" in Argentina, according to which God was in fact an electromagnetic force produced by all living beings, which at once bound them together and catalyzed their creative organizing activities (Hodges 1986).

Defeat of this movement meant that development in Nicaragua would continue along peripheral capitalist lines. As the market for coffee became saturated during the 1940s, cotton production gradually increased, its share of Nicaraguan exports growing from 5% in 1950 to 45% in 1965 (Vilas 1986: 50).

The best cotton lands were found in the northeastern zone ... Expansion forcibly displaced the farmers occupying these lands and producing foodstuffs, principally basic grains. Insofar as the growth in production was done by cultivating new lands, rather than increasing yields, cotton cultivation generated a massive population push toward the agricultural frontier ... as well as to the urban centers (Vilas 1986: 50).

A similar expansion of beef production for export took place in the 1960s (1986: 51).

Creation of the Central American Common Market in the 1960s opened Nicaragua to foreign investment. Much of this investment (54.4% in 1969) went into the industrial sector, but Nicaraguan industry remained heavily dependent on both imported inputs and on foreign markets (Vilas 1986: 52-53).

The result of this process was the emergence, by the middle of the 1970s, of an economy which was fully incorporated into the world capitalist system, with a relatively high level of proletarianization (Vilas 1986: 66-67), but with production almost entirely dedicated to exports. What little investment there was in increasing productivity, went into the agro-export sector (Vilas 1986:57).

There was, however, a certain uniqueness to the Nicaraguan social formation. Victory over the *Sandinismo* of the 1920s and 1930s had been guaranteed largely by the *Guardia Nacional* built and led by the Somoza family. This family, together with its retainers, developed over the course of the next several decades an effective stranglehold on the Nicaraguan economy. Control of the state apparatus permitted the Somocistas to secure for their own benefit most of the foreign financing and state credit advanced to support development of infrastructure, modernization of industry, construction etc. As a result of this process, by the end of the 1970s, the Somoza family and its allies controlled roughly 25% of the country's productive capacity (Vilas 1986: 88).

Since its Sixth Congress in 1928, the Communist International had counseled a rather conservative application of the popular front strategy in Latin America. Communists were instructed to build an alliance with the progressive, national, industrial bourgeoisie as well as with the petty bourgeoisie and the peasantry in order to overthrow the hegemony of large landed property, and lay the groundwork for industrialization and the development of the conditions for socialist construction. Armed struggle was, in most cases, rejected. While credible in a country such as Mexico, which had experienced a successful progressive bourgeois revolution with significant populist participation, such a strategy made little sense in a country like Nicaragua, where the indigenous bourgeoisie was largely backward and agrarian, and where what little industrial development existed was subordinated to foreign capital. Indeed, the anti-Somocista bourgeoisie in Nicaragua was primarily agrarian and Conservative rather than industrial and Liberal in character.

The *Frente Sandinista de Liberacion Nacional* (FSLN) formed in resistance to this strategy. Initially the FSLN was attracted to the *foco* strategy developed by Che Guevara. This theory assumes that the effective brake on development imposed by incorporation into the world capitalist system had rendered the peasantry and proletariat fully revolutionary. A small group of intellectuals establishes a presence in the countryside. Their bold actions against the landowners and their state ignites the revolutionary potential of the peasantry, and sets off a *guerilla* which eventually becomes a peoples' war and is able to mount an effective contest for power. Soon, however, the specificity of the Cuban experience became apparent. The conditions which had led the Cuban peasants to respond to Castro's *focos* did not exist in Nicaragua. Under the leadership of Carlos Fonseca Amador, Ricardo Morales Aviles, and Tomas Borge, the FSLN began to implement a strategy of prolonged popular war, building up base areas among the peasants in the remote mountainous regions of the country, preparing eventually to "surround the cities with the countryside," much as

Mao and Ho had done in the very different societies of Asia. Central to this strategy was the conviction that it was among the peasantry, which had largely been protected from the disintegrating effects of the market economy, and which had preserved its traditional communal institutions and popular religious traditions, that the seeds of the new society would be nourished (Hodges 1986). The *guerra popular prolongada* dominated the FSLN politically and organizationally until the end of the 1970s.

Sandinistas based in the urban areas soon became frustrated with this strategy. The Nicaraguan population was increasingly urban and proletarian, they argued. Conditions existed for building a proletarian party among these urban workers, dedicated not simply to national liberation, but to socialism. The base areas being organized by the leadership, they argued were too far removed from the real centers of power to pose a serious threat to the regime and the oligarchy. For the *proletarios*, as they were called, as well, popular religion played a critical role in the development of socialist consciousness.

Roger Lancaster's study of the role of popular religion in the Nicaraguan revolution documents this process quite well. Nicaraguan popular religion, like popular religion generally, reflects a kind of pre-political class consciousness.

> From the point of view of these non-elites, the practice of religion ... may be a means of manipulating nature and the supernatural; it may be seen as a means of manipulating elites; it may act as a means of social and economic leveling at the lowest level of social organization ... In its own quasi-ethical sphere, popular religion may foster and consolidate the class consciousness of the poor by embodying in the rich all that is evil and full of vice, while at the same time interpreting in the poor all that is just and good and finds favor with God. Even a purely escapist religion may achieve the effect of producing a powerful, if restrained class consciousness through its very escapism and indirection, so long as the poor can find an interpretation that offers redemption for those who suffer in this life and punishment for their oppressors in the next ... (Lancaster 1988: 30).

The difficulty, of course, is that this sort of popular religion is essentially passive in character.

> The role of the ordinary person is to be as good a Christian as possible; to keep the religious observances; to see to it that the children are baptized ... to avoid the mortal sins; and if lax, then not to be *too* lax. It is not so much the role of ordinary people to *represent* the religious ideals, then as to *endorse* and *recognize* them (Lancaster 1988: 34).

When the popular religion *has* led to action, it has usually taken the form of the millenarian revolt, in which the peasants take action because they believe that their redeemer has already arrived, or will arrive shortly, to win what would otherwise certainly be a losing struggle on their behalf.

The task, of course is to rationalize this class consciousness so that it becomes a catalyst for the participation of the masses in the reorganization of society. The FSLN was able to tap into a process of rationalization which was already underway within the Roman Catholic Church——a process which we examined in some detail earlier in this chapter.

The result of this process was the emergence of a revolutionary ideology in which the general line of the FSLN——its understanding of the "line of march, conditions, and ultimate general result" of the Nicaraguan struggle was mediated to the people through a partially rationalized Christianity. Theologians working in or with the FSLN elaborated this partially rationalized discourse——the theology of liberation——and pastoral agents organized the people around its symbols, discourse, etc. This had the effect, on the one hand of vastly increasing popular support for the revolutionary project, since it came to be seen as an extension of popular traditions, but also of enlisting the people in the work of their own cultural transformation, so that rather than waiting for the messiah, they took up the work of actually reorganizing their society.

The gradual development of base areas, and the organization of the urban proletariat carried out by the *guerra popular prolongada* and *proletarista* factions respectively was cut short by the crisis of the late 1970s. Growing contradictions within the bourgeoisie coupled with pressure from progressive sectors of the U.S. bourgeoisie temporarily in power, made it impossible for the Somocista regime to continue to rule in the old way. Elements within the FSLN——the *tercercista* faction grouped around Daniel and Humberto Ortega——grasped the depth of this crisis and argued for the formation of a broad front in opposition to the Somoza regime, coupled with increasingly bold military action to demonstrate the weakness of the regime, and the growing strength of the FSLN. The strategy produced almost immediate results, and by the middle of 1979, the Somoza regime had been toppled by a massive popular insurrection. The FSLN emerged as the only force capable of governing.

The Sandinista strategy for the reorganization of Nicaraguan society developed within the limits imposed by the broad alliance which brought the front to power. Nationalization was restricted to the assets of the Somoza family and the National Guard, to banking, foreign commerce, natural resources——and to those capitalists who intentionally produced below capacity (Vilas 1986: 154). At the same time, state enterprises which were unproductive were threatened with privatization. The result was a public sector controlling roughly 41 % of the country's productive capacity: less than in Latin American countries which had experienced effective populist revolutions (Mexico, Peru) or in France or Italy (Vilas 1986: 154).

Sandinista development strategy was heavily influenced by the thinking of Agrarian Reform minister and *proletarista* leader Jaime Wheelock. Land reform affected only idle, underexploited, or leased land, and land belonging to owners with more than 350 or 700 hectares, depending on the region (Vilas 1986: 159). Emphasis continued to be placed on production of agricultural exports in order to generate the foreign exchange necessary to centralize the capital required to promote industrialization. Wheelock envisioned a gradual development up the technological ladder from agricultural production, to agricultural processing, to the production of the equipment necessary for agricultural processing, etc. Continued emphasis on production of agroexports necessarily entailed some tension with the parts of the peasantry, especially in the frontier regions, which wanted to withdraw from the marketplace and engage in subsistence farming, and with the working class, whose demands for cheap and plentiful food could only partially be satisfied. Wage demands were severely restricted, in favor of rapid increases in the *social wage*: i.e. investment in housing, health care, public transportation, and especially in education (Vilas 1986: 180).

During its early stages this strategy was relatively successful in increasing production. Nicaragua's Gross Domestic Product increased 10 % in 1980, 8.5 % in 1981, fell 1.2 % in 1982, and increased 5.3 % in 1983, for an overall increase of 5.6 % between 1980-1983. During the same period the GDP for Central America as a whole fell by 1.4 % (Vilas 1986). Investment in sectors central to the Sandinista development strategy soared.

Assigning 1978 the base value 100, the coefficient of agricultural investment reached the level 274 in 1981 and 400 in 1982 ... In Costa Rica ... the coefficient of agricultural investment was reduced from 100 to 27 between 1978 and 1982 and in Guatemala it fell from 100 to 43 in the same period (Vilas 1986: 234).

Public consumption (a measure of investment in development of human social capacities) increased from 11 % of total consumption in 1977 to 33 % in 1983. Wage consumption increased from 49 % to 86 % and luxury consumption fell from 51 % to 14 % (Vilas 1986: 235).

At the political level, the FSLN rejected the Leninist model of dictatorship of the proletariat in favor of a relatively open political process. Only former Somocistas and others suspected of compromising the project of national liberation were excluded from the political process. The productive bourgeoisie in the political as in the economic arena was given considerable freedom. The leading role of the FSLN was guaranteed on the one hand by the Sandinista Popular Army, and, on the other hand by the developing cultural hegemony of the revolutionary forces.

Of central importance in this regard was the New Education process. In five months shortly after the revolution, the National literacy crusade reduced the rate of illiteracy from 50.4% to 12.9%, teaching half a million adults to read (Vilas 1986: 214). Total school enrollment increased from 501,600 in 1978 to 1,127,000- in 1984. Of these, 194,800 were adults. Prior to the revolution there was no popular adult education (1986: 215). This means that between one fifth and one quarter of the population was participating in education of some kind. The educational process, in the meantime, was reorganized to stress the values of the revolution.

In short, the Sandinista development strategy centered on involving the masses in their own development. Popular communal institutions and popular religious traditions were to serve not simply as a basis for leveraging support for the socialist project on the basis of a vague nationalism or collectivism, but were, rather, actually reorganized and transformed to become catalysts for the development of human social capacities.

Why, if the Sandinista strategy was so sophisticated, were the Sandinistas unable to maintain the support of the masses? Any answer to this question must, of course, take as its point of departure the changing international situation: the crisis of socialism, on the one hand, which weakened international support for the revolution, and, on the other hand, the success of the U.S. backed contra war which imposed such high levels of military spending that the Sandinistas were unable to sustain the gains of the early years of the revolution. Second, because of the way in which the FSLN came to power, the strategy which it implemented did not really reflect the most intense (or advanced) expressions of Sandinista thinking. Dominated by the *tercercista* faction, or rather the Ortega wing of that faction, FSLN policy increasingly came to resemble a very liberal interpretation of Leninist popular front theory, influenced to be sure by Sandinista motifs, but geared primarily to maintaining broad based support rather than to carrying out an in-depth transformation of the whole way of life of the population.

There were, however, some very real contradictions built into the process from the very beginning——contradictions which reflect the limitations of the Sandinista fusion of populist and Leninist strategic elements. On the one hand, the theology of liberation, on which the FSLN relied to galvanize popular support, failed to carry out a really comprehensive rationalization of the popular religious traditions. The result was the persistence of essentially regressive elements within the very core of Sandinismo. Of critical importance here is the popular cult of the *guerillero*.

> Like the priest, the saint, and the martyr, the guerilla's life is one of deprivation and sacrifice. His passion is distinctly spiritual, not physical ...
>
> ... Indeed, if we see a continuous Judeo-Christian (sic) religion of *sanctification through sacrifice*, the new constellation appears as a reconstellation of the ancient messianism. Like the spiritual example of Christ the Redeemer and the moral authority of the priest, the social authority of the FSLN rests on its exemplary action through self-sacrifice (Lancaster 1988: 133).

Liberation theology thus conserves precisely that element of Christianity (and of the religions inherited from tributary social formations generally) which we identified as the foundation of otherworldliness and passivity.[3] The problematic operates on two levels. On the one hand it locates the source of salvation not in labor——not in participation in the self-organizing activity of the cosmos——but rather in warfare and the self-sacrifice which it entails. Second, the demand for saving action is displaced from the peasant and worker to another: ultimately onto the martyred *guerillero* who dies in the mountains to liberate us, and immediately on to the FSLN which is the institutionalized form of the *guerillero*'s charisma. While the ideology legitimates the leading role of the FSLN, it simultaneously undermines the ideological conditions for the participation of the masses in the process of socialist construction. What the FSLN required above all was not the uncritical support of the masses, but rather their willingness to *labor* tirelessly, at every increasing levels of effort, without prospect of immediate prosperity, and to develop themselves in every arena, so as the create the conditions for the new society.

At the same time, the marketplace remained intact, and its alienating dynamic made itself felt with increasing intensity, generating alongside this ideology of salvation through self-sacrifice the consumerist pressures characteristic of every market society. In this sense, it was already too late for Nicaragua to attempt to follow a strictly noncapitalist road to

---

[3]. It is important to note in this context that the most important Nicaraguan religious thinker, Ernesto Cardenal, in fact transcends most of the limitations of liberation theology. We will have an opportunity to analyze his work in some detail later on.

development. Unlike the China of 1949, the Nicaragua of 1979 was already a fully capitalist society with a large, thriving urban sector.

The operation of these tendencies was intensified, rather than being mitigated, by the contra war and by the dominance of the Ortega wing of the *tercercista* faction in the FSLN leadership. The Ortegas were concerned above all with maintaining the unity of the popular front, and thus hesitated to risk too much in the way of internal political and ideological struggle. We will have an opportunity to explore this question in greater detail later on, when we examine the struggle around the poetry workshops initiated by Ernesto Cardenal's Ministry of Culture.

In the end, the FSLN ran up against the same contradiction as the Soviet and Chinese parties before them. They were unable to overcome the consumerism of the masses and unite the Nicaraguan people around the task of socialist construction. This said it must be recognized that, the limitations we have noted notwithstanding, their strategy took them further in this direction than the strategy of either the Soviet or the Chinese parties, and would have taken them further had the global crisis of socialism not intervened. There is good reason to believe that, as it reflects on the defeat of 1990, the FSLN may be able to overcome its current difficulties and chart a course which will lead Nicaragua to socialism——and beyond.

C. The Crisis of Dialectical Materialism

1. Postmodernism

It should, in any case, be apparent at this point that none of the principal tendencies of the socialist movement have actually grasped the "line of march, conditions, and ultimate general result" of the struggle to transcend market relations. On the contrary, from the standpoint of the early 1990s, it is difficult not to conclude that the socialist movement as we have understood it is dead, and that the historical aspirations of this socialism can be carried forward only by a movement of a very different nature.

As we noted in the first chapter, the crisis of socialism has issued in a larger crisis of the whole philosophical tradition of which dialectical materialism is a part. The failure of socialism to prove itself "the solution to the riddle of history" has led to a rejection of all doctrines which attempt to grasp the cosmos as an organized totality, and draw on this knowledge to carry forward the cosmohistorical evolutionary process. This is the phenomenon known as postmodernism.

We have already developed a philosophical critique of postmodernism.

Our purpose here is to show how postmodernism develops logically out of dialectical materialism, and to analyze postmodernist strategies for social transformation. The key figure in this regard is, without question, Louis Althusser. Althusser began his work in the late 1950s and early 1960s as a critic of the "socialist humanist" current which was then gaining strength, and of the larger trend towards what his Maoist and semi-Maoist allies called "modern revisionism." At the theoretical level, Althusser rejected all attempts to read dialectical and historical materialism as what he called a form of "historicism," a teleological doctrine which purported to grasp, and somehow to realize historically, the underlying essence of human nature, of which all historical forms were ultimately just an expression. Practically, he saw in this doctrine the theoretical basis for the political line of the Twentieth Congress of the Communist Party of the Soviet Union, which called for creating a "party of the whole people" and a "state of the whole people," and for a strategy of "peaceful coexistence, peaceful competition, and peaceful transition to socialism."

In response to the humanist offensive, Althusser advanced a complex and highly controversial reading of Marx, according to which all of Marx's earlier works——those before the *German Ideology* and the *Theses on Feuerbach*——were to be rejected as still fundamentally Hegelian in character. Not only Marx's political economy, but also his true philosophy was to be found in *Capital* and the later writings. Here, humanistic historicism gives way to a new understanding of the dialectic, according to which reality is constituted by "complex structured totalities," which, far from being the expression of any single organizing principle, are in fact always internally unstable and contradictory.

The continuity with the ideas of Lenin and especially Mao should be apparent. All three regard the dialectic as first and foremost a doctrine of contradiction and struggle. But Althusser went further even than Mao, charging that Stalin, because of his focus on the development of the productive forces, and his conviction that socialism was at once the necessary expression and the condition for the full development of human productive capacities, had conserved and reproduced elements of a humanistic and historicist interpretation of dialectical materialism and prepared the way for the humanist offensive of the 1950s and 1960s.

Althusser's interpretation of the dialectic had definite sociological implications. Neither socialist nor capitalist relations of production are the necessary "expression" of industrial forces of production. Nor are specific political parties, ideologies, etc. the necessary expression of specific classes, class fractions, etc. Rather, each specific social formation is a complex and contradictory whole which integrates a mixture of different technological forms, relations of production, political structures, ideological structures, etc. Nor is there ever any one single "resolution" to the contradictions of a specific social formation or political conjuncture. On the contrary, everything depends on the relative effectiveness of the political and ideological strategies developed by the leading elements of the classes and class fractions involved.

For sociologists, historians, and political strategists working in the dialectical and historical materialist tradition, Althusser's work was an enormous liberation. Althusser set socialist theory free to analyze the actual structure of economic, political, and cultural relations within capitalist society. And the results were often impressive, particularly in the fields of class theory and political analysis. Poulantzas, Laclau, Therborn, Anderson, and others produced sophisticated analyses of the origins, development, and current contradictions of the bourgeois state.

And yet Althusserian theory never produced a coherent political strategy. Many of Althusser's students were drawn towards Maoism, though Althusser himself remained within the fold of the *Partie Communiste Francaise* (PCF) throughout most of his life, and some of his followers, such as Poulantzas, were ultimately drawn towards variants of Eurocommunism. This lack of strategic coherence should not surprise us. From an Althusserian perspective, while everything was possible, nothing was necessary or even likely. The materialist teleology which had been the hallmark of socialist strategic thinking, and with it the struggle to *discover* the "line of march, conditions, and ultimate general result" of human history, were gone. Indeed, socialism itself no longer appeared to be necessary.

The underlying pessimism of the Althusserian trend soon took its toll. Poulantzas committed suicide. Althusser murdered his wife, trying to flush her down the toilet. And the more creative thinkers within the trend soon made explicit the rupture with dialectical materialism which was already implicit in Althusser's own writings.

The most important thinkers in this regard are Ernesto Laclau and Chantal Mouffe. Laclau and Mouffe locate the roots of the crisis of Marxism in the "essentialism" and "class reductionism" of Marx himself. We have already noted that Marx believed that the unity and solidarity of the working class would be constituted at the economic level, on the basis of labor's participation in the self-organizing activity of the cosmos, and of the material interdependence created by industrial forces of production. Most Marxist theory, beginning with Kautsky has recognized that this is not true ——that the unity, and indeed the very identity of the working class must be constituted though political organization and ideological struggle. Even so, for the entire Leninist tradition, and even for Gramsci, the class struggle in the political and ideological spheres carries out historic tasks which are defined, in the last instance, at the economic level. The Communist Party does not struggle for the hegemony of just any revolutionary subject, but specifically for that of the working class, however complex and subtle the system of alliances it constructs, and however important the political and cultural, as well as the economic bases of those alliances.

> Whether the working class is considered as the political leader in a class alliance (Lenin) or as the articulatory core of a historical bloc (Gramsci) its fundamental identity is constituted in a terrain different from that in which the hegemonic practices operate. Thus there is a threshold which none of the strategic-hegemonic conceptions manages to cross. If the validity of the economist paradigm is maintained in a certain instance ——last though decisive—— it is accorded a necessity such that hegemonic articulations can be conceived only as mere contingency. This final rational stratum, which gives a tendential sense to all historical processes, has a specific location in the topography of the social: at the economic level (1985: 76).

Laclau and Mouffe reject the very existence of such a "final rational stratum." The historic failure of the economistic teleology, they argue, calls radically into question the project of the dialectical tradition ——the whole effort to understand society as an organized totality which develops according to definite laws, independent of human will.

> The symbolic ... character of social relations implies that they lack any ultimate literality which would reduce them to necessary moments of an immanent law. There are not two planes, one of essences and the other of appearances, since there is no possibility of fixing an ultimate plane of signification. Society and social agents lack any essence, and their regularities merely consist of the relative and precarious forms of fixation which accompany the establishment of a certain order (98).

> If society is not sutured by a single unitary and positive logic, our understanding of it cannot provide that logic (143).

The attempt to impose such a logic, they argue, leads inevitably to authoritarianism.

> Leninism makes no attempt to construct, through struggle, a mass identity

418   *Towards Synergism*

not predetermined by any necessary law of history. On the contrary, it
maintains that there is a 'for itself' of the class accessible only to the
enlightened vanguard, whose attitude towards the working class is purely
pedagogical. The roots of authoritarian politics lie in this interweaving of
science and politics (59).

Laclau and Mouffe thus break with the dialectical vision of society as a
totality which develops according to an internal unitary logic, in favor of
a view of society as a crystallization of symbolic and power relations.
Within this context antagonisms may develop along relatively uniform fault
lines, as in the Third World where imperialist exploitation and the
predominance of brutal and centralized forms of domination tend from the
beginning to endow the popular struggle with a center, with a single and
clearly defined enemy (131). This kind of antagonism, Laclau and Mouffe
maintain, leads to what they call a "popular" struggle. On the other hand,
in

> advanced industrial societies ... the proliferation of points of antagonism
> permits the multiplication of democratic struggles, but these struggles,
> given their diversity do not tend to constitute a people ... and to divide the
> political sphere into two antagonistic fields (131).

"Hegemony" (134ff) in this context takes on a radically new meaning.
Where for Gramsci, hegemony was essentially a form of class alliance
based on the identification of the task of the fundamental social class whose
identity was already defined at the level of the economic, with broad
national, popular, democratic and religious identities shared by the people
as a whole, for Laclau and Mouffe hegemony is "a political type of
relation," in which complex patterns of antagonism and equivalence
transform the identities of the various social elements which form them,
and in which no single element is necessarily dominant.

This is really just a theoretically elegant (or obtuse) way of saying that
politics, far from being an expression of pre-defined social identities or
social laws, is instead an open-ended process which is at once defined by,
and which at the same time redefines, the identities of the various forces
which participate within it.

This focus on the open-ended character of the political process leads
Laclau and Mouffe to replace the Leninist distinction between democratic
and socialist tasks, with a new understanding of the modern world centered
on the concept of the "democratic revolution." They argue that "there is
nothing inevitable or natural in the different struggles against power," and
that it is, therefore,

necessary to explain in each case the reasons for their emergence and the different modulations they may adopt. The struggle against subordination cannot be the result of the situation of subordination itself (152).

They thus make a distinction between subordination and oppression.

> We shall understand by a relation of subordination that in which an agent is subject to the decisions of another ... We shall call relations of oppression, in contrast, those relations of subordination which have transformed themselves into sites of antagonisms ... The problem is ... to explain how relations of oppression are constituted out of relations of subordination (154).

Laclau and Mouffe argue that this shift began to take place about two hundred years ago with the French revolution. As the liberal democratic discourse which emerged from this revolution became hegemonic, it set in motion a proliferation of democratic struggles which challenged not only the relations of subordination characteristic of the *ancien regime* ——to the nobility and the clergy—— but also relations of subordination of worker to capitalist, women to men, colonized to colonizer. It is in this context that the socialist, feminist, and anticolonial struggles of the past century and a half should be understood. For Laclau and Mouffe socialism is just one dimension of a larger democratic revolution (154-158).

The present period is, in this view, characterized, on the one hand by the continued proliferation of democratic struggles, and on the other hand by the articulation of these new forms of resistance into an antidemocratic discourse. Popular support for the neoliberal project of dismantling the Welfare State is explained by the fact that the neoliberal right has succeeded in mobilizing against the latter a whole new series of resistances to the bureaucratic character of the new forms of state organization (169). In the light of this situation, they argue that

> the alternative of the Left should consist in locating itself fully in the field of the democratic revolution and expanding the chains of equivalence between the different struggles against oppression. The task of the Left therefore cannot be to renounce liberal democratic ideology, but on the contrary to deepen and expand it in the direction of a radical and plural democracy. (176)

There are a number of difficulties with the theoretical perspective and political strategy advanced by Laclau and Mouffe. First of all, Laclau and Mouffe seem to take for granted that a recognition of the mediation of human social relations through language necessarily implies the rejection of an "essentialist" epistemology, ontology, and ethics. This is not so. It is equally possible to regard human language and communication as simply a complex manifestation of the relationality which characterizes all matter, and which is, if you will, its very "essence"——a view which, I have argued above, is increasingly born out by scientific experiment (Davies 1988: passim). Without any analysis of the logic of cosmos or human development, Laclau and Mouffe can provide us with no objective ethical ground for the democratic values they claim to uphold. If there is no underlying logic of society, no *telos* towards which human society, and the cosmos generally, are developing, or if this *telos* is not knowable to human reason, then moral norms are purely and simply a crystallization of power relations. On what grounds should we prefer a hegemonic formation which unites antiracist, antisexist and anticapitalist practices, over one which is racist, sexist, and capitalist?

Second, their analysis of the historical sources of resistance to oppression leaves a great deal to be desired. We will see that human beings have resisted oppression since the very dawn of civilization. Recent research (Gottwald 1979) has documented peasant revolts as early as the Late Bronze Age. Furthermore, this resistance has been legitimated historically not by a conviction that human society is without remainder a human product, but rather by a profound sense that existing social arrangements are out of accord with an objective moral order sanctioned either by divine will or natural law. And the historic success of social movements, in the long run at least, has depended more on the extent to which their goals corresponded with objective historical imperatives, than on their ability to stitch together short term political alliances. Social movements, furthermore, are not simply movements *against* oppression, but movements *for* the development of human social capacities.

More specifically, Laclau and Mouffe seem to have a very flat understanding of the radical democratic impulse which is so central to their analysis. The great aspiration of radical democracy, as opposed to classical liberalism, is to simultaneously expand the human capacity to organize and reorganize social institutions, and to include ever larger numbers of people in this reorganizing process——not, simply, to resist subordination. Indeed, the exercise of power, if it is to be relational, reciprocal, and democratic, requires a willingness to be affected as well as to affect others. It is at this, deeper level——the level of the developing human capacity to organize and reorganize human social relations——and not at the more superficial level of "resistance to oppression" that the radical democratic impulse flows into socialism.

Finally, as important as this radical democratic impulse is to the socialist movement, it is only one stream among the many that flow into socialism. The scientific impulse to know the cosmos, the technological impulse to organize matter in ever more complex ways, the religious project of developing human spirituality to the point where men and women are actually interested in the development of the cosmos as a whole——these projects as well are all part of the larger socialist process.

2. A Synergistic Analysis of the Crisis of Socialism

We are now in a position to summarize our own analysis of the crisis, which is already implicit in our analysis of the history of the socialist movement. The socialist movement has failed for two reasons. On the one hand, human organizing capacities have not developed to the point required for communism. Industrial forces of production provide an insufficient material basis for the realization of the historic aims of the socialist project. Industrial production exploits matter, physical, chemical, biological and social, merely displacing complex organization from one locus to another. The industrial worker is never a fully conscious participant in the self-organizing activity of the cosmos. Even under the best conditions he carries out production plans developed by scientists and engineers who understand far better than he the imperatives of the production process. It is hardly surprising that it should have proved so difficult to make the industrial working class the conscious subject of the historical process.

Second, the penetration of market relations into every sphere of life gradually erodes the social fabric, and makes every activity simply a means of realizing individual consumer interests. This means that, as Lenin himself pointed out, the working class never develops spontaneously beyond the point of trade unionism. Indeed, as capitalism develops, even trade union solidarities become difficult to sustain.

Taken together these two dynamics lead to serious economic, political, and ideological deformations. At the economic level, socialism has tended, implicitly, if not explicitly, to set for itself the goal of maximizing the production of value (complexity understood only in its quantitative aspect) rather than promoting the all sided development of human social capacities. Concretely this has taken the form either of social democratic "collective consumerism" or of Stalinist "production for its own sake." While both of these goals are more advanced than the capitalist aim of maximizing profit, they are at best rather distant approximations of the authentic *telos* of human productive activity, i.e. contributing to the development of the cosmos to increasingly complex levels of organization.

At the political level, socialism has tended to regard the masses from an industrial point of view, as an inert mass without any dynamism of its own, which must, and can, be molded and shaped into a new socialist humanity. Like the industrial mode of production generally, this approach can provide the basis for extraordinary advances in the level of human social organization. The Communist Parties represent humanity's highest organizational achievement to date, and the Communist International the planet's first authentically global political organization. At the same time, this linear approach to power makes it impossible to tap into, and thus to actually develop, the organizing capacities of the masses themselves, who, not surprisingly, never advance to a fully communist level of development.

At the ideological level, socialism has been characterized by an underlying contradiction between its atheistic ontology and its dialectical, teleological sociology. The claim that human society can be grasped as a whole, and that human history develops necessarily towards an ultimately meaningful *telos* is central to the entire socialist project. At the same time, dialectical materialism has taken a rigorously atheistic position. We have seen in earlier chapters that "God" is essentially a representational form of the complex, self-organizing character of the cosmos as a whole. To deny God, therefore, is to deny the self-organizing character of the cosmos——i.e. to deny that the cosmos is an organized totality developing necessarily towards an ultimately meaningful *telos*. To uphold a dialectical sociology, while rejecting a dialectical ontology effectively reduces humanity to an isolated island of meaning in a larger ocean of chaos. Human organizing activity ultimately serves only human consumption interests. It is not an investment in the future of the cosmos.[4]

Three points are in order here. First of all, this orientation is a reflex of the larger industrial problematic of which socialism is a part, for industry always and only breaks down and displaces complex organization, releasing energy which allows humanity to reorder nature in accord with its own advanced but still limited and particular imperatives. And it reflects the meditation of industrial production through the marketplace, which makes all activity into simply a means of realizing consumption interests. Second, this orientation in fact reinforces the industrial-market problematic, by requiring all organizing activity to be legitimated in merely human, consumer terms. Finally, atheism, by denying the ultimate meaningfulness of the cosmos, makes it increasingly difficult to sustain faith in the

---

[4]. This is true even if we accept a "dialectics of nature" which regards human labor power as the grand culmination of millennia of cosmic evolution, but which denies the underlying ontological necessity of this cosmic evolutionary dynamic.

meaningfulness of human history. In practice, as we have seen, historic socialism has either reimported religious motifs covertly into its linking ideologies (the dominant pattern in successful socialist movements), or else drifted towards postmodernist nihilism. Neither approach is adequate to the task of cultivating among the people a rational grasp, and thus a rational love, of the self-organizing cosmos, and the capacity to contribute actively to its development.

The history of the socialist movement is largely the history of an unsuccessful effort to overcome these obstacles. Social democracy looked simply to the quantitative growth and/or the intellectual and moral development of the working class to generate socialist consciousness——an expectation which was disappointed as the corrosive impact of generalized market relations became clear. Leninism grasped more clearly the obstacles to the development of socialist consciousness among the broad layers of the working class. But neither Lenin nor any of his interpreters were able to develop an effective strategy for actually raising the masses to a fully communist level of development——either before or after the revolution. Instead, they developed a cluster of related strategies which permitted a core of professional revolutionaries to secure the support of the masses, and control of the "commanding heights of society" without effecting such a transformation. The various tendencies within what we have called the "third force"——populism, theomachy, and socialist humanism——were able to recover Marx's insights into humanity's drive for holism and the corrosive impact of market relations, and were able to develop a strategy for actually organizing and developing the working classes. But in the absence of an adequate ontological ground, even the powerful insights of the third force remained vulnerable to the disintegrating forces of the marketplace and to attack by nihilistic postmodernism.

III. Beyond Idealism and Materialism: Towards a Synergistic Synthesis

It should be apparent by now that neither objective idealism nor dialectical materialism represents a fully adequate "solution to the riddle of history." On the one hand, objective idealism, while providing a more or less adequate ontological ground for organization, was unable to grasp either the latent potential or the internal contradictions of the material world generally, and of the social form of matter in particular. Dialectical materialism, on the other hand, while grasping precisely the latent potential and internal contradictions of human society, left the process of organization ungrounded, and thus inadequately theorized. The result has been the disintegration of both trends into various forms of "postmodernism," authoritarian and fideistic in the case of objective idealism, nihilistic and deconstructive in the case of dialectical materialism.

At the same time, it is important to recognize that it is not the principal positive claims of these two traditions which have been called into question, but rather their deficiencies and limitations. Specifically, it is still possible to affirm with objective idealism that there is an underlying organizing principle of which the cosmos as a whole is the expression. Human society is a relatively advanced expression of this principle, but by no means its final realization. This organizing principle, furthermore, provides the basis for defining principles of value on the basis of which we can make judgements regarding the beautiful, the true, and the good, and which guide us in our struggle to know and love the complex synergistic integrity of the universe.

Similarly, we can affirm with dialectical materialism that the potential for organization is latent in matter, and that matter realizes this potential through real material interactions, not by being formed from the outside. The social form of matter represents a real participation in the self-organizing activity of the cosmos, to which it contributes through the processes of socialization, production, political organization, and artistic, scientific, and philosophical creativity. While the cosmic potential for organization far transcends the capacities of finite social matter, it is, nonetheless possible to construct a social formation which realizes fully that potential to the extent possible given the limits of the social form of matter——i.e. it is possible to actually build a just social order. Certain social structures, specifically patriarchy, the warlord state, and the marketplace, constitute real obstacles to the full development of human social capacities and can and must be transcended.

Taken together these principles constitute the outlines of a synthesis which has existed for some time *in potentia* in the various "dialogues" between objective idealists and dialectical materialists, and in the strategic alliances which have emerged from these dialogues. Only rarely, however, has this synthesis been formulated explicitly, in a way which brings together the insights of both traditions.

There are, however, at least three figures who have done precisely this, and who have exercised a profound influence on the development of our own synergistic theory. Nicaraguan poet Ernesto Cardenal completes the process of religious rationalization which we have been tracing throughout this work, transforming the objective idealism of Social Catholicism into a materialist spirituality. Buckminister Fuller, a North American mathematician and engineer, has theorized the development of human productive capacities to an implicitly synergistic level, providing the real technological basis for the development of a postindustrial, postmarket society, something to which dialectical materialism aspired, but which it could never achieve. Feminist philosopher Mary Daly has liberated the scholastic tradition from its patriarchal deformations, and in the process laid the groundwork for transcending the contradiction between idealist and materialist dialectics. Let us examine each in turn.

A. The Cosmohistorical Vision of Ernesto Cardenal

Cardenal was born in Granada in Nicaragua in 1925. After studying literature at the *Universidad Nacional Autonoma de Mexico* and at Columbia University, he returned to Nicaragua and became actively involved in the struggle against the elder Somoza, a struggle which issued in the abortive rebellion of April 1956. Several of his comrades were captured and killed. Later in this same year he experienced a religious conversion. Cardenal had had a long series of love affairs, which, however, left him unsatisfied. "God simply showed himself to me as love ... In reality this was a thirst for the absolute, which human love could not satiate, but I didn't know it. (Cardenal 1992: xvi)" He entered the Trappist monastery at Gesthsemani, in the United States, where poet Thomas Merton was novice master, beginning several years of close collaboration. After two years he left, partly due to health reasons, partly because Merton was urging him on to a deeper understanding of his religious vocation. He spent two more years at the Benedictine Monastery of *Santa Maria de la Ressureccion* in Cuernavaca, Mexico, and finally completed his studies for the priesthood at the *Seminario de Cristo Sacerdote* in La Ceja, Colombia. The Benedictine tradition, with its vision of all human activity, from simple manual labor through the most complex forms of artistic, scientific, and philosophical creativity, as a kind of prayer, and its historically positive outlook on the human civilizational project, clearly left its mark on Cardenal's development.

After his ordination, he returned once again to Nicaragua, establishing a small religious community, *Nuestra Señora de Solentiname*, on the island of Mancarron in Lake Nicaragua. Cardenal concentrated his efforts on developing the creative capacity of the campesinos who lived around the community. Solentiname soon became a center for artists and writers——and also a center of resistance to the Somoza regime. Gradually Cardenal left behind Merton's commitment to nonviolence, and became convinced that armed struggle was the only road to liberation for his beloved Nicaragua. He made contact with, and eventually joined the *Frente Sandinista de Liberacion Nacional* (FSLN), carrying out important diplomatic tasks during the final stages of the struggle, and serving as Minister of Culture in the Sandinista government from 1979-1989. Solentiname itself was destroyed by Somoza's *Guardia Nacional.*

Cardenal's poetry represents an enormous achievement, both because of its formal innovations and because of the powerful vision which it articulates. On the formal side, Cardenal's greatest contribution has been the revitalization, one might even say the recreation, of epic poetry. The epic, of course, was the principal literary form of precapitalist societies. The subject of the epic is not (for the most part) some one individual hero, but rather a people as a whole, or even the whole universe, in the case of certain cosmogonic epics. In an epic, the cosmos appears as an organized totality, in the context of which individual and collective action is endowed with a definite meaning and value. At the same time, human action appears to be relatively insignificant, against the larger background of divine action or cosmic evolution which frames the epic.

In a market society, it was no longer possible to write epics of this kind. On the one hand, the industrial, democratic, and scientific revolutions vastly increased human organizing capacities, making humanity, for the first time, a real participant in the cosmic evolutionary process. At the same time, the emergence of a market economy undermined humanity's sense of the underlying unity of the cosmos, which appeared increasingly as a fragmented aggregate of only externally related atoms. Order and meaning, in a market society, are purely and simply an individual human product, ultimately doomed to be swept away by the tides of history. This is reflected in the form of the novel, which is the bourgeois literary form *par excellence.* The novel is first and foremost the story of an individual hero who attempts to make of his world an organized totality, whether through concrete social historical action or through some process of interior, psychological development. (On the social basis and political valence of the epic and the novel cf. Lukacs 1971).

Cardenal's poetry, especially his "documentary poems" such as *Hora Zero* and *Canto Nacional*, and his master work, *Cantico Cosmico*, mark a return to the epic form, but with a difference. Like the classical epics, Cardenal's poems are not about an individual hero, but rather about the history of his people, the history of humanity——indeed, about the history of the cosmos. But unlike the classical epic they do not push human action——even individual human action——to the margins. Cardenal's universe is one in which human civilization plays a critical, leading role, as a center for the creation of dynamic, complex organization. The worker, the organizer, the *guerilla* leader, the artist, the scientist, and the philosopher all contribute to this great cosmic drama. But then so, too, do two lovers embracing beneath the palm trees in a city park in Bangkok or Brasilia, or the mother nursing her child in a tar paper hut in Lagos or Lima. For Cardenal's epics are also, in their own way, lyrics and love songs.

In this sense Cardenal's poetry represents our first authentically postcapitalist literature. The great realist novels, even those which fully implemented the program of socialist realism, such as the novels of Gorky or Silone or Malraux, are still irreducibly bourgeois in form. They are still stories about individual heros. In Cardenal we find once again the story of peoples, of planets, of the universe, enriched, however, by the recognition, which was absent from the classical epic, of the vital importance of human creativity to the cosmohistorical evolutionary process.

This brings us to the question of Cardenal's specific vision. Ernesto Cardenal is generally classed with the theologians of liberation, largely because like them he arrived at a position of support for the national liberation movements and socialism from within the Christian tradition. This assessment of Cardenal is not, however, really accurate, and fails to grasp the full significance of his achievement. Liberation theology, even when it fully embraces Marxism as a method of social analysis and opts for social revolution and armed struggle, remains largely within a still fundamentally Christocentric problematic. The struggle for national liberation and socialism is affirmed as an integral part of salvation history——i.e. an extension of Christ's saving work on the cross——not as a stage in the realization of the self-organizing dynamic embodied in creation itself. The progressive political valence of liberation theology notwithstanding, this leads to serious errors in revolutionary practice: a tendency to lapse into a glorification of redemptive suffering, and a tendency to invest the party, revolutionary heroes or martyrs, with substitutionary messianic charisma.

While Christocentric motifs are not entirely absent from Cardenal's work, his underlying political-theological problematic is quite different. Cardenal recognizes the universe as a complex, self-organizing totality. This totality is present at first only in abstract form.

In the beginning
    --before space time--
  was the Word
Everything which is, therefore, is truth.
            Poem.
Things exist in the form of the word (Cardenal 1989: 25).

Matter *is* organization, and develops towards increasing levels of complexity, physical:

> The universe is made of union.
>   The universe is condensation ...
> Condensation, union, this is the stuff of which the stars are made.
> The Law of Gravity
>   *che muove il sole e l'altre stelle*
> is an attraction between bodies, and this attraction
> increases when the bodies move closer together...(Cardenal 1989: 253)

chemical:

> No electron wants to be alone (Cardenal 1989: 253)

and biological:

> It rained during the night and the toads are croaking in the moonlight,
>   singing for the females, calling them to copulation (Cardenal 1989: 254).

> What a beautiful species, how I love it
>   every individual born from copulation
> born for love (Cardenal 1989: 257).

The social form of matter represents a particularly advanced manifestation of this process.

> Conditions emerged which permitted the development of organisms
> and then of organisms with consciousness, persons; and then
> an organism which is at once community and persons (Cardenal 1989: 109).

Life generally, and human beings in particular, play an active role in the development of the cosmos towards increasing levels of complex organization.

> Life evolves not by adapting itself to its environment as Darwin thought,
> but by creating the environment ...

> ... human beings emerged from their caves
> and started to build cities.
> Always superior forms after inferior ones
> in organization and structure (Cardenal 1989: 111-112).

The difficulty is that human social structure, and capitalism in particular, has become an obstacle to this process of development.

> Competition impedes cooperation.
> There is a separation between one person and another.
> A broken humanity.

> A riot in Malaysia over devaluation, buses burned
> and blood running in the streets like water from a hydrant.
> At the hour when the stars shine over Wall Street,
> At the hour when the banks open in London (Cardenal 1989: 254-255).

Social revolution thus represents an integral part of the cosmic evolutionary process.

> The more violent the perturbation,
> the more intense will be the condensations
> but even the most insignificant one develops
> condensations even though they are of extremely weak intensity ...

> However weak their original intensity may have been
> the great condensations gradually become greater
> and greater, and the small ones disappear, absorbed
> by the greater ones, and finally there is left only a collection
> of enormous condensations. Like them are the phenomena we call
> socializations, and like them is
>   the Revolution ...

> Capitalism will pass away. You will no longer see the Stock Market.
> "As sure as spring follows winter ..." (Cardenal 1989: 255-256).

Ultimately, this revolutionary process issues not only in a new society, "Communal and personal, classless and stateless," but "a new humanity, with new chromosomes, (Cardenal 1989: 256)." This new humanity participates at a still higher level in the ongoing evolution of the cosmos towards its infinitely self-organizing Omega point.

We should, perhaps, note in closing, that while Cardenal has been a loyal and committed member of the FSLN, his vision in many ways over-reaches even the advanced political line of the *frente*. During the middle and late 1980s there was growing political controversy over the poetic workshops organized by Cardenal's Ministry of Culture. These workshops, which grew out of Cardenal's experience at Solentiname before the revolution, attempted to tap into the experience and creative potential of the workers and peasants to create a popular revolutionary culture. They formed part of a larger effort to tap into the communitarian religious traditions of the peasantry and proletariat in order to catalyze the formation of a new revolutionary consciousness. The program fell out of favor with Rosario Murillo, wife of Nicaraguan President Daniel Ortega's and leader of the FSLN cultural apparatus, who felt that the poetry being produced was insufficiently cosmopolitan, too political, and did not help the FSLN in its struggle to present itself as a mainstream social democratic party and to garner international support (Zimmerman 1990).

Important strategic questions were at stake in this struggle: i.e. the relative importance of building relationships with allies who can provide needed financial and political support and the task of developing human social capacities, and more specifically the task of developing the ability of ordinary workers and peasants to exercise the poetic/prophetic office with which socialism endows them——the ability of young men and women to dream dreams and of old men and women to see visions. Eventually these and related contradictions led Cardenal to leave the FSLN.

Cardenal's work clearly represents an enormous poetic and theological achievement. Cardenal has recreated, on a new and higher level, the ancient form of the epic poem. He has carried out a complete rationalization of the religious tradition, leading to a grasp of and love for the cosmos as self-organizing totality, and, at the same time, has retheologized the socialist project as an integral part of a much larger process of cosmohistorical evolution leading beyond socialism, and even communism, to the emergence of new species more beautiful than any we can now imagine, and ultimately to consummation in an infinitely self-organizing Omega. In this sense, Cardenal transcends both Christianity and socialism and achieves an implicitly synergistic perspective.

B. Buckminster Fuller and Synergetics

It may seem a bit peculiar to include Buckminster Fuller in a discussion of tendencies within the socialist movement which point implicitly towards synergism. Fuller, after all, does not seem to have regarded himself as a socialist, and, on the contrary, adopted an explicitly apolitical stance. Perhaps more than any self-conscious Marxist, however, Fuller reproduces and raises to a higher level the historic aspiration of the social democratic trend to set in motion a process of human technological development which will, apart from any political action, lead humanity beyond capitalism to a new, higher form of social organization.

At the core of Fuller's vision, like Cardenal's, is the conviction that the cosmos is an organized, evolving totality. Fuller believes that he has grasped the organizing principle of this totality and rendered it mathematically transparent in his two volume *Synergetics*. Specifically, he argues that the structure of the cosmos can be resolved into a complex system of interrelated tetrahedrons, the inter-relationship of which defines the various forms of matter, physical, chemical, biological, etc. This organizational principle, implicit in the universe, Fuller identifies with God.

> I constantly ask myself "Do you have any experientially evidenced reason to assume a greater intellect to be operating in Universe than that of humans?" ... I found that I was overwhelmed by the experiential evidence of a cosmic intellectual integrity at work in the design of Universe. Thus, when I said in 1927 that I was going to try to find out and support what the great cosmic intellectual integrity was trying to do, I committed myself as completely as humans can to absolute faith in the wisdom of the eternal intellectual integrity we speak of as God (Fuller 1992: 259).

Human beings play a very specific role in the Universe.

> That the human mind has been designed to apprehend, to comprehend mathematically, and to express intellectually eternal-Universe design interrelationships and——even more——to employ these interrelationship principles in specially formulated objective use cases as micro-macrostructures and mechanisms informs us that human beings have indeed been designed and developed for cosmic magnitude functioning. To discover whether this terrestrial installation of humans and their minds will lead to the fulfillment of this cosmic functioning, all human individuals are now entered upon their final examination (Fuller 1992: 40).

Cosmic evolution is

> irrevocably intent upon completely transforming omnidisintegrated humanity from a complex of around the world, remotely deployed-from-one-another, differently colored, differently creedoed, differently cultured, differently communicating and differently competing entities into a completely integrated, comprehensively interconsiderate, harmonious whole.

> ... cosmic evolution is also irrevocably intent upon making omni integrated humanity omnisuccessful, able to live sustainingly at an unprecedentedly higher standard of living for all Earthians than has every been experienced

by any; able to live entirely within its cosmic-energy income instead of spending its cosmic-energy savings account (i.e. the fossil fuels) or spending its cosmic-capital planned and equipment account (i.e. atomic energy ... a spending folly no less illogical than burning your house-and-home to keep the family warm on an unprecendentedly cold midwinter night (Fuller 1981: xvii).

Realization of the full potential of humanity requires what Fuller calls a "design science revolution," centered on a transition to solar energy and the adoption of his "synergetic" technology. Humanity is being prevented from taking these steps by what Fuller calls the "power structure."

Tax hungry government and profit-hungry business, for the moment, find it insurmountably difficult to arrange to put meters between humanity and its cosmic-energy income, and thus they do nothing realistic to help humanity enjoy its fabulous energy-income wealth (Fuller 1981: xvii).

Fuller charts in considerable detail the role of the power structure in obstructing human progress from the time of early navigator-priests, who first began to grasp the organizing principle of the universe, only to find their knowledge co-opted by strong men and kings, up through the present period. One of the most potent tools in the arsenal of the power structure is specialization.

The way the power structure keeps the wit and cunning of the intelligentsia——who are not musclemen, who cannot do the physical fighting——from making trouble for the power structure (if the intelligentsia are too broadly informed, unwatched, and with time of their own in which to think) is to make each one a specialist with tools and an office or lab. That is exactly why bright people have become streamlined into specialists (Fuller 1981: 62).

In spite of his recognition of the obstacles presented by the power structure, Fuller argues against political action.

In reviewing the full range of humans' presence on Earth we discover two main evolutionary trendings.

Class-two evolutionary trendings are all those events that seem to be resultant upon human initiative-taking or political reforms that adjust to the changes wrought by the progressive introduction of environment-altering artifacts. All the class-two evolutionary events tend to flatter human ego and persuade humanity to deceive itself by taking credit for favorable changes ... It therefore assumes that humanity is running the Universe wherefore, if its power-structure leaders decide that it is valid to cash in all of nature's available riches to further enrich the present rich ... that is the power-structure leaders' divine privilege.

All the class-one evolutionary trending is utterly transcendental to any human vision, planning manipulation, and corruption. Class-one evolution accounts for humans' presence on Earth. It accounts for ... humanity's all-unexpected invention of verbal communication, and thereby the integration of the experience-won information of the many ... (Fuller 1981: 229-230).

Fuller seems to expect that technical innovation will eventually triumph over the small mindedness of the power structure, and render them socially irrelevant.

The result of this process looks quite a bit like communism. In place of the marketplace Fuller proposes a systems of "cosmic cost accounting" based on energy use.

Only cosmic costing accounts for the entirely interdependent electrochemical and ecological relationship of Earth's biological evolution and cosmic intertransformative regeneration in general. Cosmic costing accounts as well for the parts played gravitationally and rationally in the totality within which our minuscule planet Earth and its minuscule Sun are interfunctionally secreted. Cosmic costing makes utterly ludicrous the selfish and fearfully contrived "wealth" games being ... played by humanity aboard Earth (Fuller 1981: 119).

If humanity realizes its potential in time to exercise this vital option, we shall witness strong competition among individuals to be allowed to serve on humanity's research, development, and production teams. Never again will what one does creatively, productively, and unselfishly be equated with earning a living. People's sense of accomplishment will derive from showing their peers and demonstrating to the great intellectual integrity of Universe, which we speak of as God, that they vastly enjoy doing their best in the unselfish production of service for others rather than just for the survival needs of themselves and their families (Fuller 1992: 252-253).

Fuller presents a vision of human social evolution as part of a larger cosmic process. His understanding of social evolution, while clearly centered on technological evolution, is at the same time clearly postindustrial. Fuller's synergetic technology does not merely break up various forms of matter in to their component parts and then reorganize them in accord with some design idea. Rather, it grasps, and taps into, the organizing principle of matter itself, and uses this principle to catalyze the development of increasingly complex forms of organization. This is a critical contribution to the concept of synergism.

At the same time, it is necessary to note that Fuller's confidence in the spontaneous process of social evolution is naive, and reflects a failure to take into account the disintegrating influence of the marketplace. This failure should not, however prevent us, from drawing on his other contributions.

C. The Feminist Philosophy of Mary Daly

We noted at the beginning of this work the role of patriarchy in undermining humanity's archaic intuition of the universe as a unified, self-organizing, and teleological totality. The subjugation and eventual destruction of the matriarchal village community, the matrix through which humanity grasped the universe as an organized totality, undermined humanity's perception of organization as a potential latent in matter, and opened the way for the emergence of idealistic theories which regard order as something always and only imposed from the outside. It should thus come as no surprise that feminist theory has developed powerful tools for overcoming the contradiction between idealist and materialist dialectics.

Historically the socialist movements, following Engels, expected the advent of socialism and the abolition of private property to be simultaneously the liberation of women and the advent of a new era of equality between the sexes. This has not, alas, been the case. While the socialist countries posted notable gains in the area of women's liberation, they have fallen far short of really undermining the basic structures of patriarchy. Indeed, most of the gains in question——an improvement in the legal position of women vis-a-vis their fathers, husbands, etc., and a massive expansion of their role in production and public life——can be attributed to the dynamic of industrial rationalization. Developing countries need the skill and labor of their women and cannot afford to pander to antiquated prejudices. In this sense, the achievements of the socialist countries in the arena of women's liberation represent simply an extension or completion of the gains posted in the advanced capitalist countries.

The radical transcendence of patriarchy can only be the product of an independent women's liberation movement. The womens' movement which emerged in the nineteenth century and which is now in its "third wave" is a complex phenomenon and embraces a wide range of different tendencies: liberal feminists who are seeking simply to apply consistently the prevailing norms of bourgeois society, radical feminists who reject both liberalism and socialism and argue that the oppression of women is not only the primordial form of oppression, but also the principal contradiction of human society in the present period, and socialist feminists who seek to integrate the insights of radical feminism with a commitment to the socialist vision.

Liberal feminism doesn't really attack the underlying roots of patriarchy——indeed it lacks any notion of structural oppression. Among feminists, "radical" and "socialist," who are attempting to get at the underlying structures, three broad trends have emerged. The first, which is often termed "essentialist," but which might better be called "idealist," because of its tendency to uphold an idea, however revalued, of the "eternal feminine," draws an a growing body of archeological evidence, which we discussed in some detail in an earlier chapter, that at least some archaic societies, from the late paleolithic through the early bronze age, were in fact matricentric or even matriarchal in the strong sense of being ruled by women. These societies were economically egalitarian, or at least free of gross exploitation, and had a highly developed spirituality centered on an intuitive grasp of the underlying unity of the cosmos. This archaic holism was destroyed by invading patriarchal tribes who, over a period of thousands of years, overthrew the matriarchy and established the militaristic structures which eventually gave birth to the great empires of the ancient world. Archaic, matriarchal communitarianism, together with its religious expression in the cult of the goddess, is taken to provide the basis for constructing a new postpatriarchal social order in which not only the oppression of women, but also ecological rape and economic exploitation have been overcome (Stone 1976, Christ 1987, 1989, Gimbutas 1989a,b,).

Idealist feminism has been criticized on a number of grounds. First, questions remain about the character of the archeological evidence, though most of the counter arguments seem to me to be less arguments against archaic matriarchy as such than they are arguments against the idea that this matriarchy was some kind of primordial paradise wholly free of conflict and tension. Second, there is a tendency for idealists to rest a bit too easy with accounts of invasions by warlike patriarchal pastoralists or "Indo-Europeans" without giving any account as to how these invaders became patriarchal, or why, unlike the agrarian societies of Old Europe or the Mediterranean Basin, they never experienced archaic matriarchy. I have attempted to remedy this deficiency in my own account by stressing the role of ecological factors in explaining which societies turn towards warfare, and by stressing the role of warfare in the patriarchal-tributary transition.

The most telling critique of idealist feminism has to do with the way in which certain idealists move from the archaic matriarchal past to the postpatriarchal *telos* towards which we are called to evolve. Some idealists, especially but not only those influenced by Jung, tend to overlook the patriarchal distortions embedded in the very feminine images they wish to resurrect. As Rosemary Ruether quite correctly points out, matricentric religions which cast men in the role of the divine son of the *Magna Mater* are hardly an adequate basis for constructing a society in which men and women treat each other as mature adults. Indeed, this idealized mother/son relationship is just one of the many male fantasies generated by patriarchal society. And neo-Jungian idealism is far worse, resurrecting what amount to stale patriarchal stereotypes which cast women as "connected to the earth," and "nurturing," and men as essentially "rational," "oriented towards power," while simply reversing the values attached to these stereotypes (Ruether 1992).

This said, it is, however, important to point out that many idealist feminists are engaged in important and useful historical and socioreligious research, and that unlike certain other trends, they take seriously the need to develop an ontologically and cosmologically grounded moral and spiritual vision for the future.

The second trend, which has its roots in psychoanalytic and socialist theory, has tended to locate the origins of women's oppression very early in human history. Shulamith Firestone (1971), for example, has argued that it is childbearing itself which made women vulnerable to male domination——and that authentic liberation is possible only on the basis of artificial reproduction. Gayle Rubin (1975) locates the origin of women's oppression in the "traffic in women" which characterized the kinship structures of tribal societies——even those which are matriarchal. This traffic required internalization of certain specific taboos——the incest taboo, but also the taboo on lesbianism which makes women "available" to men and thus transforms them into potential objects of exchange. While complex industrial societies are no longer regulated by kinship rules, we nonetheless bear the traces of socialization patterns which linger from our tribal past. The liberation of women presupposes the abolition of the taboo on lesbianism and indeed of the entire "sex-gender system."

This kind of feminism, which employs an essentially dialectical and materialist method, even when it is profoundly critical of actually existing socialism, has made important contributions to our understanding of the dynamics of patriarchy. Like most dialectical materialist theory, however, it lacks a firm ontological ground, something which is expressed at the sociological level in the absence of any clear conception of what women and men are like (logically and historically) "prior" to patriarchy. The result is an inability to develop a morally compelling vision of a postpatriarchal future. Indeed, like dialectical materialism generally, much radical and socialist feminism has gradually degenerated into a deconstructive postmodernism which is entirely without moral compass, and which lacks any ground for condemning patriarchy in the first place.

The third form of feminism is associated specifically with the work of Mary Daly. Often incorrectly classed with the idealists, Daly has in fact developed both an analysis of patriarchy and a compelling vision for the future which transcends the limitations of both idealism and materialism——and thus contributes powerfully to the emergence of a new synergistic vision.

At the very core of Mary Daly's system is the concept of "active potency."

> In Aristotelian and medieval philosophy, *potency* can mean either passive potency, or capacity to receive a perfection (called "act"), that is capacity to be acted upon, or it can mean active potency, meaning capacity to act. To put it another way, whereas passive potency is a capacity to receive something from something else, active potency is the ability to effect change. It is power (Daly 1984: 166).

Active potency, in other words, is nothing more or less than the self-organizing capacity of matter——matter realized as the potential for organization.

Patriarchy *consists* for Daly first and foremost in the repression of active potency and its displacement by its passive cognate——though "advanced" patriarchy declines to allow women to even receive perfections.

> Clearly phallic lust assigns/confines women to the arena of passive potency, as vessels/vehicles for males, who have reserved what they believe to be active potency for themselves. Moreover, its intent is the castration of women both on the level of passive potency and of active potency (1984: 166).

The repression of active potency is effected by the whole complex of methods which Daly analyzed in her *Gyn/ecology*: Chinese footbinding, Indian suttee, African genital mutilation, European witchburnings, and American Gynecology. All of these practices at once deform and stunt the growth and development of women in the name of a dead ritualized orderliness——something quite different from a living organization rooted in active potency——and promote a necrophilic aesthetic and spirituality

(Daly 1990).

The repression of active potency has both intellectual and affective effects. On the intellectual side the principal effect is the breakdown of "elemental reasoning" which can grasp the underlying organizing principle, the structure, and the final cause or *telos* of its objects.

> Crone-logically it is clear that the breakdown of Elemental reason is connected with the breakdown/cover-up of female Presence ....

> This removal from the philosophical enterprise of intuitive/imaginative reasoning ... is connected ... with the "philosophical" discrediting and erasure of "final causality" (purpose) and with the reduction of the profound meaning of final causality——which I would describe as unfolding of be-ing——to mere reified "goals" and "objectives" (1984: 154-155).

The imposition of "goals and objectives," whether by the late bronze age warlord or the modern corporate executive is, of course, not the same thing as realizing the latent potential of human society. That is why Daly says that men reserve to themselves "what they imagine to be active potency." In fact, however, patriarchal power is a mere counterfeit.

On the affective side, the repression of active potency undermines the operation of the passions. Historically

> passions were understood to be *movements* of a faculty known as the "sensitive appetite," which tends towards the good and shrinks from the evil as perceived by the senses. Essential to this analysis is the idea that passions are *caused* by something that is perceived. They are movements rooted in knowledge and not static, inexplicable blobs of "feeling" (1984: 198).

With the repression of reasoning about final cause, however, the passions are cut off from their *telos* and become "mere emotion": something to be "dealt with" rather than something which moves us towards the good, and helps us to avoid evil.

Repression of active potency, by undermining our ability to reason about final causes and thus cutting our passions off from their objects, makes authentic virtue impossible. Drawing again on the Thomistic tradition, Daly notes that

> a virtue is understood to be a good operative habit that exists in a power of the soul ...

According to this philosophical tradition, moral virtues are deeply connected with the passions. Virtues are not identical with passions, however, for virtues are habits, not movements. They are 'principles of movements' (1984: 217).

Apart from reasoning about final causes it is impossible for these principles to emerge. What takes their place are pseudovirtues, which impose order from the outside.

Liberation from patriarchy, for Daly, is first and foremost a matter of recovering and realizing active potency.

In this true and radical sense feminism is a verb; it is female be-ing ... feminism as Realizing is constant unfolding process ... the radical ontological process of Realizing female Elemental potency ... the Moving that is *feminism* requires overcoming the possessed powers——engaging in battles of principalities and powers. This Battling is our great exorcism, and it is also our ecstasy, for it is the claiming of our principalities, the Naming/Actualizing of our powers. That is, it is creation ... (1984: 194).

What Daly has achieved is nothing short of a recovery and reconstruction of the whole philosophical tradition——one which conserves the gains which were made, even under patriarchy, by thinkers such as Aristotle and Aquinas, but which, at the same time, reaches back and undoes the ancient patriarchal distortions. And this reconstruction at the same time transcends the contradiction between idealism and materialism. She conserves idealism's doctrine of an underlying ontological ground, its doctrine of essence, and its capacity to reason about final cause. At the same time, Daly rejects the idealist claim that organization comes to matter from the outside, and she clearly names——even if she does not really analyze historically or sociologically——the social basis of this idealist deformation. Daly conserves——and in fact develops in powerful new ways——the materialist insight regarding the self-organizing character of matter itself, while rejecting sharply the materialist divorce from the ontological ground, a divorce which leads inevitably into postmodern nihilism.

There are, to be sure, limitations. Daly engages the Thomistic tradition brilliantly, and one wishes she had done the same with the Hegelian-Marxist tradition. And her innocence of Marxism means that some of the more obvious links——between patriarchy and the warlord state, or later patriarchy and capitalism——links which Engels already sketched, but which Daly's work enables us to understand so much better——are never made. Indeed, there is very little here that can really be called sociological.

These limitations notwithstanding, Daly has clearly laid the groundwork for a new philosophical synthesis——a synthesis which synergism hopes to provide.

***

It should be apparent at this point that the crisis of both objective idealist and dialectical materialist movements notwithstanding, the potential of the dialectical tradition as a whole is far from spent. On the contrary, we stand on the threshold of a new era. We need now to draw the lessons of the current crisis and chart the next steps in the human civilizational project.

CHAPTER SEVEN:
The Next Steps in the Human Civilizational Project

The crisis of socialism, in both its idealist and its materialist forms, has created a fundamentally new vantage point from which to comprehend the "line of march, conditions, and ultimate general result" (Marx 1848/1978:484) of the cosmohistorical evolutionary process. Far from demonstrating that the universe is a chaos, into which order can be introduced only by means the decrees of a transcendent God, inaccessible to human reason, or through the historically conditioned activity of human societies, this crisis in fact has shown the fatal limitations of all movements which lack an adequate link to the ontological ground——to the self-organizing cosmos in which we live and think and move and have our being. It is only on the basis of rational knowledge of the universe as a self-organizing totality that humanity can discover and carry out its own vocation: to serve as a center for the creation of organized complexity in a general system which includes but also transcends the human historical process and the social form of matter generally. It is just precisely this kind of perspective which we outlined in the first chapter. We are now in a position to look at the historical process as a whole, to situate it in the context of our larger philosophical system, and to draw out the implications of our analysis for the present period.

Being itself is fundamentally relational in character. This underlying relationality means that not only the behavior, but the very essence of things, and not only their essence but their very existence, is dependent on and determined by their relationship to everything else. The universe is a general system of relations from which particular subsystems can be abstracted, but which exists, and can be truly known, only as a totality.

The essence or nature of each particular subsystem, as well as that of the system as a whole, is determined by the structure of its internal and external relationships, which makes it what it is. Each of these particular subsystems expresses more or less adequately the relationality which defines the system as a whole, but none can do so perfectly, and none can, by itself, account for the fact that there is something rather than nothing, cosmos rather than chaos. Only the structure of the general system can account for this, because only the structure of the general system can ground organization.

By organization we mean a system of relationships structured in a purposeful manner. Particular systems may be structured purposefully, but they have their purpose outside of themselves——the cell in the organ, the organ in the body, the individual body in the larger ecosystem or community. Only the general system has its purpose in itself, because its purpose *is* organization. The general system is the *ens realissimum* of the scholastics, that being whose essence it is to be.

The activity of this system is nothing other than the organizing process which brings all particular systems into being, and which thus realizes the principle of organization (its own principle) in ever more complex ways. The end or *telos* of this process is the whole, but since authentic holism by its very nature comprehends its own opposite, the process continually gives rise to new, more complex and more highly organized forms, so that the general system is at once always whole and complete in itself, but at the same time ever changing, growing, becoming...

Now in order to *know* being we must grasp its structure. This structure is, first of all, the purely formal structure of the logical system which governs the relations between all logically possible propositions. This formal system, is, however, predicated on the existence of propositions and thus of universals, the conditions for the existence of which can be comprehended only on the basis of a transcendental reflection. This transcendental reflection, in turn, demonstrates that universals are possible only if, and to the extent to which, the particulars which they cover are not really particular at all, but rather specific determinations of an architectonic organizing principle, the idea of organization itself, from which all forms of organization can be derived.

Matter is organization *in potentia*. This latent potential for organization is realized not by *taxis*, i.e. by receiving form imposed from the outside, nor through the process of exchange or *katallaxis* favored by the neoliberals, but rather through a complex process of *cosmogenesis* which involves logical implication, material interaction, and teleological attraction. Logically possible forms of organization, however simple, aspire necessarily to wholeness and totality. And yet as they try to realize this aspiration they come up against their own limited nature. This is most fully apparent in their interactions with other systems which differ from them, at once demonstrating their inadequacy and opening up the possibility of growth and symbiosis. Spirit is nothing other than the realized potential for organization.

At the simplest level matter is the metric field, the space-time continuum. But space-time necessarily implies gravity, the strong and weak nuclear forces and electromagnetism. This latter force, in turn, implies chemical organization, chemical organization biology, and biology the social form of matter. Each new level of organization is more complex than that which preceded it. Physical matter is merely ordered but lacks the more complex attributes of holism, organization, and creativity. Chemical matter is distinguished from the merely physical by its holism. The molecule has distinct properties which make it more than merely the sum of its parts. Biological matter is distinguished from the merely chemical by the appearance of organization in the narrower sense of structures which serve specific purposes.

The social form of matter is distinguished from merely biological systems by its capacity to develop increasingly complex organizational forms. Where biological organisms merely reproduce themselves in forms which are qualitatively identical, social formations continuously bring into being new personalities, create new technological artifacts and new kinds of technology, build new organizations and new forms of organization, develop works of art and new artistic styles, new scientific knowledge and new theoretical problematics, new philosophical knowledge and new philosophical systems. The social form of matter, therefore, unlike lower levels of organization, by its very nature continuously adds something qualitatively new to the cosmos. Human society is a center for the creation of dynamic, organized complexity. This is the cosmic significance of the human civilizational project.

The driving force behind this process of development is human labor, understood as the capacity to reorganize matter in increasingly complex ways. But the social form of matter is capable of reorganizing not only physical, chemical, and biological matter (technology), but also of reorganizing itself, that is reorganizing human social relationships (the exercise of power), of organizing information in accord with its own immanent principles (producing art, science, and philosophy). Taken together these processes represent a real participation in the self-organizing activity of the cosmos, and thus a real process of spiritualization.

Social systems are regulated by various structural rules which determine just how these basic organizing processes unfold. Out of the primordial, undifferentiated unity of the band, certain distinctions gradually emerged——distinctions which became the basis of tribes, clans, and lineages——together with rules governing the relations between these groups. The result was the kinship system which regulated the organizing activities of tribal societies. Kinship relations proved inadequate to regulate the increasingly complex productive activities of advanced horticultural societies, dependent as they were on the knowledge of the motions of the heavenly bodies, of climate and weather, of the cultivation of plants and the domestication of animals, and thus gave way to village community structures which could support an emerging stratum of elders, teachers, and priests. The resulting communitarian social formations, many if not all of which seem to have been matriarchal, catalyzed rapid development of human social capacities, and provided a basis in experience for humanity to begin to comprehend the cosmos as an organized, meaningful totality.

At a certain point, however, probably around the time of the discovery of metals technology, some communities discovered that they could augment their consumption levels more easily by conquering other communities, and extracting tribute from them in the form of rents, taxes, and forced labor, than they could by investing their small surplus in scientific and technological research and gradually raising the level of productivity. The result was the emergence of the warlord state. At the same time, the advent of warfare vastly increased the power of men in human society, and led to the overthrow of the archaic matriarchy and the establishment of a patriarchal regime in which order was no longer regarded as an immanent potential of matter but rather as something imposed on matter from the outside. The warlord state held back the further development of human social capacities not only because it squandered resources on warfare and luxury consumption, but also because it promoted a distorted understanding of the organizing process itself. Increasingly organization was regarded as something imposed on matter from the outside, as by a conquering king. The people themselves participated in this process not through their creative labor, but rather through suffering and self-sacrifice——through the payment of rents, taxes, and forced labor to warlord "organizers." Increasingly human society began to seem out of harmony with the divine cosmic order of the universe.

The great salvation religions were fundamentally an attempt to restore these ancient harmonies. Some, such as Taoism, the Hopi religion, and the cult of the *Magna Mater* simply attempted to recall people to the lost "way" of the village community. While in a certain sense the purest expression of humanity's drive to "restore the mandate of heaven," these movements lacked either the will or the capacity to mount a real contest for power with the warlord state, and thus survived only in remote regions inaccessible to the predations of the warlords (the Hopi) or only as an undercurrent of resistance within civilizations which continued to be dominated by the warlords (Taoism, the cult of the *Magna Mater*). Other religious movements, such as Hinduism and Buddhism, sought to achieve harmony with the cosmic order by means of withdrawal from the world, which was regarded as a realm of illusion, into the undifferentiated unity of the One (*Brahman*), or into that enlightened state in which the self realizes its own radical dependence on all other things, and thus its own ultimate unreality (*Nirvana*). Such movements offered a path to salvation for individuals, but could not sustain a consistent regime of world building or world transformation. Finally, a third cluster of movements (Hellenic philosophy, Judaism, Christianity, and Islam), sought wholeness in knowledge of and obedience to the law immanent in the universe itself and accessible to reason, or revealed by a transcendent God, creator of heaven and earth. Knowledge of the law provided a basis for reorganizing human society and thus restoring the ancient harmony.

This last cluster of movements was characterized from the very beginning by a profound ambiguity. On the one hand, Hellenic philosophy and Judaism integrated both a belief in the meaningfulness of the material world, and a profound sense of the chaos and disintegration at work within it. The result was a uniquely powerful dynamic of social transformation. At the same time, precisely because they felt so intensely the contradiction between the law in all its beauty, and the world in all its pain, these movements had a tendency to regard organization as something imposed on matter from the outside, by an extramaterial prime mover, or by a transcendent creator God. They thus reproduced within themselves the patriarchal idealism characteristic of the religions of the warlord states. This tendency was exacerbated, in the case of Islam and Christianity, by a tendency to conceive of the organizing process as first and foremost a process of military conquest or expiatory self-sacrifice.

These limitations notwithstanding, the salvation religions collectively were able to contain the predations of the warlord states sufficiently to once again unleash the development of human social capacities. Gradually people began to unlock the laws which governed the organization of matter, and learned how to use this knowledge to reorganize matter in new and increasingly complex ways. The industrial, democratic, and scientific revolutions represented an enormous leap forward in human organizing capacities, making humanity for the first time an authentic participant in the self-organizing activity of the cosmos.

Once again, however, human civilization ran up against serious obstacles. Mediation of human organizing activity through the marketplace makes all relationships simply a means of realizing individual consumer interests, concealing the social nature of humanity, which thus appears to be radically depraved, or at least irreducibly individualistic. Individuals feel themselves to be at the mercy of forces beyond their control. The self-organizing activity of the cosmos appears under the guise of market imperatives, as the inscrutable plan of a transcendent sovereign God. Eventually the operation of market forces begins to erode the social fabric and to undermine the social basis for humanity's understanding of the cosmos as an organized totality. The sovereign God of the Reformed tradition gives way to the "process without a subject," the spontaneous process of evolution "more transcendent than God," posited by contemporary postmodern and neoliberal theory (Lyotard 1984, Hayek 1988).

It was the great contribution of the dialectical tradition (Hegel 1967a,b, 1969, 1971, Marx 1844/1978) to grasp the industrial, democratic, and scientific revolutions as themselves simply the highest expression of the self-organizing dynamic which had been inherent in being from the very beginning, and to recognize the alienation generated by the marketplace as merely a vanishing moment which would ultimately be transcended. The low level of development of human organizing capacities, however, continued to hold back the realization of humanity's drive towards synergism, and in fact led to serious distortions in the movements which flowed out of the dialectical tradition. The science of the nineteenth century was dominated by an atomistic paradigm which did not support Hegel's dialectical intuition regarding the self-organizing character of the cosmos as a whole. This meant that human society was increasingly regarded as merely an island of meaning in an ultimately meaningless chaos. This tended to undermine the axiological authority of socialism, which was increasingly regarded as the meaning humanity gave to its own history rather than as the expression of an underlying drive towards organized complexity written into the fabric of being itself. Some sought an ontological ground outside of matter, and reverted to one or another form of idealism. Others attempted to rescue the ontological foundations of socialism by endowing humanity or "the historical process" with the full burden of authority historically carried by God, with results which were philosophically absurd and historically tragic.

Neither religious idealism nor dialectical materialism was able to grasp the self-organizing character of matter itself, and thus neither was able to develop a coherent strategy for transcending the contradictions of market society. The crisis of socialism, in both its "idealist" and its "materialist" forms, has nothing to do with the fundamental unworkability of the socialist vision, or with the tendency of "totalizing" ideologies to produce authoritarian regimes. It is, rather, a result of relatively minor deficiencies in philosophical perspective, deriving from the limitations of atomistic science, and therefore ultimately from the alienating impact of market relations, which in turn produced far more significant errors at the levels of social analysis, strategy, tactics, etc.

And so what remains? The answer is, quite literally, everything.

For the truth is that the socialist period[1] has been enormously productive for the development of basic human organizing capacities. Just as the salvation religions, their many limitations notwithstanding, unleashed human creativity and created the conditions for the industrial, democratic, and scientific revolutions, and the Reformed tradition restricted the tendency of capitalists to squander surplus on luxury consumption, ensuring that, the limitations of the market system not withstanding, human civilization continued to progress under capitalism, so too the socialist movements of the past 150 years have directly and indirectly created the conditions for the emergence of a new synergistic mode of social organization. For it is the centralization of capital by the socialist state, by capitalist states acting under the pressure of the international workers movement, and by nonmarket nonstate institutions such as the university which has made possible the tremendous progress in the sciences during the past one hundred years. And it is this scientific progress——the emergence of relativity and quantum mechanics, of complex systems theory and developmental and evolutionary biology, and finally of dialectical sociology——which has laid the groundwork for the philosophical synthesis set forth in this work, and ultimately for the emergence of a political-theological movement which will permit humanity to authentically transcend market relations, which thus once again unleash the development of human social capacities.

At the core of these new developments, of course, is the shift away from an atomistic scientific paradigm towards an understanding of the cosmos as a relational, self-organizing, and teleological system. As we argued in the first chapter, this new scientific paradigm is gradually giving birth to a new picture of the universe. Relationships and fields have replaced fundamental particles. Fractal self-similarity, the emergence of coherent structures, and progressive adaptation to the environment have displaced entropic disintegration as the dominant characteristics of our picture of even nonliving matter. Self-organization and symbiosis rather than random variation and natural selection now appear to be the driving forces behind biological evolution. And intelligence, or the social form of matter, appears increasingly to be a necessary implication of the existence

---

[1]. It is entirely proper to refer to the past 150 years as the socialist period, in that the vast majority of the great advances of this epoch were made either under socialist regimes (idealist or materialist), or under the active pressure of the international workers movement, which forced bourgeois regimes to restrict the destructive activity of the marketplace and to use state power to centralize resources for investment in infrastructure, education, research, and development.

of the cosmos.

But if matter itself is self-organizing, then we must take an entirely different attitude towards human organizing activity, at every level, from the design of production processes and the socialization of children, through the building an exercising of power, to the arts, science, philosophy, and religion. We are developing a new mode of organization which we will call "synergistic."[2] Where earlier modes (agrarian-tributary, industrial-capitalist) treated organization as something external to matter, imposed from the outside by an organizer, or emerging spontaneously out of the interaction of only-externally-related atoms, synergism recognizes that matter itself *is* organization. The task of the organizer is to grasp the principle of organization already implicit in matter, draw out the implications and contradictions implicit in the present structure, and drive towards a higher synthesis.

We are only beginning to explore the implications of this new paradigm. But it is possible to identify some important developments. First, human beings have become increasingly aware of their interdependence with other members of the ecosystem which they inhabit, and of the increasingly fragile integrity of the ecosystem as a whole. Where industrial capitalist and to a lesser extent socialist systems understood humanity as the creator of order in an essentially chaotic natural world, synergism recognizes humanity as an active participant in an organizing process much larger than itself, the nature of which it must respect and understand if it hopes to contribute to, rather than degrading the organization of the system as a whole.

Second, humanity is beginning to rethink the way in which it *produces* the social form of matter itself——i.e. the way in which we socialize our children.[3] Historically, strategies for socialization have centered on repression, and especially *sexual repression* and *repression of the child's bond with the mother*. Such strategies are based on the notion that humanity's underlying biological drives are fundamentally antisocial ——a perception, which, as Erich Fromm (1941) points out, derives from the way people experience each other in a market society, or, more generally, in societies in which the spontaneous social bonds of family, village community, etc., have been disrupted by market or state. In so far

---

[2]. The word "synergism" derives from the Hellenic "*syn-*" (with) *"-ergon "*(work) and implies an optimal and ever increasing level of complexity, integration, and creativity.

[3]. I am indebted, in this discussion, to the as yet unpublished, but very important work, of Maggie Mansueto.

as the child's capacity to be social derives in part from its biological constitution, and in part from its early bond with the mother, such childrearing strategies in fact lead to the *repression of the child's capacity to be social.* The resulting antisocial personality then requires the constraints imposed by the range of authoritarian movements which have emerged in market societies. Synergist childrearing strategies point at least incipiently towards greater reliance on cultivation of the child's social capacities, rather than on repression, as the basis for development of a complex, autonomous, relational personality structure.

Closely related is the growing emphasis on incorporating men fully into the childrearing process. Nancy Chodorow, for example, argues that the exclusion of men from the childrearing process not only imposes special burdens on women, but in fact creates a situation where human males, in particular, are forced, at the age of five or six to reject their relationship with their mother and to opt for identification with a father who, because he has little role in the childrearing process, remains rather distant, abstract, and more than a little threatening. The result is an inability to form deep and lasting emotional bonds, and to regard the work of civilization as confined exclusively to the conflictual and competitive realm which exists outside the household. Women, on the other hand, because they remain in a close relationship with their mothers, who act as powerful role models, have an easier time forming intimate emotional bonds, but more difficulty establishing a clearly defined autonomous identity. Incorporating men more fully into the childrearing process would lead to a softening of the role of the rejection of the mother in the socialization process, while freeing women to become more active in the public arena.

There are also important changes at work in the way in which we approach the task of production: of reorganizing physical, chemical, and biological matter. Industrial organization is based fundamentally on breaking down the pre-existing organization of matter in order to release energy and then use that energy to reorganize matter in new, presumably more complex forms. This is true at all levels. Industrial production burns fossil fuels (or smashes atomic nuclei) in order to generate energy, which is then used to break raw materials down into their constituent elements, transforming ores into metal, wood into pulp, etc., and then to reorganize these elements as machines, paper, etc. Similarly, industrial production requires that communitarian forms such as the family and the village community be broken down, and that workers be reorganized under the discipline of the factory system.

The new synergistic forms of organization are very different.

Synergistic technology is based on an understanding of production as a conscious participation in the self-organizing activity of the cosmos. From the ecological side this means that it must be based not on the exploitation of natural resources, but rather on the channelling of latent potential for organization into new and more complex forms. This means using sources of energy and raw materials which are renewable, the "capture" of which does not reduce the energy or raw material available for other uses. Production must actually increase the level of organization of the ecosystem, and not simply displace complexity from carbon based organisms to steel and silicon based machines. Production must also be clean enough that valuable species of plants and animals, which make up an irreplaceable part of the genetic endowment of this planet, are not lost.

Because of this it is particularly important to identify critical ecosystem capacities which will permit us to access the large quantities of energy necessary to produce complex forms of social organization without exporting ever greater entropy into the ecosystem and eventually undermining the conditions of our own existence. Clearly the most important of these ecosystem capacities is the sun, which is the original source of all of the energy utilized by human societies. Solar energy is a resource both directly, as we learn to develop a variety of active and passive solar energy technologies, and indirectly, as it is embodied in plant and animal life used for food, the production of alcohol fuel, etc., wind, hydraulic and geothermal energy, nuclear and fossil fuels, etc.

At the level of production processes themselves there are also important untapped ecosystem capacities. This is an area which we are only beginning to explore, in large part because we are only beginning to understand the creative self-organizing character of matter itself. There are, however, some hopeful signs.

* The "synergetic" engineering principles of R. Buckminster Fuller (1981, 1992) promise to permit the development of energy efficient dwellings, manufacturing facilities, and other structures which capture as much as possible of the sun's energy, and which recycle as much as possible of their own waste.

* Growing attention has been focused in recent years on the pharmaceutical applications of various plants, especially those growing in endangered rain forest ecosystems. Biotechnological processes generally are synergistic in that they tap into the self-organizing properties of matter rather than breaking down raw materials and reassembling them in accord with a definite plan.

* Artificial living systems and artificial intelligence research has succeeded in creating self-reproducing electronic organisms which evolve and adapt in creative ways to various electronic, computational environments, and which have actually succeeded in evolving solutions to complex computational problems which had eluded human software engineers (Langton 1989, Rothschild 1991).[4]

In a sense, synergistic production processes represent a return, albeit at a much more advanced scientific level, to the careful, intensive, cultivation of the productive capacities of matter, and especially living matter, which characterized horticultural societies.

This new, synergistic perspective is also increasingly affecting the way in which we go about building and exercising power. All power is fundamentally relational in character, and is based on the intersection of the complex interests of diverse individuals, which creates the basis for collaborative activity centered on a common plan which promises to realize as many of the interests as possible. This means that power is to be sharply distinguished from control, which merely prevents things (including bad things) from happening. Power is not something imposed from the outside on a disorganized, atomized mass of human individuals, but rather expresses a potential for organization which is already latent in the network of interests which binds human beings to each other and to the natural world. It should also be increasingly apparent, however, that power does not emerge spontaneously. It involves, by its very nature, a bringing to consciousness of the previously unconscious and implicit shared interests of the group in formation. It is also apparent that the wider the self-interest of an individual or group, the greater its potential to build and exercise

---

[4]. It is necessary to remember, of course, that there are real dangers involved in both biotechnological and artificial life and intelligence research. Scientists who conduct such research must have a profound grasp of ecosystem and evolutionary dynamics, and of the nature of human society, which is the complex system par excellence. This means not only training by, and ongoing interaction with ecosystem and evolutionary biologists and social scientists, but day to day practical involvement in diverse human activities and interaction with diverse human cultures. It is this sort of interaction precisely which seems to me to be lacking in many of the current centers of biotechnology and artificial life and intelligence research, and which leads to such phenomena as Tipler's reduction of human intelligence to information processing and his consequent project to ensure the survival of intelligence into the far future of the universe by building self-reproducing intelligent computers which, if he has his way, will effectively eat the universe.

power. This means that "organizers" such as the marketplace which simply link people up on the basis of their existing understanding of their self-interest cannot build as much power as organizers which actually expand the sphere of human self-interest. The synergistic organizer utilizes a kind of Socratic dialogue, which begins with existing networks, interests, and aspirations, draws out implications and contradictions, and drives towards a higher synthesis which realizes ever more fully the potential latent in the group and its members.

The implications of the new dialectical science are most striking, however, when we begin to look at the way in which human beings organize their experience and define their relationship to the cosmos as a whole. We have already seen in the last chapter, in our analysis of the works of Ernesto Cardenal, that in the artistic arena we are witnessing the rebirth of the epic poem, the subject of which is not the individual hero, but the whole cosmohistorical process. Doris Lessing's five part series *Canopus in Argos* represents a similar trend. Unified field theory, complex systems theory, postdarwinian evolutionary biology, and anthropic cosmology, we have seen, contain an implicit ontology which both grasps the cosmos as a relational, self-organizing, and teleological totality, comprehending in a new and qualitatively deeper way the reality of God, and validates the recognition that humanity is a real participant in (though by no means the sole organizer of) the cosmic evolutionary process. This new standpoint overcomes the spiritual limitations, and rectifies the religious errors of the great world religions, of capitalism, and of socialism. Humanity's drive towards synergism has a firm ontological foundation. The human civilizational project makes a real contribution to the process of cosmic evolution. But our aim, our *telos*, which is the same as that of the cosmos as a whole, lies beyond any one single form of social organization——indeed beyond the merely social form of matter——in a self-organizing Omega the character of which we cannot yet even begin to comprehend.

The result, of course, is something much more sweeping, and much more profound than any mere "rectification" of "errors" in the socialist movements. Just as it was necessary to transcend the purely religious standpoint in order to realize the values of the great religious movements which emerged out of the crisis of tributary civilization, and to transcend the purely technological and economic standpoint in order to realize the values implicit in capitalism, so to it is necessary to transcend the merely political standpoint in order realize the values nurtured and cultivated by socialism. Humanity is now ready to take conscious responsibility for its role in the cosmos. But in order to do this we must understand our current tasks in the light of their relationship to the human civilizational project as a whole, and understand human civilization in the light of its authentic place in the process of cosmic evolution.

Synergism has, to be sure, been born in the most difficult of times. The penetration of market relations into every sphere of life, in every corner of the planet, is rapidly destroying the ecosystem and the social fabric on which human development and social progress depend, and undermining efforts to centralize the surplus necessary for technological, economic, political, social, and cultural progress. The result is an unprecedented social crisis, in which the vast majority of humanity is rapidly losing hope in its future, and ever broader sectors of the population are resorting to random violence or self destruction.

This crisis does take different forms in different regions of the planet. Vast regions of Latin America, Asia, and especially Africa continue to stagnate, their populations plunging into the depths of a misery which would have been inconceivable during most periods in this planet's history. In parts of the Third World, neoliberalism and export oriented industrialization strategies have mobilized vast armies of human labor to produce cheap consumer goods to satisfy the whims of affluent Japanese, European, and North American consumers. In some cases this rapid industrialization has even raised living standards. But it has also completely unravelled the social fabric, causing the disintegration of families, the decay of national, popular, and religious traditions, and undermining the networks of resistance which have historically enabled the peasantry and the working class to dream of, and struggle effectively for, a new social order. The corollary of this social disintegration has been a resurgent religious fundamentalism——Hindu, Christian, but above all Islamic——which seeks to impose order from the outside on what it cannot help but see as a hopelessly chaotic social world (Mansueto unpublished).

In the socialist and "formerly socialist" countries, the story is much the same. Only a handful of religious socialist and populist regimes, such as Libya, Syria, Iraq, and Myanmar, together with Cuba and North Korea among the socialist countries, have resisted the temptations of economic liberalization altogether, and none of these countries has developed a credible strategy for transcending the limitations of bureaucratic organization. The remainder of the socialist countries have undergone a more or less dramatic process of marketization. Some, such as Serbia, Romania, and Vietnam, have held on to a broadly socialist vision, but only by imposing sharp restrictions on democratic participation and political liberalization. Most of the countries of the former Soviet bloc, on the other hand, have plunged into an orgy of self-destruction, dismantling the world's premiere educational, artistic, scientific, and philosophical establishment in order to build luxury housing and luxury hotels, and dissolving the world's largest (albeit imperfect) complex of multinational states in the name of ethnic particularism (and this in a period celebrated as an age of globalization and world unity). Not surprisingly, the former Soviet bloc has also given birth to its own species of fundamentalism, Orthodox Nationalist and National Bolshevik, which promises to make order out of the chaos engendered by the marketplace (Mansueto 1993).

The situation in the advanced capitalist countries also continues to deteriorate. If its absolute economic deprivation is not nearly as bad as in most parts of the Third World, the disintegration of the social fabric is probably worse. In the United States, at least, people who are not able to sell their labor power for a decent wage quite simply have nowhere to turn. The minimal "social safety net" left in place after the neoliberal counterrevolution of the 1980s simply does not compare with the support which was offered to previous generations of immigrant workers (and which many urban workers in the Third World still enjoy) by extended family, neighborhood, and ethnic communities.

Many, to be sure, continue to live in unparalleled affluence. But they also live in constant insecurity, never knowing if they will continue to have insurance to pay for a major medical expense, or money to send their sons and daughters to a university——or for that matter if they will be able to get home safely after a night out. And they work interminably long hours (men and women, mothers and fathers both) simply for the privilege of spending a few hours each week using some new electronic device or eating at some chic restaurant catering to the latest culinary fad. Their children grow up never knowing what it is like to spend time together in a family, or to relate informally to other people in a community. In Japan and Europe there is more security, and in Europe, at least, more leisure, but the influence of nihilistic consumerism is probably even greater. And Japanese "organized capitalism" and the European "social market" are themselves both under attack from the neoliberal right.

The greatest failure of these societies, however, is their inability, indeed, their apparent lack of interest, in investing in the development of human social capacities. No matter how outrageous the level of luxury consumption, the money just can't be found to support the development of new infrastructure, the expansion and improvement of educational institutions, or research and development——especially basic research in the natural and social sciences and in the philosophical disciplines. The vision of the future with which much of the postwar generation grew up——a future in which poverty had been eliminated, universal mass transit had displaced the automobile, and space flight was commonplace——seems further away now than it did in 1960.

In all regions of the globe, however, the underlying problem remains the same. The social disintegration engendered by the marketplace is causing people to lose sight of their fundamental vocation as human beings——the call to participate actively in the self-organizing activity of the cosmos. And the marketplace, through its short-sighted allocation of resources, is holding back the implementation of the new technologies, organizing methods, etc. made possible by the scientific-technical revolution ——technologies and organizing methods which could help humanity to realize its vocation. With socialism discredited, many millions have slipped into nihilistic postmodernism which denies the meaningfulness of the universe altogether, and many millions more have embraced various fundamentalisms which seek to impose order from the outside. We face nothing less than a crisis of the human civilizational project itself.

Within this complex and contradictory situation——the emergence of incipient synergistic capacities on the one hand, and an all-sided crisis of the human civilizational project on the other hand——three broad progressive movements have emerged, movements which are not yet fully synergistic in character, but which point towards synergism, and with which synergists must ally themselves.

The first of these movements——at once the most powerful and the most ambiguous from a synergistic perspective——is directed at global unity. New methods of transportation and social communication are rapidly knitting the diverse peoples of the Earth together into a single, integrated, global civilization. This process of global integration has made possible an unprecedented level of interaction between peoples and cultures, an interaction which is pregnant with creative potential. At the same time, the mediation of this interaction through the marketplace tends to degrade the ecosystem and the social fabric of every society on the planet. And even the creation of a stable world market presupposes the establishment of a stable international political authority——something like a world state——just as the establishment of unified national markets presupposed the creation of powerful nation states. The global victory of the forces of free trade in the 1980s has thus paradoxically given birth in the 1990s to a movement to establish a global political authority of sufficient power to protect the ecosystem and the social fabric, and serve as a guarantor of world peace. This is the significance of the establishment of the European Union, the relative strengthening of the United Nations, the international conferences on ecosystem integrity and population control in Rio de Janeiro and Cairo, etc.

This movement finds its constituency among the great transnational corporations, especially those in the high technology sectors who understand the importance of ecosystem integrity and international peace, and have some interest in the further development of human social capacities, together with their operatives in government, the research institutions, etc. Politically it finds expression in such organizations as the Trilateral Commission and the Council on Foreign Relations. Ideologically it has affinities with a number of trends from the remnants of Social Christianity (which historically played such an important role in the establishment of international institutions) through the New Age.

The movement towards global unity is progressive because even through it has been partially set in motion by the emergence of a unified world market, it points necessarily beyond the marketplace, towards a unified world state and towards an expanded role for international organizations. It is ambiguous because it remains in the hands of transnational capital and the advanced elements of the state bureaucracies, and its horizons are constrained by their interest in defending the world market and their own claims on surplus for luxury consumption.

The second movement is directed at conserving the integrity of the ecosystem and the social fabric. It is first and foremost a movement of resistance to the penetration of market relations into new spheres, and to the corrosive impact of market relations on the biological and social infrastructure of human civilization. In this sense it is the authentic conservative movement of our time——the guardian of a heritage bequeathed to humanity by fifteen billion years of cosmic evolution, and by millennia of human civilizational progress.

This movement finds its constituencies among the village communities of the planet which remain in relations of close interdependence with the ecological niches which they inhabit and which conserve humanity's primordial intuition of the underlying unity and order of the cosmos; among the working class mothers and fathers and grandparents struggling to hold their families together in the hostile atmosphere of a market economy; among the leaders of religious institutions; and among those sectors of the intelligentsia (biologists and some social scientists) who have come to understand the critical importance of the ecosystem and the social fabric for humanity and for the cosmos in general. Politically it finds expression in the "green movements" and in movements of peasant communities and urban neighborhoods around the planet which are struggling to secure the resources they need in order to carry out their vital social functions. Like the movement towards global unity, the movement to conserve the integrity of the ecosystem and the social fabric has ideological affinities with a wide range of trends from authentic social conservatism (but not fundamentalism) to elements of the New Age.

The progressive character of this movement should be apparent. But it, too, has its ambiguities: first and foremost a tendency to blame scientific and technological progress, rather than the marketplace, for the disintegration of the ecosystem and the social fabric, and thus to drift ever so gradually towards Luddite antitechnologism or religious fundamentalism.

The third movement is directed at the development of human social capacities, and at securing the resources necessary to support this development. This movement has its core constituency among the most talented and creative sectors of the working class: people involved in developing new artistic styles, new theoretical problematics, new philosophical systems, and new spiritual disciplines. But it also finds support among people struggling to raise their children into creative, powerful, wise, and loving human beings, people who long to do work with their hands and minds which will make a real contribution to society, teachers dedicating their lives to work in inner city schools, organizers trying to tap into the enormous creative potential which is languishing in our ghettos. This movement *is* the movement towards synergism, which, however, has not yet become conscious of itself, and has not yet found its direction. In the neoliberal ideological milieu of the 1990s it often describes itself as a movement to "develop human capital" or "remain competitive" and is often satisfied with market based solutions which fall far short of its implicit goal: to develop human social capacities so that humanity, in turn, can fulfill its great vocation as a real participant in the cosmohistorical evolutionary process. But even so, it taps into humanity's most progressive impulse: to contribute something new to the world around us, to leave our communities, our country, our planet more productive and more powerful, wiser and more beautiful, than we found it.

What are the tasks of synergism in the context of this complex and contradictory situation? We begin by distinguishing between strategic tasks and immediate tactical imperatives. By strategy we understand a comprehensive plan for organizing, developing and deploying the basic resources we need in order to reorganize human society. By tactics, on the other hand, we understand the definition of the principal contradiction in the present period, and thus the disposition of the resources we have already organized and developed.

Synergism is just beginning to emerge as a philosophical trend out of the crisis of religious idealism and dialectical materialism. We are still in the process of developing a philosophical system and drawing out its broad practical implications. Our first task, therefore, remains theoretical. The current scientific revolution has laid the groundwork for the emergence of a new synergistic theory. But it has only laid the groundwork. Many of the results of the new sciences have themselves been framed in a way which reflects the residual influence of the old atomistic paradigm. This is because our whole scientific language, grounded as it is in a set-theoretical mathematics and formal logic, is itself atomistic. This not only makes it difficult to draw out properly the philosophical and practical implications of the new science, but holds back the progress of scientific research itself. We must continue to break down the barriers of the old atomistic paradigm and elaborate a new theory which is fully synergistic in character.

There are a number of dimensions to this task. First, we need to develop new methods of research which can grasp "whole systems" rather than merely isolating and analyzing the relationships between a limited number of variables. Second, we must apply this new paradigm to every type of system: physical, chemical, biological, social, supersocial. We must constantly observe and analyze reality in order to identify and understand organizing processes at every level on the dialectical scale——physical, chemical, biological, social——and then attempt to comprehend how these organizing processes contribute to the self-organizing activity of the cosmos as a whole. Third, we must synthesize the results of our research into a single, unified, philosophical system. We need to grasp the logic of being itself, so that we can understand how it is that complex organization develops. We must grasp the cosmos as a unified system, so that we can understand where we are going and how to get there. And we must elaborate a comprehensive theory of value.

This new theory will play a vitally important role in the synergistic mode of organization. First, even more so than earlier modes of organization, synergism recognizes that it is only by grasping the immanent, self-organizing dynamic of matter itself (including the social form of matter) that we can learn how to create still more complex organizational forms. Synergistic theory, in other words, is itself a material productive force. As it observes and analyzes the myriad organizing processes which constitute the cosmos, theory is constantly searching for new ways of bringing resources and people together, tapping into latent organizing potential, and setting into motion new and more complex activities. Second, synergistic theory makes it possible to situate our organizing activity within the larger context of human social progress and cosmic evolution, and to insure that our activity remains at once rooted in and ordered to these larger processes. An organizing process which is regulated only by the immanent dynamic of its own will to power, and which is open only to the possibilities presented to it by the resources it controls and understands, and by its present constituency, will tend ultimately to become cut off from the self-organizing activity of the cosmos. While the struggle of any particular organization to build its own power is, in and of itself, a good, the full potential of this power can be realized only if the organization's power is rooted in and ordered to the self-organizing dynamic of the cosmos as a whole. Otherwise, the organization will begin working against the trend of cosmic history, and no matter how much power the organization is able to build, its efforts will ultimately be doomed.

Our second task is organizational. It involves identifying, training, positioning, and mentoring leaders who are capable of organizing at a fully synergistic level. Our aim here is not to "infiltrate" institutions with the aim of gaining control, but rather to identify and train people who can set in motion organizing activities at every level which begin to tap into the potential of the synergistic mode. From this point of view it is just as important to identify, train, and position parents, engineers, managers, artists, scientists, philosophers, and religious leaders, as it is to identify, train, and position political leaders in the more traditional sense.

Nor is our aim to "remake" people, or "form" them in a synergistic model. Rather, we need to look for creative people at the cutting edges of their fields, who are either already beginning to work at a synergistic level, or who are wrestling with problems which make them open to a synergistic perspective. Our task is to

a) help them break with the old atomistic paradigm, and grasp synergism——i.e. comprehend the relational, self-organizing, and teleological character of matter itself,

b) help them develop basic organizational skills (relational individual and small group meetings, etc.,) and help them position themselves in strategically significant institutional sites, and

c) continue to mentor them as they set in motion new and more complex activities.

By identifying the most creative people in our society, in every field of endeavor, helping them to break through to a fully synergistic level of development, positioning them in strategic institutional locations, and mentoring them as they develop new, more complex activities, we will set in motion the creation of a new, synergistic mode of organization in the womb of the old capitalist and bureaucratic structures.

As we identify, train, and position organizers in the full range of social institutions, and as they begin to set in motion new and more complex activities, they will find themselves in an increasingly paradoxical and contradictory situation. On the one hand, they will find that they are increasingly powerful, simply because synergistic theory makes it possible to organize at a far more complex level than any previous theoretical system. The owners of capital, and those who occupy positions of bureaucratic authority, will find themselves increasingly dependent on our synergistic engineers, organizers, artists, scientists, philosophers, and spiritual leaders to tackle their most complex problems, and to set in motion the new activities which make their own prosperity and power possible. At the same time, the leaders we position will increasingly come up against the limits imposed by the marketplace. They will find themselves unable to centralize the resources which their activities require, even as vast quantities of wealth are squandered on luxury consumption.

The resolution to this contradiction lies in the synergistic network itself. We need to bring the network of organizers which we train into increasingly intense relationships with each other. First they may simply exchange information and ideas. But gradually they will begin to exchange more than information. They will begin to collaborate in their strategic planning, to pool resources for joint projects, develop nonmarket supply and distribution networks, and eventually organize their own consumption and that of the people they lead in a way which restricts the flow of resources to luxury consumption, and increases the total surplus available for investment. The result will be a new mode of centralizing and allocating resources——a mode which centralizes the largest surplus possible for investment, and allocates resources in such a way as to make the largest contribution possible to the development of human social capacities and the self-organizing activity of the cosmos. The marketplace and bureaucratic structures will become increasingly irrelevant, vestigial organs which are largely ignored, and eventually dismantled, almost as an afterthought.

It may appear that this emphasis on reorganizing institutions represents an abandonment of the commitment to revolutionary structural transformation which characterized historic socialism. This is not true. It does, however, represent a fundamentally different understanding of the nature of fundamental social change, as well as the way in which such change comes about, and the end towards which it is directed. If we examine the transition from the tributary to the capitalist modes of production, it becomes apparent that the most important factor in this transition was the gradual creation of a nontributary, i.e. market, mode of organizing and deploying social resources. The abolition of feudal and mercantilist restrictions played a secondary role. Similarly, the marketplace will be transcended only when humanity has succeeded in developing new, synergistic, postmarket modes of centralizing and allocating resources. Restriction of market forces and the centralization and allocation of capital by the state, while often necessary, and even powerfully progressive, is not by itself adequate for a transition to postmarket modes.

Synergism regards this process of developing and implementing new modes of organization as an ongoing task. There is no one regulating structure the restriction of which will unleash once and for all human creative potential. There is no one mode of social organization which by itself constitutes the solution to the riddle of history. That solution, rather, consists in embracing the riddle itself, as a unique and powerful catalyst for organization, growth, and development.

The struggle to transcend the marketplace will, furthermore, be prolonged——a process of many decades at the very least, and more likely of centuries. In all probability humanity will attempt a number of partial solutions, solutions of the kind which have already been suggested by the three great progressive movements which we identified earlier: establishment of a powerful international political authority, the use of this global state power, as well as national, regional, and local state power and the authority of nonmarket, nonstate institutions to defend the integrity of the ecosystem and the social fabric and to centralize and allocate the resources necessary to promote the development of human social capacities.[5]

Synergists need to support these movements, while pointing out clearly their limitations. We need to tap their latent potential, gradually drawing out their implications and internal contradictions, and driving towards a higher synthesis. We cannot, however, rely on the internal logic of these movements to carry the people towards synergism, either through a gradual transformation of quantitative, incremental reforms into a qualitative break with the marketplace, or through a gradual working out, under synergist leadership, of the internal contradictions of the progressive

---

[5]. Warren Wager's <u>A Short History of the Future</u> (1989) describes in considerable detail a future scenario in which, after a global holocaust in 2044, humanity undertakes each of these tasks in succession, beginning with the construction of a unified world state. This state is eventually dismantled by a "small revolution," the protagonists of which look suspiciously like today's "greens" and "ecofeminists." The decentralized society which emerges from this revolution eventually begins to stagnate, until a series of "great projects," including large scale space exploration, once again captures the imagination of the planet and catalyzes renewed social progress.

movements themselves. Rather, we need to elaborate our own vision, and carry it to the people, so that, moved by the incredible beauty of the universe, convinced by the rational power of our system, and drawn by the good towards which it points, they are actually won over to synergism——or rather raised to a level of development which permits them to become authentic participants in the synergistic organizing process.

There is no royal road to synergism, only the gradual, difficult, struggle to comprehend ever better the organizing principle of the cosmos, and to use this knowledge to develop every more complex forms of organization.[6] Our priority must remain the task of developing our theoretical system, and identifying, training, positioning, and mentoring leaders. We need to practice a tactics of alliance——of unity and struggle with the other progressive movements——but a strategy of the gradual, patient, development of human social capacities and of the essentially passive accumulation of forces. It is the development of these capacities and the accumulation of these forces which themselves carry humanity forward.

If this seems like a much more sober vision than that advanced by objective idealism and dialectical materialism, and the socialist movements which they inspired, then perhaps it is. No great revolution which might appear on our near or distant horizon is going to bring the human civilizational project suddenly to completion. Our future is one of continuous struggle to develop, to evolve, to become more than we are. But then our goal, the goal of human civilization, is not and can never be some final historical state. Human civilization plays a vitally important role in the cosmohistorical evolutionary process, as a center for the creation of dynamic, organized complexity. But human civilization——even the social form of matter generally——is not an end it itself.

We need to conclude our argument with a few brief considerations on what lies beyond history. The claims of Professor Tipler notwithstanding, this is an area in which our knowledge remains severely limited. What we do know is this. The biological infrastructure on which the social form of matter currently depends may not survive into the far future of the universe.   This does not, however, mean that complex organization, life, and intelligence, are ultimately doomed, that organization is merely *a* tendency, and not *the* tendency, of the universe. If matter is, indeed, the

---

[6]. In this sense we synergists need to reverse the decision made by humanity centuries ago, when a few men took up arms, conquered their sisters and brothers, and laid tribute on them, rather than follow along the slow and difficult path of progress through research, investment, and development.

drive towards organization, then there is nothing which, in principle, prevents the further evolution of matter into a form which allows intelligent life to survive into infinity, continuing to develop, and eventually embracing the universe as a whole, at what has come to be called the Omega point. Indeed, certain variants of the anthropic cosmological principle suggest that this *must* happen.

We have already outlined in Chapter One a preliminary argument for this position. We have also explained why we do not accept Professor Tipler's version of the Omega Point Theory, nor, in particular, his strategy for reaching Omega. In depth discussion of the nature of supersocial, transhistorical modes of cosmic organization will have to await another work. We will, however, say a few words regarding the implications of Omega for our understanding of the significance of our own individual lives, and of the life of our civilization.

Specifically, we can say with confidence that we will Be at Omega. Cosmos is not a system of atoms which are only related to each other in a merely external manner. It is, rather, a web of relationships, interconnected at the most profound levels of its being. The faintest smile of a mother who cradles her child in some barrio of Juarez or some shanty town of Bangkok is felt by the most distant star. And it is felt as a smile. Nothing which we do which contributes in some way to the sociality, the productivity, the power, the beauty, the truth, the goodness, or the complex, synergistic integrity of the cosmos will be lost. Similarly, in so far as the universe is a system of relations, and not of only externally related atoms, we *are not*, or at least are not only, our limited, empirical selves. We *are* rather, Omega itself, *in potentia*, and Omega is our own true self, conserving everything we have done, realizing everything we have left undone. In this sense, at least, we persist into Omega——even when the Earth itself has long since vanished ...

This is, to be sure, a rather different kind of hope than that offered by the salvation religions, by capitalism, by or materialist socialism. On the one hand synergism teaches us that a great deal is possible, and that our work makes a real difference to the cosmos. We are not merely the children of a benevolent God who orders all and cares for all and to whom we owe only obedience, but rather active partners in the design and creation of the cosmos. On the other hand, our activity is not directed towards some proximate social goal——even a very difficult one like communism——but rather towards an end which lies beyond history, in the distant future of the Omega point. In this sense our vision will always transcend our reach. We will always be able to intuit possibilities which we cannot prove, and prove that things are possible which we cannot realize in our lifetime————or even within the scope of human history. This is the unique grandeur, and also the terribly frustrating limitation of humanity, which is "just a little lower than the angels."

The salvation religions, capitalism, and socialism, in this sense, all asked humanity to be satisfied with far too little. The salvation religions deprived humanity of its rightful place as an authentic participant in the cosmohistorical evolutionary process. Secular systems such as capitalism and materialist socialism, on the other hand, deprive humanity of both its ontological ground in the organizing principle of the cosmos, which is the unique source of its power, and also rob humanity of its authentic, transhistorical destiny. The result is either a spiritual debasement of humanity, or else an idolatrous investment of partial totalities————human history, or worse still the party, or even its leader————with the attributes of the Beautiful, the True, the Good, and the One. And idolatry is always unsatisfying. Even the simplest peasant, after all, knows the kings die like other men (Ps 82)————and that the cult of the king is, therefore, ultimately a cult of death.

***

There can be little doubt that we live in dark times. The achievements of literally millennia of human civilization seem to be jeopardy. The marketplace undermines the integrity of the ecosystem and the social fabric, tearing apart families, depriving millions of the resources they need to make a productive contribution to society. In this environment of social disintegration it becomes ever more difficult to organize the resources necessary for investment in infrastructure, education, research, and development. The planet seems almost inadvertently to have escaped the threat of nuclear annihilation (at least for a while). But the arms race has given way to an even more ominous contest between the United States and the former Soviet bloc to see who can dismantle their artistic, scientific, educational and cultural apparatus more quickly. Sometimes it seems as if everything has been lost.

But we know that the darkness can never triumph. Matter itself is relationship, holism, organization, development. Engels said that whenever the self-organizing dynamic of the cosmos is defeated in one place, it begins anew in another. Life, love, creativity, power, and knowledge are what bring the cosmos into being in the first place, and they will not, *can not*, yield to the forces of chaos and destruction.

We have, therefore, every reason to labor unendingly, to contribute with everything we do to the construction of Omega, and no reason to weep over our own limitations. The work that we cannot finish, others will take up, knowing that we labor in them, even as we, now, make straight their path. Nor is there any excuse to rest satisfied with any partial totality, when we can grasp the concept, at least, of something which far transcends any merely human achievement.

This is the true faith of a mature humanity——a humanity which is certain of both its power and its limitations, which has grown to the point that it can feel at home in a universe of which it is not the center, in which it is not and never will be Lord, but which it nonetheless recognizes as meaningful, and to which it has something very important to contribute. Our species is being called on to grow up, to cast aside both its childish dependence and its infantile, nihilistic rebelliousness, and get on with its work. The present darkness notwithstanding, our future is brilliant.

BIBLIOGRAPHY

Abalkin, L.
1986 "The Interaction of Productive Forces and Production Relations," in *Problems of Economics 28:12*

Adams, Richard
1988 *The Eighth Day: Social Evolution as the Self-Organization of Energy*. Austin: University of Texas

Ahlstrom, Sydney
1972 *A Religious History of the United States*. New Haven: Yale Univ. Press

Aldaraca, Bridget et al
1980 *Nicaragua in Revolution: The Poets Speak*. Minneapolis: Marxist Educational Press

Alexander, Robert
1981 *The Right Opposition*. Westport: Greenwood

Alinsky, Saul
1946 *Reveille for Radicals*. Chicago: University of Chicago Press

1971 *Rules for Radicals*. New York: Random House

Alper, Harvey
1987 "Order, Chaos, and Renunciation: The Reign of Dharma in India," in Kliever, Lonnie, ed. *The Terrible Meek*. New York: Paragon

Althusser, Louis
1965/1977 *For Marx*. London: Lane

1968/1970 *Reading Capital*. London: New Left Books

1966-1969/1971 *Lenin and Philosophy*. New York: Monthly Review.

Amin, Samir
1975 *El capitalismo y la question campesina*. Mexico: Nuevo Tiempo

1978a *The Law of Value and Historical Materialism*. NY: Monthly Review

1976/1978b *The Arab Nation*. New York: Zed Press

1979/1980 *Class and Nation, Historically and in the Current Crisis*. New York: Monthly Review

1981/1982a *The Future of Maoism*. New York: Monthly Review

1982b *Dynamics of Global Crisis*. New York: Monthly Review

1988/1989 *Eurocentrism*. New York: Monthly Review

1990 "The Future of Socialism," in *Monthly Review* 42:3

1990 *Transforming the Revolution*. New York: Monthly Review

Anderson, Perry
1974 *Passages from Antiquity to Feudalism*. London: New Left Review

Anderson, Philip, Arrow, Kenneth, and Pines, David
1988 *The Economy as an Evolving Complex System*. NY: Addison Wesley

Aquinas, Thomas
1260/1955-1957 *Summa Contra Gentiles*, published as *On the Truth of the Catholic Faith*, trans. A.C. Pegis et al, 5 volumes

1272/1952 *Summa Theologiae*. Chicago: Encyclopaedia Britannica

Arendt, Hannah
1958 *The Human Condition*. Chicago: University of Chicago Press

1971 *The Life of the Mind*. New York: Harcourt Brace and Jovanovich

Aristotle
c 350 B.C.E./1946 *Politics*, trans. Ernest Barker. Oxford: Clardendon Press

c 350 B.C.E./1952 *Metaphysics*, trans. Richard Hope, New York: Columbia University Press

c 350 B.C.E./1973 *De Anima*, in *Introduction to Aristotle*, trans. Richard McKeon, Chicago: University of Chicago Press

c 350 B.C.E./1973 *Ethics*, in *Introduction to Aristotle*, trans. Richard McKeon, Chicago: University of Chicago Press

Augustine
426/1972 *The City of God*, trans. Henry Bettenson. New York: Penguin

Avineri, Shlomo
1972 *Hegel's Theory of the Modern State*. Cambridge, UK: Cambridge Univ. Press

1981 *The Making of Modern Zionism: The Intellectual Origins of the Jewish State*. New York: Basic

Barrow, John, and Tipler, Frank
1986 *The Anthropic Cosmological Principle*. Oxford: Oxford Univ. Press

Barton, William and Capobianco, Michael
1991 *Fellow Traveler*. New York: Bantam

Bashkar, Roy
1989 *Reclaiming Reality*. London: Verso

Bellah, Robert
1970 *Beyond Belief*. New York: Harper

1973 "Introduction," Durkheim, Emile, *On Morality and Society*. ed. Bellah, Robert, Chicago: University of Chicago Press

1975 *Broken Covenant*. New York: Seabury

1980 *Varieties of Civil Religion*. New York: Harper

1985 *Habits of the Heart*. New York: Harper

1991 *The Good Society*. New York: Knopf

Belo, Fernando
1975/1981 *A Materialist Reading of the Gospel of Mark*. Maryknoll: Orbis

Bennett, Charles
1988 "Dissipation, Information, Complexity, and Organization," in *Emerging Syntheses in Science*. ed. Pines, D., New York: Addison-Wesley

Berk, Stephen
1974 *Calvinism versus Democracy: Timothy Dwight and the Origins of American Evangelical Orthodoxy*. Connecticut: Archon

Bernstein, Eduard
1909/1961 *Evolutionary Socialism*. New York: Schocken

Berryman, John
1984 *Religious Roots of Rebellion*. Maryknoll: Orbis

Bettelheim, Charles
1976 *Class Struggles in the U.S.S.R.*. Volume One, NY: Monthly Review

1978 *Class Struggles in the U.S.S.R.*. Volume Two, NY: Monthly Review

Birnbaum, Lucia Chiavola
1980 "Earthmothers, Godmothers, and Radicals," *Marxist Perspectives* 1

1983 "Religious and Political Beliefs of Sicilian and Sicilian American Women," presentation to the Conference of the American Italian Historical Association," San Francisco, California, 30 December 1983

1985 *Liberazione della Donna*. Middletown, CT: Wesleyan University Press

Boff, Clodovis
1986 *Theology and Praxis*. Maryknoll: Orbis

Bogdanov, Alexander
1928/1980 *Tektology*. Intersystems Publishers

Boulding, Kenneth
1985 *The World as a Total System*. Beverley Hills, CA: Sage

Boyer, Paul and Nissanbaum, Stephen.
1974 *Salem Possessed: The Social Origins of Witchcraft*. Cambridge: Harvard University Press

Brundage, John
1985 *The Fifth Sun*. Austin: University of Texas Press

Budiansky, Stephen
1992 *The Covenant of the Wild*. New York: Morrow

Burbank, Garin
1976 *Grass Roots Socialism*. Baton Rouge: Louisiana State University Press

Cahoone, Lawrence
1988 *The Dilemma of Modernity*. Albany: SUNY Press

Campbell, David
1990 "Introduction to Nonlinear Phenomena," in *Lectures in the Sciences of Complexity*. ed. Erica Jen. New York: Addison-Wesley

Cardenal, Ernesto
1980 *Antologia*. Barcelona: Editorial Laia

1989 *Cantico cosmico*. Managua: Nueva Nicaragua

1992 *Los ovnis de oro*. Bloomington. IN: Indiana University Press

Chaitin, G..
1988 *Algorithmic Information Theory*. Cambridge: Cambridge Univ. Press

Chang Kwang-chi
1963 *The Archeology of Ancient China*. New Haven, CT: Yale Univ. Press

Charlesworth, James
1988 *Jesus Within Judaism*. New York: Doubleday

Chiang Chun-chiao
1975 *On Exercising All-Around Dictatorship Over the Bourgeoisie*. Peking: Foreign Language Press

Childe, V. Gordon
1851 *Man Makes Himself*. New York: Mentor

Chodorow, Nancy
1978 *The Reproduction of Mothering*. New York: Monthly Review

Christ, Carol
1987 *The Laughter of Aphrodite: Reflections on a Journey to the Goddess*. San Francisco: Haper and Row

1989 "Rethinking Theology and Nature," in *Weaving the Visions: New Patterns in Feminist Spirituality*. ed. Christ, Carol and Plaskow, Judith. San Francisco: Harper and Row

Claudin, Fernando
1975 *The Communist Movement: Comintern to Cominform*. NY: Monthly Review

Cohen, Mitchell
1987 *Zion and State: Nation, Class, and the Shaping of Modern Israel*. New York: Blackwell

Cohen, Norman
1958 *The Pursuit of the Millennium*. New York: Oxford

Communist Party of China
1976 *A Basic Understanding of the Communist Party of China*. Toronto: Norman Bethune Institute

Compilation Group for the Modern History of China Series
1976 *The Taiping Revolt*. Peking: Foreign Languages Press

Conforti, Joseph
1981 *Samuel Hopkins and the New Divinity Movement: Calvinism, the Congregational Ministry, and Reform in New England Between the Great Awakening and the Revolution*. Grand Rapids: Christian University Press

Connolly, William
1988 *Political Theory and Modernity*. Oxford: Basil Blackwell

Costello, Paul
1982 "Reaping the Whirlwind: Soviet Economics and Politics 1928-1932," *Theoretical Review* 27

Crivelli, Gustavo Balsamo
1921 *"L'essenza di cristianesimo,"* La Parola del Popolo. 23 Dec 1921

1922 *"Socialismo e religione,"* La Parola del Popolo. 28 Jan 1922

Crivello, Antonino
n.d. *"Natale,"* poem, undated clipping from unidentified magazine
n.d. *"Polemica Poetica,"* with Giuseppe Luongo, undated clipping from unidentified magazine

Crowley, J.E.
1974 *This Sheba Self: The Conceptualization of Economic Life in Eighteenth Century America.* Baltimore: Johns Hopkins Press

Cunningham, Agnes, ed. and trans.
1982 *The Early Church and the State.* Philadelphia: Fortress

Cuomo, Mario
1984 "Speech to Democratic National Convention," *New York Times.* 17 July 1984

Dahl, Robert
1989 *Democracy and its Critics.* New Haven: Yale University Press

Daly, Mary
1978/1990 *Gyn/Ecology: The Metaethics of Radical Feminism.* Boston: Beacon

1984 *Pure Lust: Elemental Feminist Philosophy.* Boston: Beacon

Dahm, Helmut
1987 "The Philosophical-Sovietological Work of Gustav Andreas Wetter, S.J.," in *Philosophical Sovietology: The Pursuit of a Science.* Dordrecht: Reidel

Davies, Paul
1988 *The Cosmic Blueprint.* New York: Simon and Schuster

1991 *The Mind of God.* New York: Simon and Schuster

Davis, Mike
1986 *Prisoners of the American Dream.* London: Verso

Dayton, Donald
1983 "Social Concern in Nineteenth Century Evangelicalism," in *The Coming Kingdom.* New York: Paragon

de Janvry, Alain
1981 *Reformism and the Agrarian Question in Latin America.* Baltimore: Johns Hopkins

de Ste. Croix, C. E. M
1981 *The Class Struggle in the Ancient Greek World: From the Archaic Age to the Arab Conquests*. London: Duckworth

Descartes, Rene
1637/1975 *Discourse on Method*, trans. Laurence J. Lafleur. Cambridge, UK: Cambridge University Press

1641/1975 *Meditations*, trans. Laurence J. Lafleur. Cambridge, UK: Cambridge University Press

del Carria, R.
1966 *Proletari senza rivoluzione*

Deleuze, Gilles
1988 *Spinoza: Practical Philosophy*. San Francisco: City Lights

Deng Ming-Dao
1990 *Scholar Warrior: An Introduction to the Tao in Everyday Life*. SF: Harper

Denton, Michael
1985 *Evolution: A Theory in Crisis*. New York: Burnett Books

Diaz Polanco, Hector
1977 *Teoria marxista de la economia campesina*. Mexico: Juan Pablo

Dobb, Maurice
1948 *Soviet Economic Development Since 1917*. New York: International

Dumenil, Gerard, Glick, Mark, and Rangel, Jose
1984 "The Tendency of the Rate of Profit to Fall in the United States, Part I," *Contemporary Marxism* 9

Dumont, Luis
1970 *Homo Hierarchicus*. Chicago: University of Chicago Press

Dunn, Stephen and Ethel
1967 *The Peasants of Central Russia*. NY: Holt, Rinehart and Winston

Durkheim, Emile
1893/1964 *The Division of Labor in Society*. New York: Free Press

1897/1951 *Suicide*. New York: Free Press

1911/1965 *Elementary Forms of Religious Life*. New York: Free Press

Dussel, Enrique
1974/1981 *History of the Church in Latin America*. Grand Rapids, MI: Eerdmans

Eco, Umberto
1976 *A Theory of Semiotics*. Bloomington, IN: Indiana University Press

1988 *The Aesthetics of Thomas Aquinas*. Cambridge, MA: Harvard University Press

Edwards, Jonathan
1746/1957f *A Treatise Concerning Religious Affections*, in *Works*. New Haven, CT: Yale University Press

1750/1957f *Qualifications Requisite for Full Communion*, in *Works*. New Haven, CT: Yale University Press

1754/1957f *Treatise on the Will*, in *Works*. New Haven, CT: Yale University Press

1765/1957f *The Nature of True Virtue*, in *Works*. New Haven, CT: Yale University Press

Emmanuel, A.
1972 *Unequal Exchange*. New York: Monthly Review

Engels, Frederick
1880/1940 *The Dialectics of Nature*. New York: International

1880/1978 *Socialism: Utopian and Scientific*. in *Marx-Engels Reader*. New York: Norton

1884/1948 *The Origins of the Family, Private Property, and the State*. Moscow: Progress

1895/1978 "Introduction to Marx's *Class Struggles in France. 1848-1858*," in *Marx-Engels Reader*. New York: Norton

Esposito, John
1984 *Islam and Politics*, Syracuse, NY: Syracuse University Press.

Esteva, Gustavo
1978 "Y si los campesinos existen?" *Commercio Exterior* XXVIII

Ferguson, Thomas
1981 "The Reagan Victory: Corporate Coalitions in the 1980 Presidential Campaign," in *Hidden Election*. Ferguson, T. and Rogers, Joel, eds., New York: Pantheon

1986 *Right Turn*. New York: Hill and Wang

1989 "By Invitation Only: Party Competition and Industrial Structure in the 1988 Election," *Socialist Review* 89:4

1992a "Who Bought Your Candidate and Why?" in *The Nation*. April 6/13, 1992

1992b "The Lost Crusade of Ross Perot," in *The Nation*. August 17/24, 1992

Feuerbach, Ludwig
1841/1957 *The Essence of Christianity*. New York: Harper

Finks, P. David
1984 *The Radical Vision of Saul Alinsky*. Ramsey, N.J.: Paulist Press

Firestone, Shulamith
1971 *The Dialectic of Sex*. New York: Morrow

F.M.
1922 *"Cristo,"* La Parola del Popolo, 14 October 1922 *"dalla Difesa della Lavoratrice*

Footman, David
1962 *Civil War in Russia*. New York: Praeger

Foster, Stephen
1971 *Their Solitary Way: The Puritan Social Ethic During the First 100 Years of Settlement*. New Haven, CT: Yale University Press

Foster, W.Z.
1968 *History of the Communist Party of the U.S.A.*. NY: International

Foucault, Michel
1966/1970 *The Order of Things*. New York: Random House

Frank, Andre Gunder
1975 *On Capitalist Underdevelopment*. Oxford: Oxford University Press

1978 *Dependent Accumulation and Underdevelopment*. London: MacMillan

Frend, W.
1957 *The Donatist Church*. Oxford: Clarendon

Freud, Sigmund
1930/1961 *Civilization and its Discontents*. London: Hogarth

1940/1969 *An Outline of Psychoanalysis*, New York: Norton

Freyne, Sean
1980 *Galilee From Alexander the Great to Hadrian*. South Bend: NotreDame

Fromm, Erich
1941 *Escape from Freedom*. New York: Holt Rinehart and Winston

1947 *Man for Himself*. New York: Holt Rinehart and Winston

1955 *The Sane Society*. New York: Holt Rinehart and Winston

1963 *The Dogma of Christ*. New York: Holt Rinehart and Winston

1973 *The Anatomy of Human Destructiveness*. NY: Holt, Rinehart, and Winston

Fuller, Buckminster
1975-1979 *Synergetics*. New York: MacMillan

1981 *Critical Path*. New York: St. Martin's Press

1992 *Cosmography*. New York: Macmillan

Galbraith, James
1988 *Balancing Acts*. New York: Basic

Geissler, Suzanne
1981 *Jonathan Edwards to Aaron Burr: From the Great Awakening to Democratic Politics*. Lewiston, New York: Edwin Mellon

Geyer, R.F. and van der Zouwen, J.
1982 *Dependence and Inequality: A Systems Approach to the Problems of Mexico and Other Developing Countries*. Elmsford, NY: Pergamon Press

Gimbutas, Marija
1989a *Goddesses and Gods of Old Europe: Myths and Cult Images*. Berkeley: University of California Press

1989b *The Language of the Goddess: Unearthing Hidden Symbols in Western Civilization*. San Francisco: Harper and Row

Girardi, Giulio
1973a *Marxismo e cristianesimo*. Assisi: Cittadella

1973b *Cristianesimo, liberazione humana, lotta di classe*. Assisi: Cittadella

1976 *Credenti e non credenti per un mondo nuovo*. Assisi: Cittadella

Glazer, Nathan
1961 *Social Basis of American Communism*. Westport: Greenwood

Gleick, James
1987 *Chaos*. London: Penguin

Gorbachev, Mihail
1987 *Perestroika*. London: Collins

Gordon, Manya
1941 *Workers Before and After Lenin.* New York: Dutton

Gottwald, Norman
1979 *The Tribes of Yahweh.* Maryknoll: Orbis

Gough, Kathleen
1973 *Imperialism and Revolution in South Asia.* NY: Monthly Review

1981 *Rural Society in Southeast India.* NY: Cambridge University Press

1989 *Rural Change in Southeast India.* New York: Oxford University Press

Gramsci, Antonio
1948 *Il materialismo storico e la filosofia di Benedetto Croce.* Torino: Einaudi

1949a *Il Risorgimento.* Torino: Einaudi

1949b *Note sul Macchiavelli, sulla politica, e sullo Stato Moderno.* Torino: Einaudi

1949c *Gli intelletualli e l'organizzazione di cultura.* Torino: Einaudi

1950 *Letteratura e vita nazionale.* Torino: Einaudi

1951 *Passato e presente.* Torino: Einaudi

1954 *L'Ordine Nuovo.* Torino: Einaudi

1966 *La questione meridionale.* Roma: Riuniti

Greven, Philip
1977 *The Protestant Temperament.* New York: Knopf

Gribben, John and Rees, Martin
1989 *Cosmic Coincidences.* New York: Bantam

Gundle, Stephen
1987 "The PCI and the Historic Compromise," *New Left Review* 163

Gutierriez, Ramon
1991 *When Jesus Came the Corn Mothers Went Away.* Stanford: Stanford Univ. Press

Haken, H.
1988 *Information and Self-Organization.* New York: Springer-Verlag

Hall, Gus
1991 "The Crisis in the Soviet Union," pamphlet published by the Communist Party, U.S.A.

Hatch, Nathan
1977   *The Sacred Cause of Liberty: Republican Thought and the Millennium in Revolutionary New England.* New Haven: Yale University of Press

Harris, Errol
1965   *Foundations of Metaphysics in Science.* London: Allen and Unwin

1987   *Formal, Transcendental, and Dialectical Thinking.* Albany: SUNY Press

1991   *Cosmos and Anthropos.* Atlantic Highlands, NJ: Humanities International

1992   *Cosmos and Theos.* Atlantic Highlands, NJ: Humanities International

Harris, Marvin
1979   *Cultural Materialism.* New York: Random House

Hayek, Frederick
1973   *Law, Liberty, and Legislation, Volume One: Rules and Order.* Chicago: University of Chicago Press

1988   *The Fatal Conceit.* Chicago: University of Chicago Press

Hegel, G.W.F.
1807/1967b *Phenomenology of Mind*, trans J.B. Baillie. New York: Harper

1812/1969   *Science of Logic*, trans. A.V. Miller. NJ: Humanities Press Int'l

1820/1942   *Philosophy of Right*, trans. T.M. Knox. Oxford: Oxford University Press

1830/1971   *Encyclopaedia of the Philosophical Sciences*, trans. William Wallace. Oxford, UK: Oxford University Press

1831/1956   *Philosophy of History*, trans. J. Sibree. New York: Dover

Heimart, Alan
1966   *Religion and the American Mind: From the Great Awakening to the Revolution.* Cambridge: Harvard University Press

Heller, Agnes, ed.
1983   *Lukacs Revalued.* Oxford: Basil Blackwell

Hill, Christopher
1972   *The World Turned Upside Down.* London: Temple and Smith

Hinnebusch, Raymond
1982   "The Islamic Movement in Syria" Sectarian Conflict and Urban Rebellion in an Authoritarian-Populist Regime," Dessouki, Ali E. Hillal. *Islamic Resurgence in the Arab World.* New York: Praeger Publishers.

Hinton, William
1966 *Fanshen*. New York: Vintage

1983 *Shenfan*. New York: Random House

1990 *The Great Reversal*. New York: Monthly Review

Hobbes, Thomas
1962 *Leviathan*. New York: Macmillan

Hobsbawm, Eric
1959 *Primitive Rebels*. New York: Norton

Hodges, Geoffrey
1986 *The Intellectual Foundations of the Nicaraguan Revolution*. Austin: University of Texas Press

Horsely, Richard and Hanson, Paul
1985 *Bandits, Prophets, and Messiahs*. Maryknoll: Orbis

Howe, Daniel Walker
1979 *The Political Culture of the American Whigs*. Chicago: Univ. of Chicago Press

Howe, Irving
1976 *World of Our Fathers*. New York: Simon and Schuster

Howe, Irving, and Coser, Lewis
1957 *The American Communist Party*. Boston: Beacon

Huai-nan tzu
d 122 B.C.E./1990 *Huainanzi*, in Thomas Cleary, trans. and ed. *The Tao of Politics*, Boston: Shambala

Industrial Areas Foundation
1979 *Organizing for Family and Congregation*. New York: Industrial Areas Foundation

1990 *Organizing for Change: The IAF at 50 Years*. New York: Industrial Areas Foundation

Ingrahm, R.L.
1992 *A Survey of Nonlinear Dynamics: "Chaos Theory"*. Singapore: World Scientific

Ismael, Tareq
1976 *The Arab Left*. Syracuse, NY: Syracuse University Press.

Jensen, J.
1984 "The Great Uprisings," in Davidson, S. and Weiler, N. Sue, eds., *A Needle, a Bobbin, a Strike*. Philadelphia: Temple University Press

Johnson, Paul
1978 *The Shopkeepers' Millennium: Society and the Revivals in Rochester 1815-1837*. New York: Hill and Wang

Jonas, Suzanne
1981 "An Overview: Fifty Years of Revolution and Intervention in Latin America," in *Contemporary Marxism* 3

Jones, A.H.M.
1974 *The Roman Economy*. Oxford: Blackwell

Kagarlitsky, Boris
1990 *The Dialectic of Change*. London: Verso

Kant, Immanuel
1781/1969 *Foundations of the Metaphysics of Morals*, trans. Lewis White Beck. Indianapolis: Bobbs-Merrill

Kautsky, Karl
1972 *Foundations of Christianity*. New York: Harper

Kelly, Kevin
1992 "Deep Evolution: The Emergence of Postdarwinism," in *Whole Earth Review* 76

Kipnis, Ira
1952 *The American Socialist Movement*. New York: Monthly Review

Knoke, David
1982 *Network Analysis*. Beverly Hills: Sage

1990 *Political Networks*. Cambridge: Cambridge University Press

Kohlberg, Lawrence, Levine, C., and Hewer A.
1983 *Moral Stages: a Current Formulation and A Response to Critics*. Basel: Karger

Konrad, George and Szelenyi, Ivan
1979 *Intellectuals on the Road to Class Power*. NY: Harcourt Brace Jovanovich

Kramer, Samuel Noah
1963 *The Sumerians*. Chicago: University of Chicago Press

Kyrtatas, Dimitris
1987 *The Social Structure of Early Christian Communities*. London: Verso

La Parola del Popolo. publication of the *Federazione Socialista Italiana*

Laclau, Ernesto
1977 *Politics and Ideology in Marxist Theory.* London: Verso

Laclau, Ernesto and Mouffe, Chantal
1985 *Hegemony and Socialist Strategy.* London: Verso

Lancaster, Roger
1988 *Thanks to God and the Revolution.* Berkeley: Univ. of California Press

Langton, Christopher
1989 *Artificial Life: Proceedings of an Interdisciplinary Workshop on the Synthesis and Simulation of Living Systems.* New York: Addison-Wesley

Lanternari, Vittorio
1965 *The Religions of the Oppressed: A Study of Modern Messianic Movements.* New York: New American Library

Lapointe, Archie, et al
1992 *Learning Mathematics.* Report #22. Washington, D.C.: National Center for Education Statistics

Lasch, Christopher
1977 *Haven in a Heartless World.* New York: Basic Books

1979 *The Culture of Narcissism.* New York: Norton

1981 "The Freudian Left and the Cultural Revolution," *New Left Review* 129

1990 *The True and Only Heaven.* New York: Norton

Leff, Gordon
1967 *Heresy in the Late Middle Ages.* New York: Barnes and Noble

Lenat, Douglas B.
1980 *The Heuristics of Nature.* Report HPP-80-27, Stanford Heuristic Programming Project

Lenin, V. I.
1894/1959 *The Development of Capitalism in Russia.* Moscow: Progress

1902/1929 *What is to Be Done?* New York: International

1905/1971 *Two Tactics of Social Democracy.* in *Selected Works*, NY: Int'l

1908/1970 *Materialism and Empiriocriticism.* Moscow: Progress

1913/1971 "The Three Sources and the Three Component Parts of Marxism," in *Selected Works*. New York: International

1920/1971 *Left-Wing Communism, an Infantile Disorder*. in *Selected Works*. New York: International

Lenski, Gerhard and Jean
1982 *Human Societies*. New York: McGraw Hill

Leo XIII
1891/1943 *Rerum Novarum*. in *Two Basic Social Encyclicals*. Washington, D.C.: Catholic University of America Press

Lessing, Doris
1979 *Shikasta*. New York: Knopf

1980 *The Marriages Between Zones Three, Four, and Five*, NY: Knopf

1980 *The Sirian Experiments*. New York: Knopf

1982 *The Making of the Representative for Planet 8*. New York: Knopf

1983 *The Sentimental Agents in the Volyen Empire*. New York: Knopf

Levi-Strauss, Claude
1949/1969 *The Elementary Forms of Kinship*. Boston: Beacon

1958/1963 *Structural Anthropology*. New York: Basic

Lewin, Moshe
1968 *Russian Peasants and Soviet Power*. Evanston: Northwestern University Press

Lin Biao
1965 "Long Live the Victory of the People's War," Speech delivered 9-3-65

Lipow, Arthur
1982 *Authoritarian Socialism in America* Berkeley: Univ. of California Press

Locke, John
1690/1967 *Two Treatises on Government*. London: Cambridge Univ. Press

Lockridge, Kenneth
1970 *A New England Town: The First Hundred Years*. New York: Norton

Longworth, R.C.
1994 "Job Fears From Cyprus to Sri Lanka," *Chicago Tribune* 4 Sept 1994

Lovett, Clara
1982 *The Democratic Movement in Italy*. Cambridge: Harvard

Lowy, Michel
1991 "The Crisis of Really Existing Socialism," Monthly Review 43:1

Lukacs, Georgi
1971 *History and Class Consciousness*. Cambridge: M.I.T. Press

1976 *The Ontology of Social Being*. Budapest: Magveto

Lukes, Stephen
1973 *Emile Durkheim*. London: Lane

Lyotard, Jean-Francois
1984 *The Postmodern Condition*. Minneapolis: Univ. of Minnesota Press

Maduro, Otto
1982 *Religion and Social Conflict*. Maryknoll: Orbis

Mandel, Ernest
1968 *Marxist Economic Theory*. New York: Monthly Review

1978 *Late Capitalism*. London: Verso

Mansueto, Anthony
1985 "Religion and Socialism in Italian American History," *Proceedings of the American Italian Historical Association*

1988 "Religion, Solidarity, and Class Struggle," in *Social Compass* XXXV:2-3

1990 "The Role of Religion in the Socialist Transition," in *North Star Review* 3

1992a "The Industrial Areas Foundation: A Preliminary Analysis of its Social Basis and Political Valence," in *Dialectic, Cosmos, and Society* 1:1

1992b "Synergism," in *Dialectic, Cosmos, and Society* 1:2

1992c "The Crisis of Neoliberalism and the Emergence of a Progressive-Institutionalist Bloc," *Dialectic, Cosmos, and Society* 1:3

1993a "The Current Situation in the European Countries of the Former Soviet Bloc," in *Dialectic, Cosmos, and Society* 1:4, reprinted as a chapter in *Russia and the West: A Dialogue of Cultures*, Tver State University Press

1993b "The Contributions of Complex Systems Theory to Ethics," *Proceedings of the Third Annual Conference of the Chaos Network*

1994a "The Next Steps in the Human Civilizational Project," *Dialectic, Cosmos, and Society* 2:1

1994b "Beyond Postmodernism: The Contributions of Anthropic Cosmology and Complex Systems Theory to the Social Sciences," *Filosofskie nauki*

1994c "The Cosmohistorical Vision of Ernesto Cardenal," *Dialectic, Cosmos, and Society* 2:2

1994d "Towards Synergism: A Personal Journey," *DCS* 2:2

1994e "Visions of Cosmopolis: The Religious Dimensions of the UFO Phenomenon," *OMNI* October 1994

1995a "In These Dark Times ...," *Dialectic, Cosmos, and Society* 7

forthcoming "From Dialectic to Organization: Bogdanov's Contribution to Social Theory," *Studies in Soviet and East European Thought*

Mao Zedong
1926/1971 "Analysis of the Classes in Chinese Society," in *Selected Works*. Peking: Foreign Languages Press

1927/1935 "Report on the Peasant Movement in Hunan," in *Selected Works*. Peking: Foreign Languages Press

1937a/1971 "On Contradiction," in *Selected Works*. Peking: Foreign Languages Press

1937b/1971 "On Practice," in *Selected Works*. Peking: Foreign Languages

1937c/1971 "Combat Liberalism," in *Selected Works*. Peking: Foreign Languages Press

1938/1971 "The Role of the Chinese Communist Party in the National War," in *Selected Works*. Peking: Foreign Languages Press

1940/1971 "Current Problems of Tactics in the Anti-Japanese United Front," in *Selected Works*. Peking: Foreign Languages Press

1945/1971 "The Foolish Old Man Who Removed the Mountains," in *Selected Works*. Peking: Foreign Languages Press

1949/1971 "On the People's Democratic Dictatorship," in *Selected Works*. Peking: Foreign Languages Press

1957a/1971 "On the correct Handling of Contradictions Among the People," in *Selected Works*. Peking: Foreign Languages Press

1957b/1971 "Speech at the Chinese Communist Party's National Conference on Propaganda Work," in *Selected Works*. Peking: Foreign Languages Press

1963/1971 Where Do Correct Ideas Come From?" in *Selected Works*. Peking: Foreign Languages Press

Marcuse, Herbert
1955 *Eros and Civilization*. Boston: Beacon

1964 *One Dimensional Man*. Boston: Beacon

Marsden, George
1980 *Fundamentalism in American Culture*. New York: Oxford

Margulis, Lynn and Fester, Rene, eds.
1991 *Symbiosis as a Source of Evolutionary Innovation*. Cambridge, MA: MIT Press

Mariategui, Jose Carlos
1979 *Siete ensayos de interpretacion de la realidad peruana*. Mexico: Ediciones Era

Maritain, Jacques
1951/1973 *Integral Humanism*. Notre Dame, IN: Univ. of Notre Dame Press

Marx, Karl and Engels, Frederick
1843/1978 *Contribution to Hegel's Philosophy of Right: Introduction*. in *Marx-Engels Reader*. New York: Norton

1844/1978 *Economic and Philosophical Manuscripts*. in *Marx-Engels Reader*. New York: Norton

1846/1978 *The German Ideology*. in *Marx-Engels Reader*. NY: Norton

1849/1978 *Wage Labor and Capital*, in *Marx-Engels Reader*. NY: Norton

1848/1978 *The Communist Manifesto*. in *Marx-Engels Reader*. NY: Norton

1857-1858/1973 *The Grundrisse*. New York: Vintage

1859/1966 "Preface," to *Contribution to a Critique of Political Economy*. in Fromm, Erich, *Marx's Concept of Man*. New York: Continuum

1867/1977 *Capital*, Volume One. New York: Vintage

1881/1978 "Letter to Vera Zasulich," in *Marx-Engels Reader*. NY: Norton

Matthews, Caitlin
1991 *Sophia: Goddess of Wisdom*. London: Mandala

McGinn, Bernard
1979 *Apocalyptic Spirituality*. New York: Paulist

McKenna, Dennis
1994 *Invisible Landscape*. San Francisco: HarperCollins

McLoughlin, William G.
1960 "Introduction," in Finney, Charles Gradison. *Lectures on the Revival of Religion*. Cambridge: Harvard University Press

1978 *Revivals, Awakening, and Reform*. Chicago: Univ. of Chicago Press

Meikle, Scott
1985 *Essentialism in the Thought of Karl Marx*. London: Duckworth

Merkel, Wolfgang
1992 "After the Golden Age: Is Social Democracy Doomed to Decline?" in Lemke, Christiane and Marks, Gary *The Crisis of Socialism in Europe*. Durham, NC: Duke University Press

Michael, Franz
1966 *The Taiping Rebellion*. Seattle, WA: University of Washing Press

Mill, J.S.
1965 *Utilitarianism*. Indianapolis: Bobbs-Merrill

Miller, Alice
1986 *Thou Shall Not Be Aware*. New York: Meridian

Mitrany, David
1961 *Marx Against the Peasants*. New York: Collier

Moore, Barrington
1966 *Social Origins of Dictatorship and Democracy*. Boston: Beacon

Moore, Stanley
1980 *Marx on the Choice Between Socialism and Communism*. Cambridge: Harvard University Press

Moss, Leonard and Campannari, Stephen
1982 "In Quest of the Black Virgin," in Preston, James, Editor, *Mother Worship*. Chapel Hill: University of North Carolina Press

Mottu, Henri
1977 *La manifestation de l'Esprit selon Joachim de Fiore*. Neuchatel

Naquin, Susan
1981 *The Shantung Rebellion*. New Haven: Yale University Press

Nash, Gary.
1970 *Class and Society in Early America.* Englewood Cliffs, NJ: Prentice-Hall

Nesti, Arnaldo
1972 "Religione e conflitto sociale," *Testimonanze*

1974 *Gesu Socialista: una tradizione popolare italiana.* Torino: Claudiana

Nicolis, G. and Prigogine, I.
1977 *Self-Organization in Non-Equilibrium Systems.* New York: Wiley

Osband, Kent
1982 "Maurice Dobb and the End of NEP: A Critique," in *Theoretical Review* 27

Osthathios, Geevarghese Mar
1979 *Theology of a Classless Society.* Maryknoll: Orbis

Ottoway, David and Marina
1981 *Afrocommunism.* New York: Homes and Meier

Owen, Launcelot
1963 *The Russian Peasant Movement 1906-1917.* London: King

Pare, Luisa
1977 *La proletariada agricola en Mexico.* Mexico: Siglo XXI

Parsons, Talcott
1957 *The Social System.* New York: Free Press

1964 *The Structure of Social Action.* New York: Free Press

Patterson, William and Campbell, Ian
1974 *Social Democracy in Post-War Europe.* London: Macmillan

Pawlikowski, John
1982 *Christ in the Light of Christian-Jewish Dialogue.* NY: Paulist Press

Piaget, Jean
1957 *Logic and Psychology.* New York: Basic

1968 *Structuralism.* New York: Basic

Pines, David, ed.
1988 *Emerging Syntheses in Science.* New York: Addison Wesley

Plato
c 385 B.C.E./1968 *Republic.* trans. Alan Bloom, New York: Basic

c 350 B.C.E./1960 *Timeaus*. New York: Penguin

Plotinus
c 270/1964 *The Enneads*, in *The Essential Plotinus*, trans. Elmer J. O'Brien, S.J. Indianapolis: Hackett

Portelli, Hughes
1974 *Gramsci et la questione religieuse*. Paris: Anthropos

1975 *Gramsci et le bloc historique*. Paris: Anthropos

Poulantzas, Nicos
1974 *Fascism and Dictatorship*. London: New Left

1975a *Classes in Contemporary Capitalism*. London: New Left

1975b *Political Power and Social Classes*. London: New Left

1978 *State Power and Socialism*.   London: New Left

Purcell, Edward
1973 *The Crisis of Democratic Theory: Scientific Naturalism and the Problem of Value*. Lexington: University of Kentucky Press

Radkey, O.
1958 *Agrarian Foes of Bolshevism*. New York: Collier

1962 *The Hammer Over the Sickle*. New York: Collier

Rapoport, A.
1985 *General Systems Theory: Essential Concepts and Applications*. Cambridge, MA: Abacus Press

Ratzinger, Joseph Cardinal
1984 "Instruction Regarding Certain Aspects of the Theology of Liberation," United States Catholic Conference

1986 "Christian Freedom and Liberation," United States Catholic Conference

Reeves, Marjorie
1969 *Prophecy in the Late Middle Ages*. New York: Oxford

1976 *The Prophetic Future in Joachim of Fiore*. New York: Oxford

Reich, Robert
1992 *The Work of Nations*. New York: Vintage

Reich, Wilhelm
1970 *The Mass Psychology of Fascism*. NY: Farrar, Strauss, and Giroux

Renda, F.
1977 *I fasci siciliani*. Torino: Einaudi

Resnik, Stephen and Wolff, Richard
1987 *Knowledge and Class: A Marxian Critique of Political Economy*. Chicago: University of Chicago Press

Romanell, Patrick
1969 *Making of the Mexican Mind*. Notre Dame: University of Notre Dame

Romano, S.F.
1959 *Storia dei fasci siciliani*. Bari: Laterza

Rousseau, Jean-Jacques
1762/1962 *Le contrat social*. Paris: Freres

Rowley, David
1987 *Millenarian Bolshevism*. New York: Garland

Rubin, Gayle
1975 "The Traffic in Women," in Rayna Reiter, *Toward an Anthropology of Women*. New York: Monthly Review Press

Ruether, Rosemary
1974 *Faith and Fratricide*. New York: Harper

1992 *Gaia and God: An Ecofeminist Theology of Earth Healing*. San Francisco: Harper San Francisco

Russell, Bertrand
1957 *Why I Am Not a Christian*. New York: Simon and Schuster

Rutman, Darret
1965 *Winthrop's Boston: Portrait of a Puritan Town*. Chapel Hill: University of North Carolina Press

Salvatore, Nick
1982 *Citizen and Socialist*. Urbana: University of Illinois Press

Sarkisyanz, E.
1965 *Buddhist Backgrounds of the Burmese Revolution*. The Hague: Nijhoff

Saudino, Domenico
1922a *"Il controllo dell nascite,"* *La Parola del Popolo*.14 January 1922

1922b *"Commemorazione per Giordano Bruno,"* *La Parola del Popolo.* 18 February 1922

n.d. *La chiesa ed i priviligi di classe,* manuscript

Saussure, Ferdinand de
1973 *Cours de linguistique generale.* Paris: Payot

Schneider, Jane and Peter
1976 *Culture and Political Economy in Western Sicily.* New York: Academic Press

Schofield, Anne
1984 "The Uprising of the 20,000," in Davidson, S. and Weiler, N. Sue, eds. *A Needle, a Bobbin, a Strike.* Philadelphia: Temple University Press

Sennet, Richard, and Cobb, Jonathan
1972 *Hidden Injuries of Class.* New York: Knopf

1976 *Fall of Public Man.* New York: Knopf

Sereni, E.
1968 *Capitalismo nelle campagne.* Torino: Einaudi

Service, Elman
1966 *The Hunters.* Englewood Cliffs, N.J.: Prentice Hall

Sewell, William
1980 *Work and Revolution in France.* NY: Cambridge University Press

Seyvastyanov, V., Ursul, A., Shkolenko, Yu.
1979 *The Universe and Civilization.* Moscow: Progress

Shannon, Claude and Weaver, Warren
1949 *The Mathematical Theory of Communication.* Urbana: University of Illinois Press

Sheldrake, Rupert
1981 *A New Science of Life.* London: Blond and Brigs

1989 *The Presence of the Past.* London: Fontana

Silone, Ignazio
1952 *Una manciata di more.* Milano: Monadori

1955 *Vino e pane.* Milano: Monadori

1968 *L'avventura d'un povero cristiano.* Milano: Monadori

Smith, Adam
1776/1976 *An Inquiry into the Nature and Causes of the Wealth of Nations*. Oxford: Oxford University Press

Spencer, Herbert
1873/1973 *The Study of Sociology*, in *On Social Evolution*, ed. Peel, J.D.Y., Chicago: University of Chicago Press.

Spinoza, Baruch
1675/1955 *Ethics*. New York: Dover

Spriano, P.
1967 *Storia del Partito communista italiana*. Roma: Riuniti

Stalin. Joseph
1939/1972 *Foundations of Leninism*. New York: International

1952/1972 *Economic Problems of Socialism in the U.S.S.R.* NY: Int'l

Stavenhagen, Rodolfo
1977 *El campesinado y las estrategias del desarollo rural*. Mexico: Centro de Estudios Sociologicos, Collegio de Mexico

Steele, Robert David
1992 "E3I: Ethics, Ecology, Evolution, and Intelligence," in *Whole Earth Review* 76

Stein, William
1987 "Ideology in Popular Struggle: Contradictions of Popular Movements in Highland Peru," in Kliever, Lonnie, ed. *The Terrible Meek*. New York: Paragon

Stone, Merlin
1976 *When God Was A Woman*. London: Dorset

Susiluoto, Ilmari
1982 *The Origins and Development of Systems Thinking in the Soviet Union*. Helsinki: Suomalinen Tiedeakatemia

Sweezey, Paul
1978 *Post Revolutionary Society*. New York: Monthly Review

Tannenbaum, Frank
1937 *Peace by Revolution: An Interpretation of Mexico*. NY: Columbia Univ. Press

Tawney, Richard
1922/1926 *Religion and the Rise of Capitalism*. New York: Harcourt, Brace

Teilhard de Chardin, Pierre
1955/1975 *The Phenomenon of Man*. New York: Harper and Row

Teran, Sylvia
1976 *"Formas de conciencia social de los trabajadores del campo,"* Cuadernos
Agricolas I

Theissen, Gerd
1982 *The Social Setting of Pauline Christianity.* Philadelphia: Fortress

Therborn, G.
1976 *Science, Class, and Society.* London: New Left

1992 "A Balance Sheet for the Left,: in *New Left Review* 194

Tipler, Frank
1989   "The Omega Point as Eschaton: Answers to Pannenberg's Questions for
Scientists," in Zygon 24:2

Tischner, Josef
1987 *Marxism and Christianity.* Washington DC: Georgetown Univ. Press

Tracy, Patricia
1980   *Jonathan Edwards, Pastor: Religion and Society in Eighteenth Century
Northampton.* New York: Hill and Wang

Trappl, Robert
1986 *Power, Autonomy, Utopia: New Approaches Toward Complex Systems.* New
York: Plenum

Trotsky, Leon
1932 *The History of the Russian Revolution.* New York: Simon and Schuster

1937/1972 *The Revolution Betrayed.* New York: Pathfinder Press

Tung, Jerry
n.d. "NDM's Strategy for a Post-Industrial Society"

Turing, Alan
1950/1981 "Mind," in Hofstadter, D.R. and Dennett, D.C. *The Mind's I.* New York:
Basic

Tyler, Hamilton
1964 *Pueblo Gods and Myths.* Norman, OK: University of Oklahoma

Unger, Roberto Mangabeira
1987   *Social Theory: Its Situation and Its Task.* Cambridge: Cambridge University
Press

van Zantwijk, Rudholph
1985 *The Aztec Arrangement.* Norman. OK: University of Oklahoma Press

Vatican II
1966 *The Documents of Vatican II*. New York: Guild Press

Vecoli, R.
1963 "Chicago Italians Prior to World War I," unpublished doctoral dissertation, University of Wisconsin

1969 "Prelates and Peasants: Italian Immigrants and the Catholic Church," *Journal of Social History* 2:3

Velona, Fort
1958 *"Genesi del movimento socialista democratico del Parola del Popolo,"* La Parola del Popolo New Series December 1958 - January 1959

Venturi, Franco
1966 *Il popolismo russo*. New York: Grossett and Dunlap

Vilas, Carlos
1986 *The Sandinista Revolution*. New York: Monthly Review Press

von Balthasar, Hans Urs
1968 *Love Alone*. London: Allen and Unwin

von Bertalanffy, Ludwig
1968 *General Systems Theory*. New York: George Braziller, Inc.

von Neumann, J.
1966 *Theory of Self-Reproducing Automata*. Urbana: University of Illinois

von Neumann, J. and Morgenstern, O.
1947 *The Theory of Games and Economic Behavior*. Princeton: Princeton University Press

Waddington, C.H.
1957 *The Strategy of the Genes*. London: George Allen and Unwin Ltd.

Wagar, Warren
1989 *A Short History of the Future*. Chicago: University of Chicago Press

Wallerstein, Immanuel
1974 *The Modern World System*. New York: Academic Press

Walzer, Michael
1965 *Revolution of the Saints*. Cambridge: Harvard University Press

1983 *Spheres of Justice*. New York: Basic

Waters, Frank
1963 *The Book of the Hopi.* New York: Viking Penguin

Watson, William
1974 *The Chinese Exhibition.* (a guide to the exhibition of archaeological finds of the People's Republic, Toronto, 1974)

Weber, Max
1920/1958 *The Protestant Ethic and the Spirit of Capitalism.* NY: Scribners

1921/1968 *Economy and Society.* New York: Bedminster

Weiler, N. Sue
1984 "The Uprising in Chicago," in Davidson, Sue, and Weiler, N. Sue, eds. *A Needle, a Bobbin, a Strike.* Philadelphia: Temple University

Weinstein, James
1967 *The Decline of Socialism in America.* New York: Monthly Review

Weisskopf, T.
1979 "Marxian Crisis Theory and the Post War Rate of Profit in the U.S.," *Cambridge Journal of Economics* 1979: 3

1981 "Current Economic Crisis in Historical Perspective," *Socialist Review* 57

Wesson, Robert
1991 *Beyond Natural Selection.* Cambridge, MA: MIT Press

Wesson, Robert G.
1963 *Soviet Communes.* New Brunswick, NJ: Rutgers University Press

Wetter, Gustav
1952/1958 *Dialectical Materialism: A historical and systematic survey of philosophy in the Soviet Union.* New York Praeger

Whetten, Nathan
1948 *Rural Mexico.* Chicago: University of Chicago Press

Whitehead, Alfred North
1929 *Process and Reality: An Essay in Cosmology.* New York: Macmillan

Whitehead, Raymond
1977 *Love and Struggle in Mao's Thought.* Maryknoll: Orbis

Wiener, Norbert
1948 *Cybernetics.* New York: Wiley

Winthrop, John
1929 *The Winthrop Papers*. Boston: The Massachusetts Historical Society

Wolf, Eric
1969 *Peasant Wars of the Twentieth Century*. New York: Harper

Yao Wen—yuan
1975 "The Social Basis of the Lin Biao Anti Party Clique," Chicago: Liberation

Yovel, Yirmiyahu
1989 *Spinoza and Other Heretics*. Princeton: Princeton University Press

Zeitlin, Irving.
1988 *Jesus and the Judaism of His Time*. New York: Polity

Zhao Ziyang
1987 "Advance Along the Road to Socialism with Chinese Characteristics," *Beijing Review* 30:45

Zimmerman, Marc
1990 *Literature and Politics in the Central American Revolutions*. Austin: University of Texas Press

Zimmermann, R.E.
1991 "The Anthropic Cosmological Principle" Philosophical Implications of Self-Reference," in Casti, John, and Karlqvist, Anders. *Beyond Belief: Randomness, Prediction, and Explanation in Science*. Boca Raton, FL: CRC

Zinn, Howard
1980 *The People's History of the United States*. New York: Harper

Zitara, Nicola
1971 *L'unita d'Italia, Nascita di una colonia*. Milano: Jaca Book

Zurek, Wojcieck Hubert
1990 *Complexity, Entropy, and the Physics of Information*. NY: Addison-Wesley

Index of Subjects

Index of Names

509

About the Author

Anthony Mansueto is an internationally recognized philosopher and social theorist whose research regarding the role of religion in social progress and the philosophical implications of the new science have attracted attention in Europe, North America, Latin America, India, and the former Soviet bloc. He has taught at colleges and universities in the United States and Mexico, and served for three years as Director of the Justice and Peace Commission for the Catholic Diocese of Dallas. He is currently President and Research Director at the Foundation for Social Progress, a nonprofit, nonpartisan research, education, and organizing institute, and Editor of *Dialectic, Cosmos, and Society*, a new journal dedicated to rethinking the next steps in the human civilizational project and the role of that project in the larger cosmo-historical evolutionary project. He lives in Chicago with his wife and collaborator, Maggie Vosburg Mansueto, and may be reached at the Foundation for Social Progress, P.O. Box 59875, Chicago, IL 60659, or by email at ircg@aol.com.